Marino Xanthos

Reactive Extrusion: Principles and Practice

Polymer Processing Institute

Series Editor: Joseph A. Biesenberger

Dr. Joseph Biesenberger is Professor of Chemical Engineering at Stevens Institute of Technology and President of the Polymer Processing Institute at Stevens, which he co-founded in 1982 together with Professor L. Pollara, Provost Emeritus of Stevens Institute. Dr. Biesenberger received his B. S. in Chemical Engineering in 1957 from New Jersey Institute of Technology, and M. S. E. and Ph. D. in Polymer Engineering and Chemical Engineering, respectively, from Princeton University. During 1962 he was a Montecatini Fellow with Professor G. Natta at the Milan Polytechnic Institute. He joined Stevens Institute in 1963 and served as Department Head of Chemistry and Chemical Engineering from 1971 to 1978. Dr. Biesenberger's areas of research are polymerization engineering (reaction kinetics, reactor design) and polymer processing (reactive extrusion, devolatilization). He is co-author of *Principles of Polymerization Engineering,* published by Wiley in 1983, and editor of *Devolatilization of Polymers,* published by Hanser in 1983. He is also author or co-author of numerous book chapters and more than 100 research papers.

Volume Editor:

Dr. Marino Xanthos is presently Director of Research of the Polymer Processing Institute with personal research interests in reactive processing, polymer blends and composites and plastics recycling. Since 1980, he has been affiliated with Stevens Institute of Technology, Hoboken, N.J., as Adjunct Professor of Chemical Engineering and Academic Advisor in its overseas Polymer Engineering programs. He received his Ph.D. in 1975 from the University of Toronto, Canada, in Chemical Engineering with graduate work in the area of polymer composites and his B.Sc. in Chemistry from the University of Salonica, Greece. As Manager of R&D and Technical Services for Marietta Resources International, Boucherville, Canada during 1975-1980 he contributed to the pioneering efforts that led to the development of mica reinforced plastics. Holder of U.S. and Canadian patents, editor-in-chief of *"Advances in Polymer Technology"* and author or co-author of more than 40 technical articles.

Marino Xanthos (Editor)

Reactive Extrusion

Principles and Practice

A Monograph with 92 Illustrations
and 17 Tables

Hanser Publishers, Munich Vienna New York Barcelona

Distributed in the United States of America and in Canada by
Oxford University Press New York

Editor:
Dr. Marino Xanthos, Polymer Processing Institute at Stevens Institute of Technology,
Castle Point on the Hudson, Hoboken, NJ 07030, USA

Distributed in USA and in Canada by
Oxford University Press
200 Madison Avenue, New York, NY 10016

Distributed in all other countries by
Carl Hanser Verlag
Kolbergerstraße 22
D-8000 München 80

Library of Congress Cataloging-in-Publication Data
Reactive extrusion : principles and practice / edited by Marino Xanthos.
p. cm. – (Polymer Processing Institute)
Includes bibliographical references and index.
ISBN 0-19-520951-6
1. Plastics-Extrusion. 2. Chemical reactors. I. Xanthos, Marino. II. Series: Polymer Processing Institute (Series)
TP1175.E9R42 1992
668.4′13–dc20 92-52945

Die Deutsche Bibliothek – CIP-Einheitsaufnahme
Reactive extrusion: principles and practice : a monograph / ed. by Marino Xanthos. – Munich ; Vienna ; New York ; Barcelona : Hanser ; New York : Oxford Univ. Press, 1992
(Polymer Processing Institute)
ISBN 3-446-15677-1
NE: Xanthos, Marino [Hrsg.]

ISBN 3-446-15677-1 Carl Hanser Verlag, Munich Vienna New York Barcelona
ISBN 0-19-520951-6 Oxford University Press New York
Library of Congress Catalog Card Number 92-052945

Printed and bound in Germany by Passavia Druckerei GmbH Passau

Prologue

This monograph represents the first in a series edited under the auspices of the Polymer Processing Institute. The topic, reactive extrusion (REX), also describes an ongoing research program at PPI.

The Polymer Processing Institute is an independent research corporation hosted by Stevens Institute of Technology, with which it maintains close ties, and is located in New Jersey on the banks of the Hudson River. Its mission is to serve industry by advancing the scientific underpinnings of polymer technology through industry-sponsored research, development and education, and to disseminate information pertaining thereto via all appropriate mechanisms for technology transfer. In addition to generic research, PPI carries out contract R&D with individual companies or with groups of companies, including deliverables, deadlines and confidentiality.

PPI's areas of expertise include development of high performance products and advanced processes; property characterization; and computer modeling. Its human resources, in addition to the excellent Stevens faculty from all academic departments to which it has access, include a technical staff of professionals and an experienced group of associated consultants, some of whom have contributed to this monograph. Its characterization and process labs, and computer center, are well equipped and professionally managed.

In pursuit of its mission to transfer technology, PPI operates its own extension center for the plastics industry, supported by the New Jersey Governor's commission on Science and Technology, edits its own journal, *Advances in Polymer Technology,* offers 4-5 advanced-level short courses annually for industrial engineers and scientists, and supports the education and research needs of graduate and undergraduate students at Stevens. In fact, this monograph is an outgrowth of a PPI short course on REX.

While not a new or novel process, per se, REX has been the subject of vigorous research activity in recent years, both in industry and academe, and has resulted in numerous commercial processes and products. The primary reason for the success of REX is the extruder's unique suitability as a vehicle for carrying out chemical reactions in the bulk phase, i.e. without the use of diluents, to produce "value-added", specialty polymers, through chemical modification of existing polymers or, when appropriate, to produce polymers from monomers. This virtue stems from its ability to pump and mix highly viscous materials and to facilitate the staging of multiple process steps in a single machine, including melting, metering, mixing, reacting, side-stream addition and venting and, under appropriate circumstances, even shaping.

The combination of chemical reaction and polymer processing, in general, remains a rich potential source for further development of new and novel products and processes. Before REX, another reactive process, viz., reactive injection molding (RIM), was the object of intense developmental activity.

My personal involvement with reactive processing began in 1980 with the organizing committee for the First International Symposium on Reactive Processing of Polymers in Pittsburgh. In 1985 and 1986, respectively, I organized a Topical Workshop on "Polymerization and Polymer Modification", held in Bermuda under the auspices of the American Chemical Society, and the first annual short course on REX for PPI, held in Hilton Head, for which this monograph is intended to serve as a text.

Future topics planned for the PPI series include polymer devolatilization, melt mixing and polymer blends, among others. The second will most likely emerge as the second edition of an existing monograph published by Hanser in 1983, entitled *Polymer Devolatilization,* which has been the text for another annual PPI short course by the same title. It is our hope that these monographs will facilitate the flow of important, timely technological information among industrial organizations and universities.

Joseph A. Biesenberger
Series Editor

Preface

The use of extruders as continuous reactors for processes such as polymerization, polymer modification or compatibilization of polymer blends involves technologies that are gaining increasing popularity and compete with conventional operations with respect to efficiency and economics. The need to analyze such technologies resulted in the introduction of an advanced course on reactive polymer processing offered repeatedly by the Polymer Processing Institute during the last few years. The objective of the three-day course was to establish an understanding of the applied and fundamental aspects of the process commonly known as "Reactive Extrusion" and present the current state-of-the-art from both chemistry and equipment aspects. To this effect, the course faculty was assembled by calling upon the talents of distinguished engineers and chemists, all pioneers in reactive extrusion, but also actively involved in programs applying reactive extrusion technology to industry needs.

It was only natural that the popularity of the "Reactive Extrusion" course led to discussions on producing a monograph that would assemble and disseminate to broader audiences the material presented during the course. In early 1991, the transparencies, slides and hand-outs were finally transformed into individual chapters by the same instructors who participated in the course. The result is the present book, the first in its kind, intended to benefit engineers, scientists and technologists involved in this industrially important sector of polymer processing technology.

Following R.C. Kowalski's *introduction* the book is divided into three major parts. The first part presents *applications* of the reactive extrusion technology. Case histories of industrial studies on polyolefin modification in extruders along with economics are discussed by R.C. Kowalski. M. Xanthos analyzes continuous reactive extrusion processes such as polymerization and controlled degradation by considering available information on the chemistry of the systems. The industrially important carboxylation reactions and the use of anhydride or acid modified polymers to prepare compatibilized polymer blends are presented by N.G. Gaylord. The second part of this monograph by S.B. Brown is a most exhaustive survey of virtually all *chemical reactions* that have been conducted in extruders including polymerization, grafting, copolymer formation, crosslinking, functionalization and controlled degradation. More than 600 reactive extrusion processes listed in the recent open and patent literature are classified according to their type and polymers involved. The *engineering fundamentals* of reactive extrusion are included in the third part of the book. D.B. Todd's chapter features a full description and comparison of available extrusion equipment as well as details on process parameters and requirements. The application of polymerization engineering principles to extruder reactors and the relative importance of mixing/reaction on the process efficiency are described by J.A. Biesenberger. Finally, W.M. Davis discusses the important subject of heat transfer in extruder reactors including temperature control and scale-up. The three different sections of the book and each of their respective chapters may be read in no particular order, depending on the reader's interest and background.

Literature *references* are included alphabetically at the end of the book in a master list that comprises all seven chapters. In the text, references are listed according to author's name and year of publication. Every effort has been made to ensure consistency throughout the book, with respect to format, terminology and abbreviations; however, the diverse styles of the contributing authors and the great variety of the sources of information would have made the task of further uniformizing extremely lengthy, and probably unnecessary. Thus, each chapter is self-contained, often with its own list of symbols and abbreviations; metric, English or S.I. units remain as they appeared in the original literature reference or in the contributing author's manuscript without any further editing.

Many thanks are due to my fellow co-authors who through their prompt response to my editorial requests helped to complete this monograph in the shortest possible time. Also, special thanks to Ms. Maribel Gonzalez of PPI whose skills in word processing transformed into a structured document the "amorphous" collection of floppy disks and type-written manuscripts that served as raw material for this book.

Hoboken, New Jersey
February 1992

Marino Xanthos
Volume Editor

Contents

Part II
Review of Reactive Extrusion Processes

Part III
Engineering Fundamentals of Reactive Extrusion

Contributors

Joseph A. Biesenberger, Polymer Processing Institute at Stevens Institute of Technology, Castle Point on the Hudson, Hoboken, NJ 07030, USA

S. Bruce Brown, Polymer Chemistry and Materials Laboratory, General Electric Research and Development Center, Schenectady, NY 12301, USA

William M. Davis, Exxon Chemical Company, Polymers Group, 1900 East Linden Avenue, P.O. Box 45, Linden, NJ 07036, USA

Norman G. Gaylord, The Charles A. Dana Research Institute for Scientists Emeriti, Drew University, Madison, NJ 07940, USA

Ronald C. Kowalski, Exxon Chemical Co., 1900 East Linden Avenue, P.O. Box 45, Linden, NJ 07036, USA; Present Address: 108 Union Avenue, New Providence, NJ 07974, USA

David B. Todd, Polymer Processing Institute at Stevens Institute of Technology, Castle Point on the Hudson, Hoboken, NJ 07030, USA

Marino Xanthos, Polymer Processing Institute at Stevens Institute of Technology, Castle Point on the Hudson, Hoboken, NJ 07030, USA

Introduction

By Ronald C. Kowalski, Exxon Chemical Co., 1900 East Linden Avenue, P.O. Box 45, Linden, NJ 07036

Recent years have seen a sharp increase in interest around the world in Reactive Extrusion. Sessions on the subject have been added to national meetings of the American Chemical Society, American Institute of Chemical Engineers, Society of Plastics Engineers, and to international meetings of the Polymer Processing Society, International Union of Pure and Applied Chemistry, and others. Recently, the Polymer Reaction Engineering Conference of the Engineering Foundation added a session to its program on Reactive Extrusion, and learned by survey that its audience's first request for future emphasis was overwhelmingly on that subject. At the 1990 39th Annual Technical meeting of the Society of Plastics Engineers, this author was asked to present a paper on "Future Trends in Reactive Extrusion". With 16 parallel sessions competing in the same time slot, that paper drew an SRO audience of 500 engineers, breaking the 39 year record for attendance at a single paper of the Extrusion Division.

The course on Reactive Extrusion from which this book derives has been conducted by the Polymer Processing Institute at Stevens Tech for the past five years. It was stimulated by the enthusiasm shown by the international audience at an ACS Topical Workshop in Bermuda in 1985. The course has attracted more than 250 students, representing nearly all the major companies active in polymer manufacture, compounding and formulation, a few having sent as many as 25 people.

At Exxon Chemical, several years ago, we measured this level of worldwide industrial interest in Reactive Extrusion via a patent survey and a literature survey for the period 1966-83. We found a total of more than 600 different patents granted to 150 companies. Those holding five or more are summarized in the following list (see next page).

In comparison, only 57 technical papers were found for the same time period, mostly by extruder vendors. Only three were from the companies in the above list! So it is clear that everyone is involved and, although technical publications have increased since that survey, there is still a lot of secrecy about what is being studied.

Why this level of commercial interest?

Simply put, the answer is that extruders uniquely can handle pure high viscosity polymers. They can melt, pump, mix, compound, and devolatilize them, and

Part I

Applications of Reactive Extrusion Technology

Chapter 1

Fit the Reactor to the Chemistry

Case Histories of Industrial Studies of Extruder Reactions

By Ronald C. Kowalski, Exxon Chemical Co., 1900 East Linden Avenue, P.O. Box 45, Linden, NJ 07036; Present Address: 108 Union Avenue, New Providence, NJ 07974

1.1 Introduction

Melt phase chemical reactions done in the extruder configuration have been studied and used commercially at Exxon Chemical Company for over 25 years. Many different chemistries have been studied during that period. Three of those have become known in the general literature because of early patent positions, publications and commercial production.

They are:

(1) Controlled Rheology, the controlled free radical degradation of polypropylene to produce extremely narrow MWD, which we pioneered and put into commercial production 25 years ago.

(2) Free radical grafting of maleic anhydride, acrylic acid and other monomers to polyolefins to modify their compatibility and other chemical properties, first studied in 1967.

(3) Low temperature halogenation (bromination and chlorination) of polyolefins developed during 1980-83.

All have been done in adaptations of extruders; for each a unique reactor configuration or design was developed. A principle learned through those years is that for any new chemistry we study, it is desirable to plan to develop a specific reactor geometry to suit it. This is in contrast to typical experience in solution reactions, in which the same stirred tank reactor can be used for a wide range of chemistries, including polymer reaction chemistries. The reasons for this difference lie in the comparative advantages and limitations of the extruder as a chemical reactor:

Advantages

- Longitudinal, (nearly) plug flow reactor.
- Multistaging capability.
- Uses drag flow to convey and mix high viscosity polymer.
- Wide ranges of pressure and temperature are available.

Limitations

- Mixing is difficult in high viscosity media (very low Reynolds number).
- Heat transfer out of the system (but not into the system) is poor (see Chapter 7).

The principle "Fit the Reactor to the Chemistry" will be illustrated here in steps of increasing sophistication of the technology, as developed during the 25-year period.

1.2 Case I – Controlled Rheology of Polypropylene

Prior to 1965, polypropylene, from the time of its commercial introduction some eight years earlier, had become known as "difficult to process". Marketing technical studies had pointed to possible correlations between that behavior in the processor's equipment and the variance in melt rheological properties of the manufactured resin. A definitive study (Kowalski, 1963) of its melt elasticity in particular had shown that the level of PP's elasticity was very high, that its decay rate (a retarded elastic recovery) was slow enough so that some of the elastic strain would be frozen into the product in high speed processes (as film and fiber extrusion), and that the elasticity level itself could be a major factor in the resin's processability.

In a paper (Kowalski, 1986) prepared as part of the American Chemical Society tribute to the 50th anniversary of the discovery of polyolefins, we showed that in 1964-65 this concept had been combined with two others, also quite new at the time, as the basis for a research project which we called "Controlled Rheology". The two other concepts were:

(a) That a specific function in PP's molecular weight distribution, its high molecular weight "tail", was responsible for the high elasticity level, and
(b) That the tail could be preferentially reduced by a random chain scission process conducted at high temperature in an extruder.

Our patents (Staton et al., 1971, Kowalski, 1971; Kowalski et al., 1971; Beauxis and Kowalski, 1971; Steinkamp and Grail, 1975) show that this process was in fact successfully developed and processability was significantly improved. At first, free radicals were generated to conduct the chain scission from oxygen in the air using very high extruder barrel temperatures (700-800 °F) and a "reverse temperature profile". By setting the highest temperatures at the feed end of the extruder, where air was still present before the melt seal was achieved, free radical generation could be maximized and easily controlled. This process was used commercially for years and included online melt rheometers which measured the desired property and served as the control parameter for a true closed loop computer control circuit (Kowalski, 1986), one of the first in our industry. The closed loop computer technology was developed early enough in computer control history so that we were required to design and build our own computer for the process (Beauxis and Kowalski, 1971).

Later, our patents (Steinkamp and Grail, 1975) show, the process was changed to use free radicals provided by peroxide initiators, which permitted the use of much lower extruder temperatures.

Some of the early extruder degradation technology is illustrated in Fig. 1. The plot is made on a "Viscoelastic Grid" (Kowalski, 1963) which maps the parameters of molecular weight (MW) and molecular weight distribution (MWD). Polymerization reactor grades could be made at a range of MW, but only at high (broad) MWD, indicated by the right-hand end of each line. The

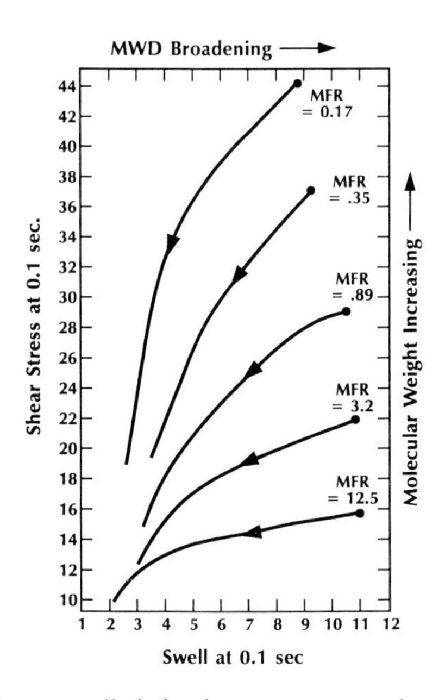

Figure 1.1 The controlled rheology process on the viscoelastic grid

lines show the degradation path of each grade fed to the extruder process. Progress down each line was made by increasing the level of the reverse temperature profile. But all the lines approached very low MWD, associated with Polydispersity Ratio (Mw/Mn) approaching 2.0. As the figure shows, the level of MW needed for a given final product application dictated the required MW of the starting material, in order to achieve very low MWD in the product.

The technological bases for the Viscoelastic Grid have been elaborated in Kowalski (1968), Kowalski (1963), and Kowalski (1986). Summarizing these:

(a) A single value for melt viscosity can be used to characterize the molecular weight of PP, more precisely than does Melt Flow Rate (MFR), and with more pertinence to high speed processing operations. We developed that to be the measured shear stress at a shear rate of 1300 sec^{-1} in a 16:1 L/D die, at fixed temperature such that die residence time was 0.1 sec. Hence, the parameter "shear stress at 0.1 sec".

(b) Analogously, "swell at 0.1 sec" represents the die swell measured in the same standardized test. The previously mentioned study (Kowalski, 1963) of PP melt elasticity had shown that die swell, measured at this short residence time, and under specific conditions, is an expression of the level of melt elasticity exhibited in a retarded elastic recovery process; passage through and out of a die.

(c) In Kowalski (1968) and again in Kowalski (1986), we showed the data in Figs. 1a and 1b, relating the above measured rheological parameters to MW and MWD (Polydispersity Ratio).

Therefore, the Viscoelastic Grid can be used as a convenient tool for representing an entire family of interrelated products, if only two molecular variables (and therefore, two rheological variables) define the family. Interrelationships between grades used for different purposes can be displayed on the Grid, as well as blending effects, or in the case of Fig. 1, the pathways for the Controlled Rheology chain scission process.

1.3 Case II – Polyolefin Free Radical Grafting of Maleic Anhydride

The Controlled Rheology work led to the conclusion that we were conducting a free radical induced chemical degradation reaction in the extruder. At the start of the research project some had assumed that if the degradation occurred, it would be by mechanical shear. That was not the case. As shown in the patents it was learned that free radicals, as from oxygen, or peroxides, were necessary. If they were excluded, very little degradation, i.e. MWD narrowing, occurred. Having learned that it was possible to run a precisely controlled chemical reaction in the extruder, it was realized that we had an opportunity to modify the concept and to run other kinds of chemical reactions. Because we had been

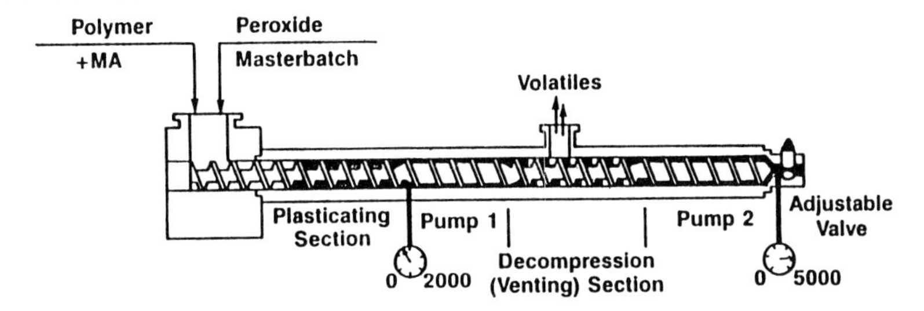

Figure 1.2 Simple maleic anhydride grafting process

able to control that degradation reaction precisely in terms of the molecular weight and the molecular weight distribution of the product, we were led into studies of free radical grafting (Steinkamp and Grail, 1975) on polyolefin backbones. Some of our early work will be described and some of the problems it revealed. Also shown will be some newly released information on the improved modern process used to do free radical grafting in Exxon Chemical's Koeln, Germany plant. And finally, we'll discuss competitive processes which have been described recently for this reaction in different equipment, with some discussion of those processes.

Fig. 2 shows our early work using a single screw extruder with two stages (Kowalski, 1989a). The first stage of pressurization leads to venting of any unreacted maleic anhydride (MA). The second stage is pressurization to the die. The first zone of the machine includes plastication combined with the beginning of reaction. The reaction would continue through the downstream portion of the machine. The sources of some of the problems with the primitive process shown can be described from this figure. The feeds consist of a peroxide masterbatch and the feedstock to be grafted, onto which MA powder has been dusted. The feedstock could have been polypropylene, polyethylene, EP rubber or some of the other polyolefins manufactured by Exxon Chemical. In this reactor, the temperature profile rises with distance down the reactor as the feeds are being melted and mixed, so it is not clear exactly where the reaction begins and how it progresses. Nor is it clear how completely the three components are mixed when the peroxide begins to decompose. If they are not thoroughly mixed, the free radicals may react only with polyolefin, or only MA, not necessarily performing the graft reaction.

Nevertheless, with such a simple system it is possible to make products which are grafted to low levels of maleic anhydride and to do some studies of the reaction as a function of temperature, peroxide level, etc.

For example, Fig. 3 (Kowalski, 1989a) is a plot of the data from that old process indicating that we can increase the amount of bonded anhydride to some rather interesting levels by increasing the amount of peroxide catalyst used. People working in the field today are aware that these levels of peroxide

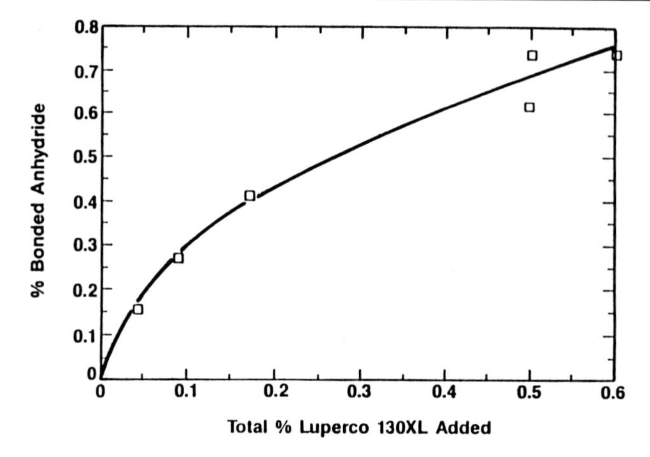

Figure 1.3 Effect of peroxide concentration on maleic anhydride grafting

are extremely high and incur consequences of color development, gelation, etc. More modern processes work with levels that are 1/10 or less than those numbers shown to get the same grafted levels of maleic anhydride.

Let us consider some of the problems encountered in trying to develop a commercial process with a reaction which is apparently successful in such a simple configuration.

(a) Incomplete mixing, undefined reaction conditions (above).

(b) Excessive peroxide usage: cost, color, gelation.

(c) Worker exposure.

Preblending the polymer with maleic anhydride permits it to sublime from the solid powdery state to the vapor. It is known that MA is quite irritating and that the irritation level varies with the individual. Most people feel irritation in the mucous membranes. Some people experience severe irritation of the skin. MA is not toxic, but it is a problem for the plant workers. It is also a problem for the customer if there is any residual unreacted maleic anhydride in the finished product.

(d) Grafting level limitation.

If 0.2% or 0.3% of maleic anhydride is grafted and some product evaluation and applications work is done, it becomes apparent that still higher levels of MA are desirable. One can do reactions in solution very easily to graft significantly more MA. When analyzed, those products show that good adhesion and allied properties require more than 0.2% or so for important market value. To increase the grafting levels significantly without product detriment it is necessary to learn more about mixing in extruders. One needs to learn how to control the internal temperature profile in the machine; increasing the level of temperature alone increases the level of grafting but also makes gels and darkens the product. In a commercial process, dark product and partly gelled product are undesirable. It is therefore necessary to increase the temperature in a precise way so that there are no hot spots, and that the average temperature profile rises uniformly throughout the melt. It is necessary to learn the limits of temperature which avoid product detriments, yet achieve higher graft level. The amount of grafting can also be increased by increasing the length of the reactor. While this is easy to do in the laboratory, commercial design will demand a reactor as short as possible, while still achieving the reaction. That requires more detailed knowledge of the reaction kinetics as a function of temperature, and as limited by mixing, as the reacting mixture proceeds down the extruder. Summarizing the requirements for increased graft level:

- Special advanced mixing technology.
- More precise internal temperature control (see especially the technology developed for Case III below).
- Optimized reaction zone length.

(e) Product quality.

For a commercial venture, one must be concerned about consistency, color, gel content, grafting uniformity and residual MA in the product. True product uniformity requires good mixing while the reaction is taking place in the reactor itself. There is a significant advantage in having all of the product made within the spec range that is set for the product, instead of using post blending as a means of blending highs and lows to achieve the product specification. For another aspect of product quality, advanced vent design technology at the extruder and further stripping downstream of the extruder are required to thoroughly remove excess MA.

Summarizing the requirements for commercial product quality:

- Improved venting technology, vent and screw design, vacuum or not.
- Precise temperature control (as above) to prevent gelation and high color.
- Thorough mixing during reaction (as above) for true product uniformity.

(f) Process engineering.

Improvements are required to avoid pre-blending and masterbatching. A commercial process should not require work for long periods of time with gas masks and other protection to prevent irritation. (It is known, however, that there are some semi-commercial ventures in the world in which that is still being done for the maleation process.) The preferred technology is to inject the reactants and the catalyst via closed systems, exactly where they are needed to control the optimum reaction. It is possible to develop metering systems that work safely with the peroxide. MA is known to have a tendency to plug small lines, valves and flow meters. But those problems can also be solved. Finally, it is necessary to provide neutralization systems for the vented material. The neutralization system should be completely enclosed and be capable of recovering all the excess MA, suitably neutralized for chemical sewer treatment.

Summarizing the process engineering requirements:

- Eliminate preblending and masterbatching.
- Good, reliable closed metering systems for direct injection of reactants and initiators.
- Neutralization systems for vents.

Exxon Chemical's commercial process for manufacturing MA modified rubber and thermoplastics has achieved the goals presented above. That process has another advantage not yet mentioned. It is capable of handling amorphous elastomers, as shown in Fig. 4 by the label "rubber bales". Customers interested in using maleic grafted polymers as impact improvers for other polymers recognize that the impact improver should usually be a rubber, very often EP rubber. Compatibilized rubber strongly influences the impact properties of the total blend. Low temperature impact properties are often best when an amorphous low MW, low T_g rubber is used. It has been learned, therefore, that MA grafted low MW amorphous EP rubber is often the most advantageous. We have therefore developed a system to feed such rubber into the reactor. The problem which was solved is that amorphous low MW grades of EP rubber are sticky and agglomerating and are normally handled only in solid bale form wrapped in plastic film. It is possible to chop the bales into pieces, but then those pieces stick to each other and to the equipment. They clog the conveying systems and the equipment used to control feed rate to the reactor. It's clear that a requirement for a continuous chemical reaction process is good control of the feed rate. Therefore, such grades of EP rubber are a special challenge. Fig. 4 indicates a capability for chopping and metering that kind of rubber. Proprietary details of how that's done are not shown.

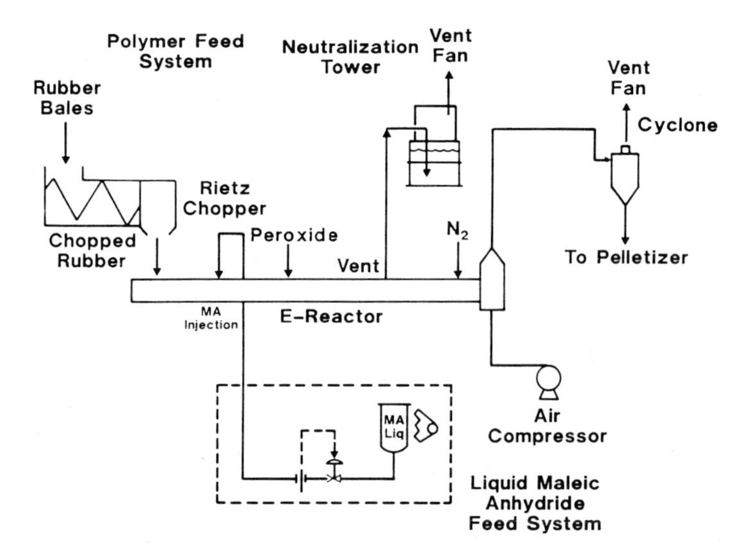

Figure 1.4 Exxon maleic anhydride graft process for baled EP rubber

Bales are fed into a device which is known in the industry as a Rietz chopper, also known not to be a rate controlling device. Technology has been developed around that machine to make it into a rate controlling device. The chopped rubber can then be metered into the reactor. For pelletized free-flowing feed of polyethylene or polypropylene, for example, other standard systems are provided.

The MA is injected via a closed liquid MA feed system, engineered to operate without plugging under continuous commercial conditions. Injection is made at a position determined by research to be the optimum one with respect to the ability to mix it into the polymer. If it is injected as liquid very close to the hopper, or into the hopper, the screw and barrel become slippery at that point, interfering with good conveying of the polymer. On the other hand, the further downstream it is injected the more extruder length is required. The figure also shows a distance between the MA injection point and the peroxide injection point. This section is designed to mix the MA well into the rubber so that when the peroxide begins to decompose the free radicals formed have an opportunity to link the two. If that mixing has not been achieved the peroxide will tend to degrade the polymer, or to react with the MA exclusively. Good mixing between the two injection points required development of special screw design technology. At the same time, the screw design also determines the temperature at the peroxide injection point. The peroxide used will have a specific half-life at that temperature. There is a complex interrelationship to be satisfied between the peroxide type, the melt temperature and the mixing level of the MA into the polymer. Again, it is not economical to delay the peroxide injection point unnecessarily, because of the cost of reactor length. Downstream of peroxide injection is the reaction zone and the vent zone. Technology around the vent is critical to strip as much as possible of the free MA without carrying particles of rubber upward. One of the reasons why MA is grafted onto polymers is to increase the adhesion of the polymer. Hot, sticky, fine particles of maleic grafted rubber or plastic passing up through a vent line will clog and plug the line. Technology has been developed around that concept to prevent plugging.

Conventional neutralization devices can be used for scrubbing MA from the gas stream. Part of the technology which is not conventional involves countercurrent stripping with nitrogen within the extruder. A twin screw, non-intermeshing machine is used which achieves good mixing, yet provides lots of internal volume and no melt blockages in the entire reactor system. Because there are no flow blockages, it is possible to induce independently moving co-continuous gas and polymer phases. Therefore, thorough countercurrent stripping is possible. The technology for good mixing in such machines, with independent continuous gas and polymer phases, is described in our halogenation patents and in the Case III discussion below. Stripping in post processing and collection devices is also possible. Fig. 4 omits that the exhaust from the

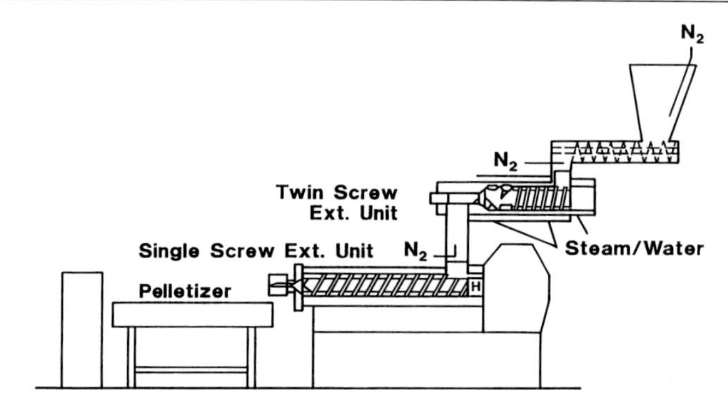

Figure 1.5 Continuous graft copolymerization equipment

cyclone actually passes through the same neutralization tower as the vent stream, and so all emissions are captured.

In light of this discussion, we can consider another technology which has been presented recently, for grafting of MA onto various polymers. Fig. 5 (Sakai, 1988) was shown as one of Japan Steel Works' reactor configurations. The machine, the basic mixer extruder, is also a non-intermeshing, counter-rotating twin screw. Added onto that is an intensive mixing zone, using convolutions on the two shafts which are not flights, but produce rather intensive mixing, like a Farrel Continuous Mixer. After the mixing section and a pump and intensive mixing, the melt is dropped into a single screw extruder to provide a significant amount of residence time to complete the reaction. Notice that the system is nitrogen blanketed at several points. A strand bath pelletizer is used which leaves the polymer open to the atmosphere at the point, with some possibility of loss of MA to the atmosphere if complete reaction has not been achieved. We would consider that a negative aspect of the process. There are no direct closed injection points into the extruders for MA or peroxide. All the feeds are injected into the first hopper. That leaves the emissions disadvantages mentioned earlier. One can see, therefore, some interesting positive and, perhaps, some negative aspects in this process.

As a final comparison, an alternative to the direct liquid injection of MA is the use of MA precoated materials in pellet form. This avoids some of the sublimation problem, but not all of it, a partial advantage. What is not avoided are the problems discussed above relative to our old process: poor control of the point of reaction and the melt temperature vs. the relative mixedness of the reactants and the peroxide. Excessive peroxide levels would tend to be required by this process, as were shown for our old process.

1.4 Case III – Halogenation of Butyl Rubber

1.4.1 General

Butyl rubber, a copolymer of isobutylene and isoprene, is used almost exclusively in tubes for tires because of its very low air permeability compared to natural rubber and other synthetic rubbers. For use in inner liners in tubeless tires, the butyl is modified through halogenation with chlorine or bromine to improve its compatibility and ensure covulcanization with the other materials used in the tubeless tire.

The manufacture of halobutyl practiced commercially by Exxon Chemical (Baldwin, 1979; Kresge et al., 1987), and others at several locations consists basically of:

1. low-temperature polymerization of isobutylene-isoprene as a slurry of particles in methyl chloride diluent, as for butyl rubber itself;
2. vaporization of the methyl chloride by hot hexane, which dissolves the particles;
3. direct halogenation of the butyl in the dilute hexane solution;
4. neutralization with dilute aqueous caustic solution;
5. precipitation of the dissolved product by steam to form a hot-water slurry of rubber particles, and simultaneously vaporizing the hexane; and
6. recovery of the rubber from the water slurry by a series of skimming, squeezing and vaporization extrusion steps.

The process is both complex and highly energy intensive.

1.4.2 Halogenation Studies

Several years ago, Exxon Chemical began to study different extruder halogenation processes on several polymers. One of these was directed to butyl rubber (Kowalski, 1989b). If successful it would eliminate the entire hexane circuit from the above sequence, with obvious projected benefits in energy, capital investment, operating manpower, environmental protection, product contamination, and considerably reduced plant complexity.

Exxon Chemical had previously conducted pioneering work in extruder degradation and free radical grafting of polymers, as described in Case 1 and 2. Halogenation of butyl in an extruder, however, was known to present a higher order of mixing and heat transfer problems:

- Nearly all of the isoprene must be halogenated.
- The reactant viscosities differ by 10^{10}.
- Butyl polymers are resistant to gaseous diffusion.
- The halogenation reaction must be conducted at low temperatures, compared with typical rubber extrusion temperatures.
- The target of the halogenation, isoprene, is only 2% of the polymer feed.

While halogenation of the isobutylene is possible, any measurable amount will produce a poorer quality product. Therefore, the halogen and polymer streams must be mixed at conditions that favor the halogenation of isoprene. Furthermore, 90-99% or so of the isoprene must be halogenated to produce an acceptable commercial product. If less is halogenated, the product will be less curable. The mixing conditions selected must bring the 2% isoprene fraction and the halogen into full contact.

It was known that the relative reactivity of isoprene and isobutylene to halogens greatly favored achieving the halogenation target. The degree of exclusivity and completeness of reaction to be achieved in the melt at a viscosity of about 10^8 cp, however, was a unique challenge.

Conventional polymer mixing technology requires that streams being mixed in the polymer melt state have closely similar rheological characteristics at the conditions of mixing. This is very far from true in the case of butyl rubber and halogen gas; for chlorine the viscosities differ by a ratio of about 10^{10}. At a commercial scale, however, we had previously demonstrated that good mixing of butyl rubber and nitrogen could be achieved (Kowalski, 1985b) in the melt in a particular extruder. While the mixing needed to dry butyl and other polymers is not on the intimate scale of halogenation, it is indeed thorough, as the product uniformity achieved showed. Moreover, it had been achieved within a short length in a twin screw extruder and at commercial rates. The configuration of that machine, a counter-rotating, non-intermeshing twin screw extruder, was the choice for beginning the development of an extruder halogenation process. But it was not the only configuration to be considered, as shown in the later discussion of the MIT work on configuration.

The commercial value of butyl and halobutyl rubber depends upon their resistance to gaseous diffusion. The need for halogen gas molecules to find (essentially) all the isoprene moieties in the residence time of an extruder (less than 2 min) is hindered by the resistance to diffusion of halogen gas through butyl. Calculation of the diffusion path length for the expected 2-min residence time showed that penetration of only about 10 μm or less could be attained in the laminar mixing mode of an extruder.

Good mixing can be achieved in extruders of any configuration by imposing a high level of shear, using high screw speed, and operating with shallow extruder channels, tight clearances, reverse flights, mixing pins, and special compounding screw sections. Using these techniques converts shear energy into sensible heat and raises the polymer temperature. In most cases, the relationship between the improvement in mixing and the level of shear energy imparted to the melt is only linear. With the melt viscosity of butyl, one would expect that good mixing would result in polymer temperatures between 200 and 250 °C.

Halogenation in the solution process is conducted at temperatures of 40 to 65 °C. While this temperature is not an absolute maximum for the chemistry, higher temperatures do favor isobutylene halogenation. The actual temperature level at which the chemistry becomes unfavorable was not known. However, some preliminary small-batch experiments in the bulk phase indicated that temperatures somewhat in excess of 100 °C would still be acceptable, still significantly below 200-250°.

For a fairly long time in the research program, the difference between the acceptable temperature range from a product quality point of view and the temperature range achievable with good mixing in a commercial, continuous extruder remained a most intractable challenge. The other problems of corrosion prevention, environmental control, the recovery of by-products, the purification of the product, etc., were rather straightforward and do not need to be reviewed. Solutions to these problems, however, were assisted by improvements in extruder stripping and devolatilization, developed during the program. These progressed to the point, as shown in a sequence of our early patents (Newman and Kowalski, 1983; Newman and Kowalski, 1984; Kowalski et al., 1985b), where it was possible to eliminate aqueous neutralization after halogenation, Step 4 in the solution sequence. Excess halogen and the halogen acid by-products could be satisfactorily stripped from the melt directly in the gas phase.

1.4.3 Chemistry

The competing reactions in the halogenation of isoprene and isobutylene are as follows:

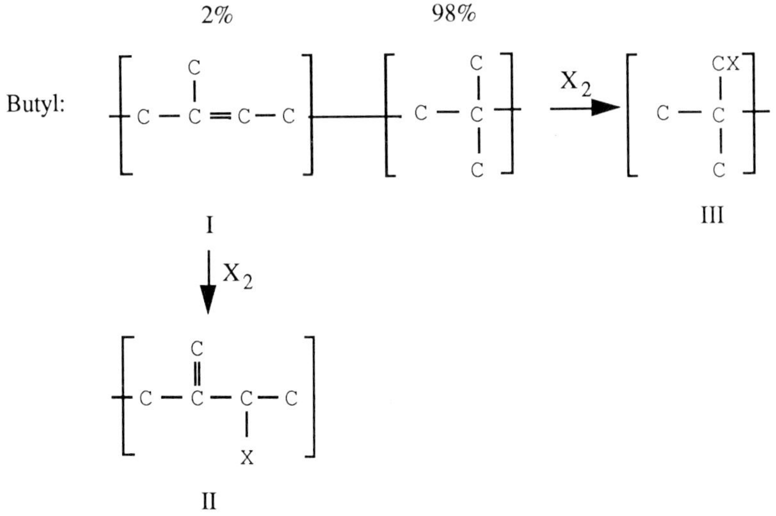

2% 98%

Butyl: [–C–C(C)=C–C–] (I) [–C–C(C)(C)–] $\xrightarrow{X_2}$ [–C–C(CX)(C)–] (III)

I $\downarrow X_2$

[–C–C(=C)–C(X)–C–] (II)

Our target was, not to produce structure III and to halogenate 90% or more of the isoprene (structure I) to structure II. Structure II is the exomethylene form of halogenated butyl, the major component of the commercial product produced in solution.

Several other isomers and dihalides are possible products of the halogenation of the isoprene moiety. Significant fractions of these isomers were, in fact, produced in the early experiments. Most, but not all, are either not curable by conventional halobutyl cure systems or respond poorly to them. An interesting exception is the halomethyl isomer:

$$\begin{array}{l} \quad\;\; X-C \\ \qquad\quad | \\ [C-C=C-C] \end{array}$$

which is curable and gives a somewhat different product when present at high levels relative to the exomethylene isomer.

We were able to develop extruder reaction processes that could control the relative level of these active isomers as well as suppress the formation of the inactive or poorly active ones (Kowalski et al., 1986; Gardner et al., 1987). Structure II is favored in general by rapid, efficient "mixing" of the reactants and by minimizing the highest temperature reached in the reaction zone. Our discussion here will emphasize the production of a product containing primarily exomethylene isomer and comparable to the historical commercial products.

1.4.4 Extruder Mixing

The challenges presented were resolved by three developments:

- Learning how to move two co-continuous phases, gas and polymer, independently through the entire reaction zone;
- Creating ways to maximize surface renewal between the two phases at the boundary, as contrasted to conventional dispersed "mixing" of one phase discontinuously in the other; and
- Using a specific fundamental extruder reactor configuration that inherently develops good "mixing" (or polymer surface and interface generation), but at a lower energy input than other reactor configurations. This meant that good mixing could be achieved at the low temperature required by the chemistry.

Work on the first two developments was conducted exclusively at Exxon Chemical. Our work on the extruder reactor configuration built upon studies at the Massachusetts Institute of Technology (MIT), which was done under the direction of Prof. L. Erwin (Erwin, 1978a; Erwin 1978b; Gailus and Erwin, 1981; Howland and Erwin, 1983; Bigio and Erwin, 1984). The three developments are, in fact, intertwined, as the following discussion shows.

Fig. 6 is a schematic representation of the mixing problem encountered in the extruder. Flow in a simple single extruder channel is shown in the upper sketch in the usual, unwrapped form. The barrel velocity vector is shown

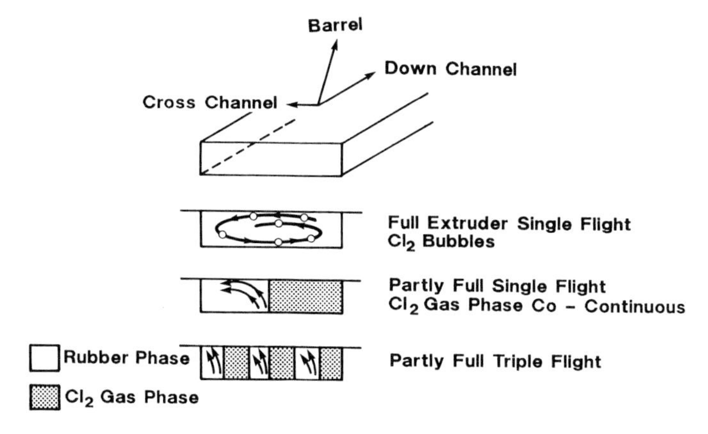

Figure 1.6 Extruder mixing mechanisms

at the screw helix angle to the down-channel component. The cross-channel component provides the means for mixing within the channel.

The other sketches are cross sections of flow patterns within the channel perpendicular to the down-channel direction. The second sketch shows the development of a creeping laminar flow vortex from the action of the cross-channel component. Many experimental studies have shown that this vorticular pattern is typically created in an extruder channel full of polymer, even being recreated downstream of a flow disturbance (Gailus and Erwin, 1981).

Also in the second sketch, the halogen vapor bubbles (for example Cl_2) are shown moving through the polymer melt with the vorticular pattern. This represents the regions in which many of our early studies were conducted in an attempt to achieve the required mixing within the constraints of the requirements previously discussed. The impact of screw speed, channel dimensions, flow restrictions, compounding sections and reverse flights were extensively studied. Some of these experiments (and equipment modifications) showed that partial conversion to the exomethylene form could be achieved, but the levels were too low.

All of the products had unacceptable levels of other isoprenyl halide isomers and of chlorinated isobutylene. Fig. 7 is typical of one of the best NMR traces obtained during this series of early studies. Compared with typical commercial chlorobutyl prepared in solution, it shows considerable quantities of by-products. (The proton nuclear magnetic trace is at 400 MHz with TMS as the reference compound.)

Cure studies of the products at this time showed them to be of little value. Temperatures in the reaction zone were typically 180 °C or higher. Under milder conditions of shear, lower temperatures were achieved. The mixing of the two phases, however, was so poor that analysis of the product showed large fractions of unreacted isoprene and very little of the target structure.

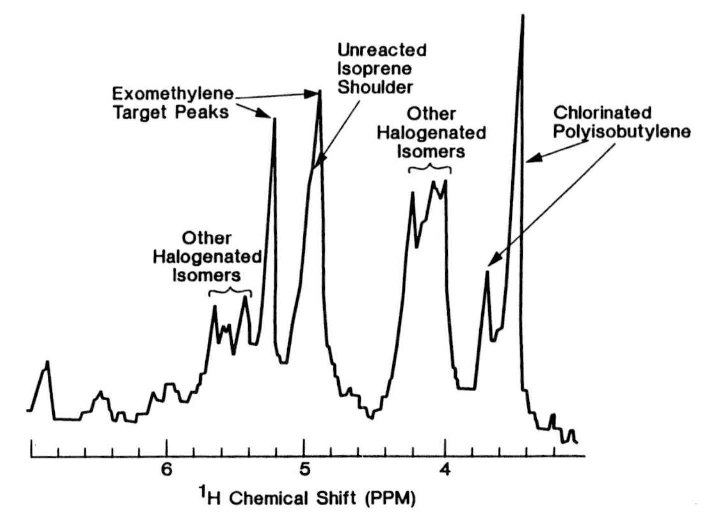

Figure 1.7 Early product NMR spectrum

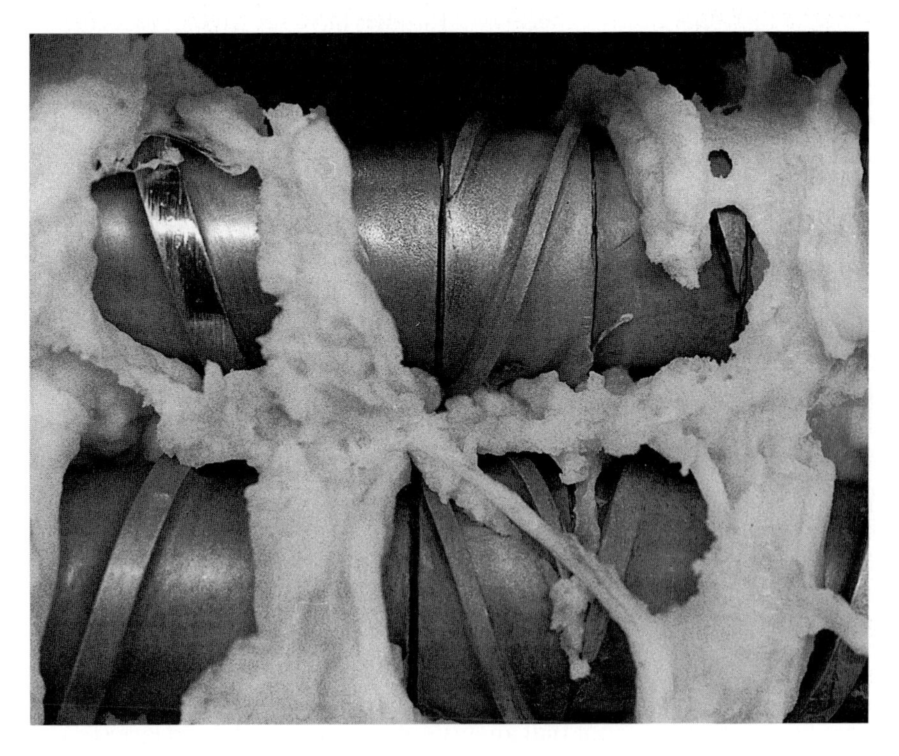

Figure 1.8 Segment of reaction zone (standard single flights without slots)

A major step toward an acceptable commercial composition came with the experiments using a reaction zone that is only partly filled throughout its entire length, i.e., two co-continuous phases. This is represented by the third sketch in Fig. 6, which shows a vortex within the polymer phase and an open down-channel passage filled with the gas phase. Contact between the two phases could take place only at the interface, which is shown schematically. Combinations of screw design, screw speed, and polymer flow rate with the counter-rotating, non-intermeshing (CNRI) configuration were found that would maximize the generation of interfacial area without *ever filling* the channel.

An example of the modifications made to improve interfacial area generation that produced an improvement in isoprene conversion is shown in the last sketch of Fig. 6. Using double- or triple-flighted sections (as shown) in the extruder multiplied the interfacial area even when all other conditions were held constant. Other modifications, including slotted flights, were also made.

Figs. 8 and 9, which show two segments of the same reaction zone, along the same screws, illustrate the desirability of using slotted flights for generating the interfacial area. (The screws were removed from the extruder at steady-state conditions by making an instantaneous shutdown.) Fig. 8 is a segment of

Figure 1.9 Segment of reaction zone (slotted flights)

standard single flights without slots; a segment containing slotted flights is shown in Fig. 9. (Flow is right to left.) The chopping effect of the slots is evident as bits of rubber fall back from a flight to its upstream neighbor. Also evident is the increased polymer surface-to-volume ratio.

When any part of the reaction zone becomes filled (which is called "choking") because of local low-screw-conveying capacity, low screw speed, a high polymer flow rate, or any flow obstruction, a "hot spot" is created at the choke point. The product then shows immediate evidence of choking by dark discolorations, which analysis shows to be a result of isobutylene halogenation. High isoprenyl halide contents are also evident.

Confirmation of the choke point-hot spot phenomenon was obtained by shutdown and quick removal of the screws on several occasions after discolored product appeared in the product. These invariably showed a segment of the screws in the reaction zone to have full channels. The dark (usually quite black) color would begin at this point.

Calibrated optical fiber probes were used in pilot-plant studies to measure temperature; as many as five were located in the reaction zone. When a probe (or probes) was positioned downstream of the choke point, the temperature trace for the probe would show a large temperature increment over the others, often 15 °C or more. If choking resulted from an operational change because of decrease in screw speed or a polymer rate increase, then the trace would show a sharp rise as a function of time.

A simplified expression for viscous energy generation, dH, by a filled screw is given by:

$$dH = C'QdT + QdP = \varrho \eta N^2 dz + \alpha N dP$$

where C' is the heat capacity per unit volume of the melt, Q is the volumetric throughput, ϱ and α are functions of machine design parameters, N represents the screw speed in RPM, and η is the viscosity, dP and dT are pressure and temperature rises over the length element dz.

The final term, αNdP, is zero if the reaction zone is only partly filled and has no choke points; therefore no pressure generation.

Another important value of partly filled (starved) operation is evident in the N^2 term. Throughput for a given extruder design is independent of N only for the starved condition, and N is then available as a process variable that is increased only enough to produce the desired interface generation, without changing the throughput. This defines the maximum end of the range of N available as a process parameter. The minimum screw speed end is determined by the devices used to generate surface, but reducing screw speed does not impact directly upon polymer throughput until a choke point is reached. Depending on the nature of the feed, throughput is controlled by an upstream metering device. Screw speed is, of course, also available to control residence time in the reaction zone.

1.4.5 Operation

A schematic representation of one of the halogenation extruders is shown in Fig. 10. The rubber feed is metered into the feed zone, passing through a dynamic seal designed to withstand the required reaction zone pressure without developing excessive temperature. Halogen gas is injected into the reaction zone, with the excess halogen and the halogen acid by-product being removed at the vent. Countercurrent stripping is possible because of the open channels existing with the co-continuous phases. Different screw configurations are shown schematically within the reaction and stripping zones.

Stripping within the reaction zone reduces the possibility of adverse reaction at the potential hot spot just upstream of the second dynamic seal at the end of the reaction zone. A second stage of countercurrent stripping is effective for additional removal of the halogen and halogen acid.

The co-continuous phases are the polymer and the low-viscosity reactant (whether liquid or gas). They are independently controllable throughout the reaction zone and offer control options that can be useful. For example, for this reaction, halogenation kinetics are sensitive to pressure on the gas side. Excellent control over the gas pressure is possible with the design in Fig. 10, via pressure control of the halogen injection and the vent systems.

The polymer phase moves by drag flow, which is determined by its high viscosity, the screw design, and the screw speed used. The low-viscosity phase moves independently of the polymer phase and of the usual extruder operating parameters - its rate is controlled by the external system for injecting the gas and by the flow characteristics and backpressure of the external vent system. The gas flow rate, its pressure, and its concentration, therefore, are all uniformly controllable and can be varied by external means (the concentration can be controlled by adding diluents). It is not influenced by the polymer phase, by its flow characteristics, by its surface generation modes, nor by product changes in rheology and temperature that result from the halogenation. Our studies show that a pressure of 10 atm (1 MPa) or more on the gas side and substantial dilution with inert gas are advantageous.

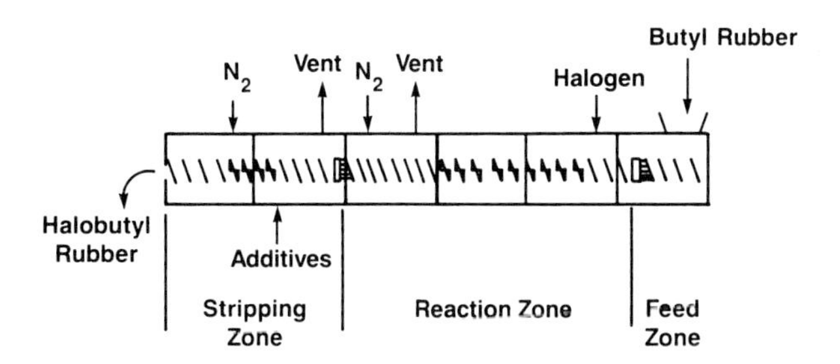

Figure 1.10 Extruder halogenation process

Our earlier commercial experience in drying rubber in the CRNI configuration (Kowalski, 1985b) had demonstrated the excellent mixing characteristics of these machines when gases are injected into a viscous polymer melt. The MIT studies, however, showed that this extruder was also an energy-efficient mixer as compared to other extruders. The polymer-gas interfacial area, therefore, could be generated at reduced temperatures. Furthermore, from those studies it was proposed that high levels of interface generation sufficient to expose all the isoprene to halogen could be achieved without blocking the channel with restrictive mixing or kneading designs.

The studies at MIT showed that channel drag flow in an ordinary extruder produces only linear shear mixing, as shown by Gailus and Erwin (1981). In a simple shear mixer the generation of the interfacial area, A_f/A_o, is proportional to the imposed strain, where A_f and A_o are the final and initial interfacial area values. If shearing continues, the interface becomes more and more oriented in the direction of the shear strain and its extension is related linearly to the extent of strain.

If the interface can be reoriented relative to the existing shear direction, then a multiplying relation of mixing as a function of strain can be achieved. Erwin's early studies (Erwin, 1978) confirmed this by actually lifting blocks of sheared black-and-white material from the shear field of an annular Couette-type instrument and orienting them 90° (a "perfect" orientation) from the strain direction before reimposing the strain. Gailus and Erwin (1981) applied these principles to an extruder channel; they showed that the flow disruption caused by a mixing pin in the extruder channel is less than perfect but that substantial reorientation and multiplication of the interfacial area does occur.

1.4.6 Configuration and Reorientation

In studying different extruder configurations, Howland and Erwin (1983) found that the reorienting feature of the CRNI extruder could be related to the "nip" region. Fig. 11, a cross section of the CRNI machine, shows the nip region

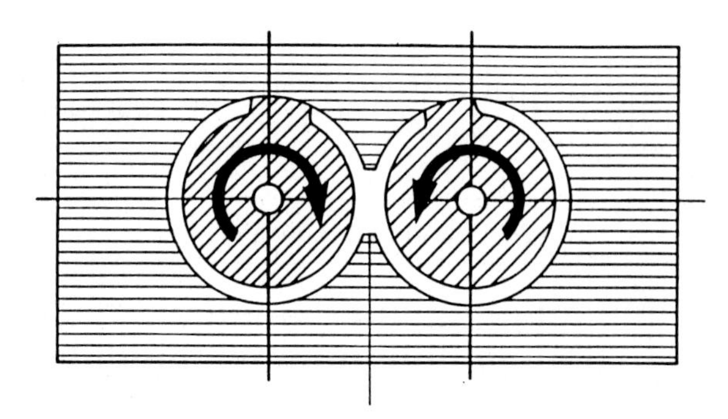

Figure 1.11 The counter-rotating non-intermeshing twin screw extruder configuration

between the screws and the open areas above and below the nip formed by truncation of the two barrel apices.

The interfacial area generated in this twin screw design is compared to a single screw in Fig. 12. "Mixing number" refers to a visual count of the number of striations, which is a measure of interfacial area. Note that the single screw relation is linear as predicted by theory, but that the CRNI twin screw mixing tends toward exponential with strain, as given by screw turns.

Fig. 13 shows a cross section of polymer taken from one of the side channels of the twin screw machine that mixed black-and-white polymer. Disruptions

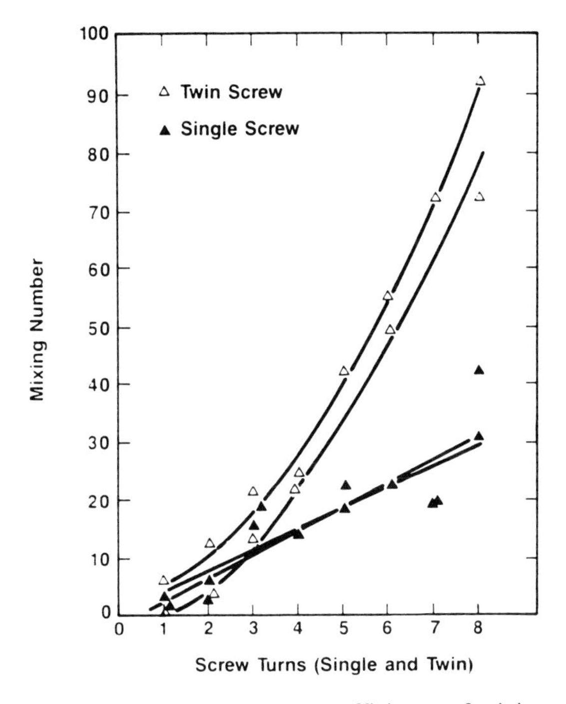

Figure 1.12 Apparent energy efficiency of mixing

Figure 1.13 Evidence of partial flow reorientations in the CRNI twin screw extruder

Acknowledgment

The extruder halogenation work described here was done by a project team consisting of the author, Dr. W. M. Davis, and Dr. N. F. Newman, the latter two also of Exxon Chemical Company at Linden, NJ.

1.5 Economic Benefits of Extruder Reactors

In this section a generic calculation is presented which shows the approximate level of economic benefit in manufacturing cost for the case of an extruder reactor process in comparison with a typical CSTR process, done in hydrocarbon (HC) solution. All the bases are clearly shown, so that the numbers are readily adjustable to fit somewhat different examples.

The benefit of reduced real estate charges is pointed out, but no value is assigned, because this value can vary enormously depending on company policy on manufacturing cost charges, and on space limitations in different locations. Notice is given so that this item is not neglected in any actual calculation made from this generic estimate.

Differential Economic Comparison
Solution vs. Extruder Reaction

Basis
Generic Polymer Modification Reaction
Solution Conc: 15% Wt.
HC Solvent: $\lambda_v = 80$ Cal/g
ΔH_{mod}: Negligible
Capacity: 20 M lb/yr
Energy: Cond. Steam, 0.7 ¢/lb
Labor Difference: (1) Operating Post
(0) MPT

Note that a modification reaction is chosen, not a polymerization, as a more usual case. Note also that labor credit is taken for only one operating post and no additional assigned technical professional (MPT). This is considered a bare minimum in view of the process differences shown in the flow sheet, Fig. 15.

The processes are compared in Fig. 15 which illustrates the commonalities and differences on the same flow sheet. Operations unique to the solution process are shown inside the solid circle. Note that this circle also includes the equipment most likely to be emissions sources. The steps up to Polymer Slurry

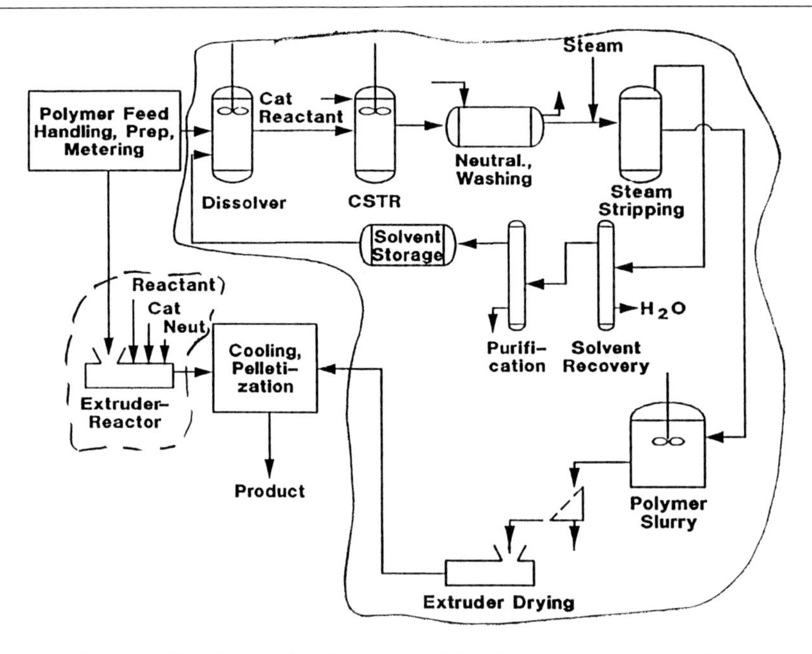

Figure 1.15 General polymer modification process comparison

Economic Differences

	$/Yr
1. Energy • Solvent Vaporization $\frac{7 \times 80 \times 1.8 \times 0.7}{1000}$ = 0.7 ¢/lb Polymer	$140,000
• Pumping, Agitation, Etc. at 10% of λ_v	14,000
2. Depreciation & Maintenance Say $\Delta I \cong \$5$ M	500,000
3. Labor 4-5 Men/Post	400,000
4. Solvent Loss 0.1%, 300 T/Yr	300,000
5. Real Estate	[ ]

are potential leak sites of volatiles to the atmosphere from all the seals in equipment flanges, instrument ports, pumps and storage tank air displacement. The steps Polymer Slurry and Extrusion Drying are sites for water contamination which require solids separation and disposal as well as water treatment before recycling or discharge.

The dashed circle shows equipment unique to the extruder process. The remainder of the equipment is common to the two processes.

Calculations are shown in the table below. The difference in capital investment between the processes is taken as a conservative $\Delta I = \$5M$, which provides the basis for Item 2, Depreciation and Maintenance at 10% of Investment.

The subtotal of $1,354,000/yr, excluding real estate charges, results in an onsite manufacturing cost saving of 6.8 ¢/lb for the 20 M lb/yr capacity in the Basis. This is a significant conservative benefit, to which must also be added the various modes of credits for emission reductions.

Chapter 2

Process Analysis from Reaction Fundamentals

Examples of Polymerization and Controlled Degradation in Extruders

By Marino Xanthos, Polymer Processing Institute at Stevens Institute of Technology, Castle Point on the Hudson, Hoboken, NJ 07030

2.1 Introduction

The present chapter considers examples of industrially important reactions that can be carried out on a large scale in single and twin screw extruders. Two general types of reactions are considered, namely polymerization and polymer modification. Polymerization examples are of the free radical and anionic types, whereas the modification example is a controlled degradation free radical reaction. For a given reaction, a methodology is presented that uses existing or specially developed information on the chemistry of the system to define the requirements of the continuous process and relate to product characteristics. Emphasis is placed on the use of laboratory data - often determined in non-extrusion equipment - on reaction mechanisms, reactants concentration, and kinetics, to select extrusion conditions, configure process equipment and predict the properties of the extruded product. A variety of reactive extrusion processes described in the recent patent and open literature are analyzed in terms of process characteristics and their relation to the chemistry of the system.

2.2 Peroxide Controlled Degradation of Polypropylene

Conventional reactor polypropylene has usually a relatively high molecular weight (MW) and broad molecular weight distribution (MWD). The resulting high melt viscosity and elasticity limit its efficient processing for some applications and can impair product quality. Post-reactor modification of MW and MWD through free radical reactions may be initiated by oxygen (See Sections 1.2 and 4.6.1), or through the action of peroxides as practiced exclusively nowadays. In this so-called controlled rheology polypropylene, (CR-PP), random chain scission reactions reduce MW and narrow MWD, thus reducing viscosity and elasticity. The CR-PP has improved processing characteristics for specialized applications. Examples of rheological advantages are lower melt processing temperatures, and higher speeds in melt spinning of very thin fibers, extrusion of thin films, as well as thin walled injection molding.

2.2.1 Reaction Characteristics

2.2.1.1 Chemistry

The degradation of PP with peroxides, ROOR, in the melt is believed to occur through a series of free radical reactions involving steps such as initiation, scission, transfer and termination (Dorn, 1985; Tzoganakis et al., 1989b).

Reduction in MW occurs through hydrogen abstraction from tertiary carbon atoms of the polymer backbone - preferably by the peroxide radicals - and the subsequent chain cleavage (beta-scission) of the formed polymer radical. The process can be terminated by recombination or disproportionation of the polymer free radicals. Oxygen and the presence of processing stabilizers and antioxidants are known to interfere with the degradation process through radical competing reactions. The following is a simplified reaction scheme:

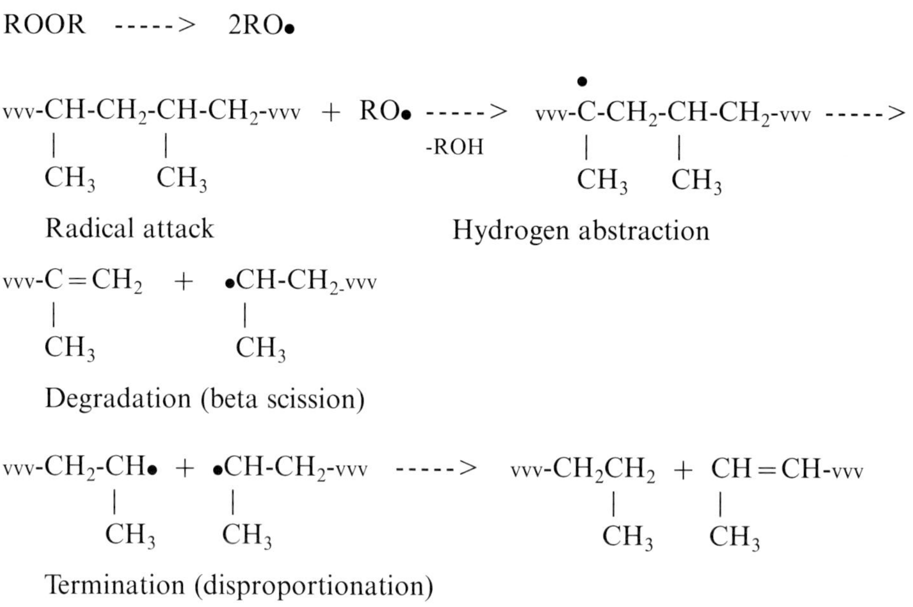

ROOR -----> 2RO•

vvv-CH-CH_2-CH-CH_2-vvv + RO• -----> vvv-Ċ-CH_2-CH-CH_2-vvv ----->
(-ROH; each CH carries a CH_3 branch)

| | | |
CH_3 CH_3 CH_3 CH_3

Radical attack Hydrogen abstraction

vvv-C=CH_2 + •CH-CH_2-vvv

| |
CH_3 CH_3

Degradation (beta scission)

vvv-CH_2-CH• + •CH-CH_2-vvv -----> vvv-CH_2CH_2 + CH=CH-vvv

| | | |
CH_3 CH_3 CH_3 CH_3

Termination (disproportionation)

2.2.1.2 Materials/Concentrations

PP homopolymers of low melt flow rate (MFR) (less than 1 g/10 min at 230 °C/ 2.16 kg) are most commonly used in the form of powder or granules obtained directly from the reactor. Organic peroxides (mostly dialkyl) are employed at concentrations ranging from 0.001 up to 1 phr, depending on the desired final MFR; they are available in liquid form (neat or diluted with an inert carrier), as crystalline solids, as free-flowing powders (liquids deposited on an inert substrate), and as polyolefin masterbatches.

The selection of the peroxide is usually based on the temperature corresponding to a given half-life, i.e. the time required for the decomposition of 50% of the peroxide. Dialkyl peroxides with 1 hr half-life temperatures ranging from 135-155 °C are commonly used. At the typical process temperatures of 180-240 °C these peroxides are capable of generating sufficiently reactive radicals to complete the process within a short residence time. Table 1 lists five widely used peroxides in order of increasing thermal stability. Peroxide **C** marketed

Table 2.1 Dialkyl Peroxides Used in the Controlled Degradation of PP

Peroxide		1 hr. Half-Life Temp.*, °C
A.	Bis(t-butylperoxy) diisopropyl benzene	137,0
B.	Dicumyl peroxide	137.0
C.	2,5-dimethyl-2,5-di(t-butylperoxy) hexane	140.3
D.	Di-t butyl peroxide	149.1
E.	2,5-dimethyl-2,5-di(t-butylperoxy) hexyne-3	151.8

* From decomposition kinetic data in dilute decane or dodecane solutions (Atochem, 1990).

as either Lupersol 101 (Atochem), Trigonox 101 (AKZO Chemie America) or Interox DHBP (Peroxide-Chemie GmbH) is listed under "Food Additives" in the U.S. Code of Federal Regulations (Callais, 1989) and in Recommendation VII of the German Health Agency (Dorn, 1985). Volatility at process temperature is another important consideration in the peroxide selection; peroxide **D**, for example, is considerably more volatile (b.p. 111 °C/760mm Hg) than either peroxide **C** or **E** (b.p. 249 °C/760 mm Hg and 243 °C/760 mm Hg, respectively), (Pennwalt,1989a).

The dilute solution half-life data of Table 2 give an indication of the relative thermal activity of the three most common dialkyl peroxides at processing temperatures. For polyolefin reactions such as the controlled degradation of PP and the cross-linking of PE, it is generally assumed that the overall rate is controlled by the peroxide decomposition rate. Thus, from first-order decomposition kinetics, processing times should ideally be equivalent to about six or seven half-lives in order to ensure 98-99% decomposition. The data of Table 2 suggest that peroxides **B** and **C** can complete the reaction in less than one minute at 200 °C, whereas peroxide **C** will be consumed in less than a minute at 220 °C. The dilute solution half-life data do not necessarily correspond to actual half-lives under extrusion conditions, and should be used only for screen-

Table 2.2 Half-Life of Dialkyl Peroxides in Dilute Solutions

Peroxide	Half-Life in Dilute Solution* seconds			
	180 °C	200 °C	220 °C	240 °C
B. Dicumyl peroxide	50	9	1.8	0.4
C. 2,5-dimethyl-2,5-di(t-butylperoxy) hexane	68	12	2.4	0.6
E. 2,5-dimethyl-2,5-di(t-butylperoxy) hexyne-3	216	36	6.8	1.4

* Estimated from decomposition kinetic data in dilute decane or dodecane solutions (Atomchem, 1990).

ing purposes. Half-lives are known to depend on type and viscosity of the medium, the distribution of the peroxide in the melt and the presence of interfering species. Manufacturers' data suggest that the half-lives of peroxides **B**, **C** and **E** in molten LDPE are two to three times longer than in decane or dodecane, and much longer than in benzene: for example, the estimated half-lives of peroxide **C** at 190 °C are 12 s in benzene, 28 s in dodecane and approximately 71 s in molten LDPE respectively (Pennwalt, 1989; Pennwalt, 1989b; Atochem, 1990).

The homolytic cleavage of dialkyl peroxides to active alkoxy radicals may be followed by a secondary reaction involving the formation of other active alkyl radicals as well as non-reactive volatile species. High temperatures and low reactivity media are known to favor secondary decomposition (Dorn, 1985) which, in turn, affect the overall peroxide efficiency. The volatile decomposition products that need to be removed during processing depend on both polymer type and temperature. t-butyl alcohol is the major product of the decomposition of peroxides **C**, **D** and **E**. Ketones, alcohols, saturated hydrocarbons, carbon dioxide and carbon monoxide are among other substances reported to be formed in inert media (Pennwalt 1989a). 2,2 di(t-amyl) peroxy propane (Lupersol 553-M75, Atochem) with 1 hr half-life temperature of 128 °C does not release t-butyl alcohol and may be used for applications where the odor of the latter is objectionable; however, at equal concentrations, it is less efficient than the common dialkyl peroxides (Ehrig and Weil, 1987).

2.2.1.3 Kinetics – Fundamental Studies

During 1988-1989, numerous studies on the fundamentals of PP degradation in laboratory size single screw extruders have been reported by Tzoganakis et al. and Suwanda et al. (See Section 4.6.1 for a complete list of references). In experiments conducted with Peroxide **C** in the temperature range from 180 to 220 °C, the experimentally determined average residence time of 1.5-3 min was mostly controlled by the single screw speed. The M_w, M_n and M_z of the product decreased with increasing peroxide concentration and could be predicted from the developed models. Residence time distribution (RTD) was found to increase with increasing peroxide concentration and decrease with temperature. In modelling, it was assumed that only the t-alkoxy radicals were active to degrade the polymer and that the rate controlling step was the first order decomposition reaction of the peroxide. Predictions in terms of mass flow rate, pressure, temperature, MWD and RTD were in satisfactory agreement with the extrusion runs suggesting that the model could find industrial use (Tzoganakis et al., 1988a).

A common finding of the above studies was that the initiator concentration and its efficiency were the most important variables affecting MW, MWD and, of course, the rheological properties of the product. (See also Tzoganakis et al., 1989c). MFR was found to be a linear function of peroxide concentration, at least up to about 0.1%. Initiator efficiency was reported to be reduced by several factors such as the presence of stabilizers, side reactions (Hudec and Obdrzalek, 1980), and as a result of insufficient mixing. Other investigators showed that the overall efficiency of the degradation process was also dependent upon the type of peroxide (Nettelnbreker and Stoehrer, 1988) and could be related to peroxide losses during processing, in particular, for volatile peroxides such as **D** (Peroxid-Chemie GmbH, 1980).

Experiments in laboratory intensive mixers (Haake or Brabender type) were shown to give valuable information on the effects of varying process parameters on the kinetics of the process, for a given resin/peroxide system. Torque vs. time curves are usually considered equivalent to curves of viscosity or MW vs. time and, as a result, they may be used to study variables such as temperature, rpm, reaction atmosphere, peroxide concentration, and method of addition. Fig. 1 shows typical torque vs. time curves obtained under a nitrogen blanket

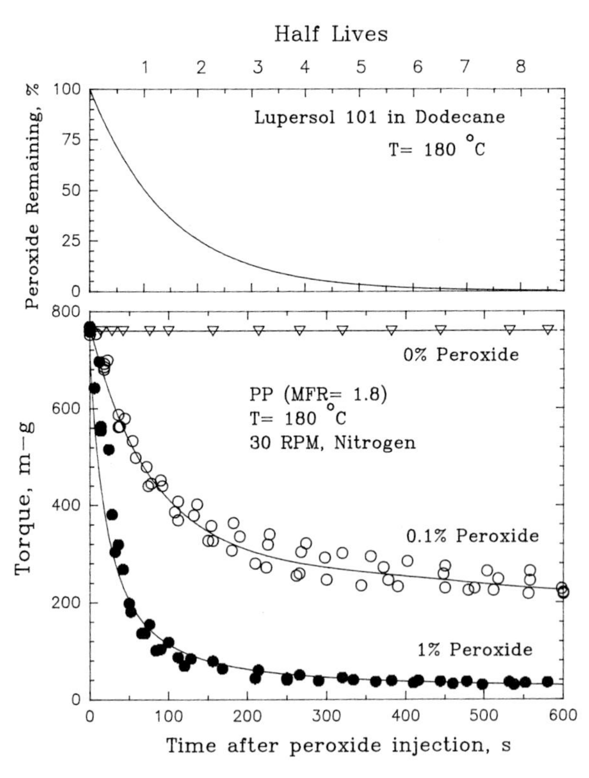

Figure 2.1 Torque vs. time curves obtained after the addition of peroxide during PP controlled degradation experiments conducted in a laboratory intensive mixer. Manufacturer's peroxide decomposition data are also shown for comparison

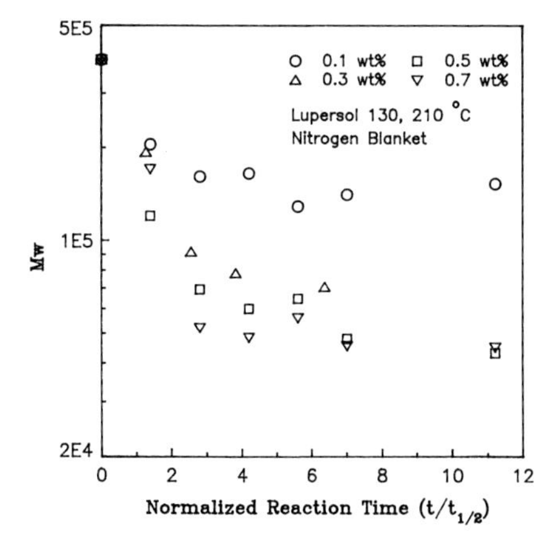

Figure 2.2 The effect of reaction time (normalized for initiator half-life) on the M_w of PP films degraded with different amounts of peroxide in an oven

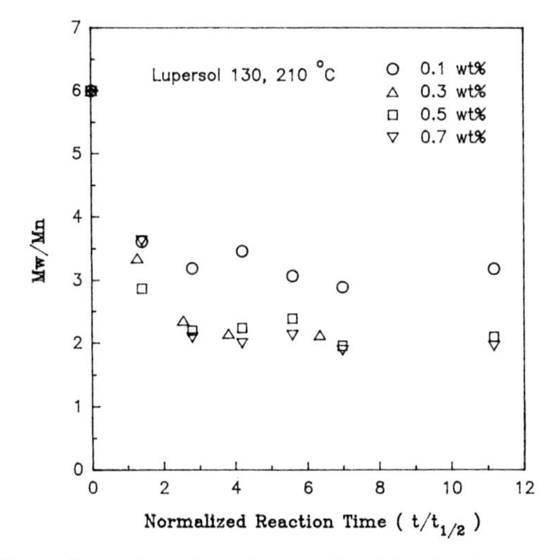

Figure 2.3 The effect of reaction time (normalized for initiator half-life) on the polydispersity of PP films degraded with different amounts of peroxide in an oven

at two different peroxide concentrations (Yu et al., 1990). The curves may be shifted toward lower or higher torque values respectively, depending on the presence of air or interfering stabilizers. The peroxide can be injected onto the melt or premixed with the powder resin. Trolez et al., (1990), have shown that the latter method yields higher peroxide **C** utilization efficiency since mixing effects are reduced; the same authors demonstrated that in the 180-210 °C range the final MW obtained through degradation with peroxide **C** was temperature independent.

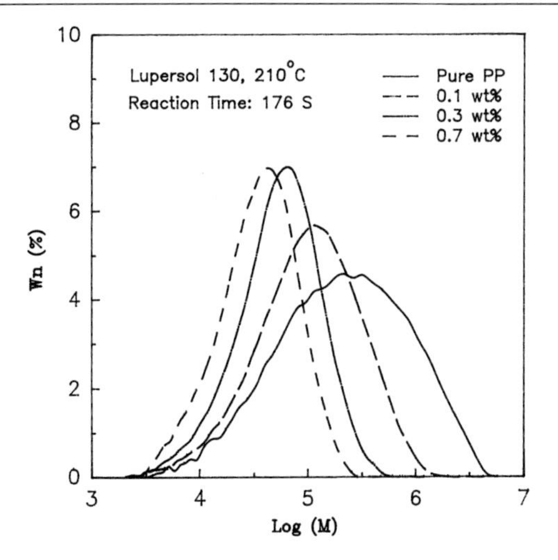

Figure 2.4 MWD curves by GPC of PP degraded with different amounts of peroxide

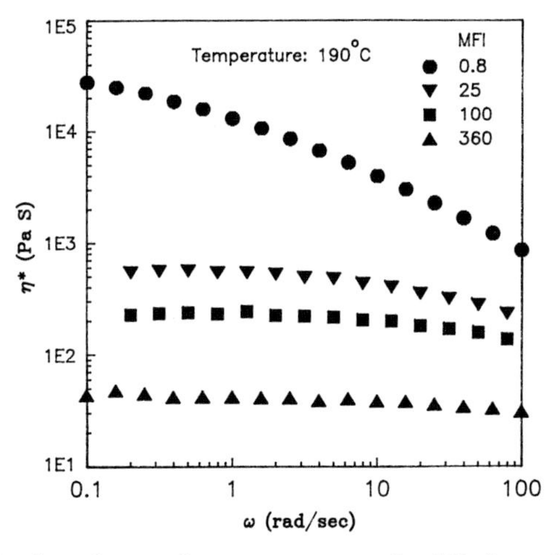

Figure 2.5 Dynamic viscosity vs. frequency curves for PP degraded in a batch mixer to different final MFR's; comparison with unmodified PP (MFR = 0.8)

Peroxide efficiency without the additional complications arising from flow and mixing effects may be assessed from kinetic data on systems containing an intimate and uniform peroxide distribution. Ryu et al. (1989) and Xanthos et al. (1989) prepared thin sintered PP films containining PP powder precoated with various amounts of peroxide **E** and followed the reaction by analyzing at various intervals for MW. The results shown in Fig. 2 suggest that the limiting value of M_w was reached at times corresponding to about 4-7 half-lives of peroxide (94-99% decomposition) as measured in dodecane dilute solution

(Atochem, 1990). The limiting M_w value was independent of initiator concentration at concentrations higher than 0.4-0.5%; MWD as measured by the ratio M_w over M_n appears to reach its limiting value of two after about 4 half-lives, at initiator concentrations higher than 0.3% (Fig. 3). These results suggest that the overall rate constant of the degradation process may be considered, for all practical purposes, equivalent to the peroxide decomposition rate provided that peroxide segregation is minimized and appropriate values for half-lives are used. Examples of data from the same work are MW curves by GPC showing the removal of the high MW tails through the peroxide initiated reaction (Fig.4). Examples of rheological curves illustrating the increasing tendency for Newtonian behavior with increasing extent of degradation are shown in Fig. 5.

2.2.2 Reactive Extrusion Process Analysis

2.2.2.1 Defining the Process Requirements

The task of defining the requirements of reactive extrusion processes is undoubtedly simplified through the understanding of the reaction characteristics and its kinetics from laboratory or pilot scale experiments. The following discussion shows how the chemistry of the PP controlled degradation may be related to the process requirements:

a) The increasing growth of CR-PP in high-volume fiber and film applications and the resulting market demand require the use of *high extrusion rates*; the fast kinetics of the reaction can, undoubtedly, contribute to such high throughput rates.
b) The length of the reaction zone, the operating temperature and the average residence times are determined by the overall rate of the degradation that, in turn, is strongly dependent upon the decomposition characteristics of the peroxide. An *estimate of residence time* could be based on the respective peroxide half-life at operating temperatures, although caution should be used in applying dilute solution data to much higher viscosity media under dynamic extrusion conditions.
c) The overall degradation is strongly dependent on the presence of oxygen as well as additives that may act as radical scavengers. Thus, the use of *non-interfering additives* under *inert atmosphere* appears to be a necessary process requirement.
d) Given the very large weight ratio of polymer/peroxide (as high as 10,000:1), *accurate metering* and *uniform distribution* of the liquid or solid peroxide in the melt is extremely important. Metering, premixing and melt mixing methods and equipment are all to influence the efficiency of the process, as well as the quality of the product.

e) The use of common peroxides results in the formation of volatile by-products such as alcohols and ketones; thus, *devolatilization* prior to pelletization would be a requirement to ensure product purity.
f) The *quality of the product* may be directly related to its MFR since this index is dependent on the initial amount of peroxide and, of course, its efficiency of utilization. Ease of pelletization would also be related to MFR, decreasing at very high MFR values due to the excessive fluidity of the extrudate.

2.2.2.2 Meeting the Process Requirements – Industrial Applications

The following section presents various industrial approaches that have been used to meet the requirements of continuous post-reactor peroxide controlled PP degradation processes. The sources of information are select patents and open literature publications, some listed also in Section 4.6.1.

- *High Conversion Rates.* High throughput rates have been achieved with large capacity equipment, such as a 12 inch diameter Egan face-cutting single screw extruder operating at 5 tons/hr (Davison, 1986), and various types of co-rotating twin screw extruders converting 2-25 tons/hr (Dorn, 1985; Nettelnbreker and Stoehrer, 1988). Increased rates may also be related to more efficient plasticization through proper screw design, melt pressurization through the addition of gear pumps and the use of high volume pelletizers (Andersen and Kenny, 1989).
- *Short Reaction Time.* Residence times of less than two minutes are feasible through the proper peroxide/ temperature combinations and the adjustment of processing conditions. Manufacturer's half-life data have been useful to predict reaction temperature and the length of the reaction zone (Greene and Pieski, 1964; Morman and Wisneski, 1984). The length of the reaction zone may be varied by varying the location of peroxide addition i.e. at the hopper, or downstream after complete melting of the polymer (Davison, 1986).
- *Accurate Metering and Intimate Mixing of Peroxide.* The relatively very small amounts of peroxide need to be metered accurately and mixed intimately in the melt. Liquid peroxides may be pumped directly into the extruder hopper (Peroxid-Chemie GmbH, 1980); however, more accurate metering and uniform distribution was achieved by masterbatching with PP powder to a 5% concentration in a high speed mixer or ribbon blender, and combine this stream in the hopper at, say, 200 kg/hr, with reactor powder at, say, 9800 kg/hr (Dorn, 1985; Andersen and Kenny, 1989). The streams were accurately metered through gravimetric feeders (Fritz and Stoehrer, 1986); provision could also exist for peroxide concentration adjustment, usually in processes equipped with on-line rheometers. Improved peroxide distribution can also be achieved by predispersing as a PP concentrate at low temperatures in a compounding extruder (Morman and Wisneski, 1984). The unreacted peroxide would then decompose in a second pass through an extruder operating

at normal process temperatures to produce directly fibers or films of very high MFR without the need for pelletization; it is understood that pellets of such high MFR would be almost impossible to produce with conventional pelletizing equipment. In one-step operations, improved distribution and mixing were claimed to be achieved by diluting the peroxide in a hydrocarbon solvent at 1:25 to 1:10 ratios (Greene and Pieski, 1964), or in mineral oil at 1:1 ratio (Davison, 1986) followed by downstream injection into the melt.

Depending on the method and location of the peroxide addition, screws in single screw machines are usually equipped with special mixing zones. In co-rotating twin extruders which handle two streams of a) PP and, b) PP/ peroxide premix simultaneously at the hopper, large pitch conveying elements are followed by a set of kneading blocks for melting and homogenization before devolatilization and, then, by elements with decreasing pitch (Dreiblatt et al., 1987). In the Farrel Continuous Mixer, a set of special counter-rotating non-intermeshing rotors have been designed to provide increased residence time in the melting/mixing/ reaction section with extensive mixing accomplished by means of optimum rotor orientation and the use of dams (Valsamis and Canedo, 1989).

- *Minimum Interference from Oxygen and Additives.* Oxygen, as an initiator of the PP degradation, does interfere with the peroxide induced reaction. Antioxidants, stabilizers and other radical scavengers usually reduce the overall degradation rate. A nitrogen blanket during premixing and metering in the hopper has been used in twin screw extrusion processes (Dorn, 1985; Fritz and Stoehrer, 1986) along with premixed long term stabilizers that were previously found to cause only a slight inhibition of the degradation; such stabilizers include Irganox 1035, Irganox 1010 (Ciba-Geigy Corp.) and their synergists at concentrations less than 0.1 phr. Oxygen interference was claimed to be reduced by injecting the peroxide downstream, after the polymer has fully melted (Greene and Pieski, 1964).
- *Removal of Decomposition Products.* Dialkyl peroxide decomposition products such as t-butyl alcohol and acetone have been devolatilized in especially configured screw sections (Nettelnbreker and Stoehrer, 1988; Curry et al., 1988; Andersen and Kenny, 1989). The design of rotor 24X of the Farrel Continuous Mixer incorporates a special second screw section for venting; the amount of residual t-butyl alcohol, ranging between 60 and 300 ppm, was found to depend on process conditions (speed, output, temperature) and initial peroxide concentration (Valsamis and Canedo, 1989).
- *Control of Product Characteristics.* M_w (affecting low shear rate viscosity and MFR) and MWD are the primary measurements of product quality. The use of melt rheometers in the early Exxon CR-PP patents is described in Section 4.6.1, along with recent attempts to develop a process control system correlating pressure drop with initiator concentration (Pabedinskas et al., 1989). In large scale production, quality control usually involves the monitoring of the

product MFR by measuring viscosity through on-line rheometers of capillary or Couette style equipped with data processing systems (Hertlein and Fritz, 1988; Dreiblatt et al., 1987; Curry et al., 1988). A single input (peroxide level) control scheme is based on the linear relationship between peroxide concentration and MFR; a multiple input system (throughput rate, rpm, peroxide level) may, in addition, control the breadth of MWD. Since MWD is dependent upon the relative characteristic times of mixing and peroxide decomposition, broad or even bimodal MWD could be produced if the peroxide decomposed before being uniformly mixed in the polymer (Andersen and Kenny, 1989). By varying cyclically the peroxide addition rate in the powder crammer feeder or, directly into the melt, bimodal materials could be produced with improved rheology for fiber spinning (Davison, 1986); the cycle period in this single screw extruder process was longer than the decomposition time of the peroxide but shorter than the dwell time.

2.3 Anionic Polymerization of Nylon 6

The continuous anionic polymerization of nylon 6 from caprolactam has been described as early as 1969 by Illig. In spite of its complexity, the process appears to offer interesting possibilities for linkage to post processing operations such as profile extrusion, blow molding, fiber spinning, film production etc. (Berghaus and Michaeli, 1990), and, also, for the preparation of compatibilized polymer blends via *in-situ* nylon 6 polymerization (Van Buskirk and Akkapeddi, 1988; Hergenrother and Greenstreet, 1988). The majority of the work reported in the patent and open literature was carried out in twin screw extruders, equipment capable of handling such a complex system; single screw extruders, in combination with gear pumps (Biensan and Potin, 1978), have also been used to ensure constant output and consistent extrudate quality.

2.3.1 Reaction Characteristics

2.3.1.1 Chemistry

The anionic polymerization of caprolactam in the presence of catalysts capable of generating lactam anions, and often in the presence of co-catalysts (activators) capable of generating the polymerization growth center, may produce high MW nylon 6 at 90-95% conversion in only a few minutes; this is in contrast to the hydrolytic polymerization process which requires up to 10 hours.

CO——NH ---- > -vvv-[-CO-$(CH_2)_5$-NH-]-vvv-
{ }
$(CH_2)_5$

The conversion to nylon 6 is an exothermic reaction with a heat of polymerization of about 125 kJ/kg (Kircher, 1987), and involves such steps as initiation, propagation and termination (Vollmert, 1973; Bartilla et al., 1986; Menges and Bartilla, 1987; Van Buskirk and Akkapeddi, 1988). In the initiation step, the ionic catalyst reacts with caprolactam to form a reactive salt that, in turn, reacts with free caprolactam to form an acyllactam with regeneration of the catalyst. In the propagation step, the acyllactam reacts rapidly with the catalyst to form a higher MW salt that reacts further with more caprolactam. The termination step usually involves the deactivation of the catalyst through the addition of reactive species. In the presence of co-catalysts such as other acyllactams, the salt resulting from the reaction between catalyst/co-catalyst reacts through fast proton exchange with the caprolactam to form the species that participate further in the propagation reactions (Van Buskirk and Akkapeddi, 1988).

2.3.1.2 Materials/Concentrations

Anhydrous caprolactam as ground powder or in its molten form is combined with various alkaline catalysts capable of forming caprolactam anions. Examples include sodium caprolactamate, alkali metal carbonates, hydrides, organometallics, etc. Co-catalysts, (activators), which are often employed to optimize the reaction rate may also lower the reaction temperature depending upon their type and concentration; acyllactams and isocyanates are suitable activators (Van Buskirk and Akkapeddi, 1988). Catalysts and co-catalysts are accurately added at concentrations ranging from 0.5-2 eq.%.; their reactivity is significantly reduced in the presence of moisture and atmospheric oxygen.

2.3.1.3 Kinetics – Fundamental Studies

In the work of Bartilla et al. (1986), preliminary tests on the reaction characteristics were performed in cast systems and in a cone and plate rheometer. The aim of the experiments was to determine suitable catalyst/co-catalyst systems at temperatures that would allow completion of the reaction in the extruder in ca. 1-3 min. Useful information was obtained from temperature rise vs. time curves, and viscosity vs. time curves obtained at different reaction temperatures. In a particularly promising system, the peak temperature of about 188 °C was reached in 1.8 min. at which time the viscosity showed a rapid exponential increase; the reaction temperature for this experiment was 160 °C and the undisclosed catalyst and co-catalyst were used at 3 and 7 phr, respectively. Polymerizations carried out in a cone and plate viscometer showed that the conversion to high MW product was strongly dependent on shear rate, increasing initially with increasing shear rate (attributed to better mixing) and decreasing thereafter, presumably the result of degradation.

In the work by Van Buskirk and Akkappedi (1988), preliminary kinetic data were obtained in ampoules under near adiabatic conditions. The time required to reach peak exotherm, (assumed to correspond to maximum conversion), was shown to be proportional to the inverse initial reaction temperature. For example, in a system containing 1 eq.% bromomagnesium caprolactam (catalyst) and 1 eq.% terephthaloylbiscaprolactam (co-catalyst) the time to reach peak temperature was 45 s at 160 °C and, by extrapolation, about 15 s at 250 °C.

2.3.2 Reactive Extrusion Process Analysis

2.3.2.1 Defining the Process Requirements

The following process requirements can be logically derived from the knowledge of the characteristics of the reaction:

a) Need for *accurate metering* of catalyst/co-catalyst, in the absence of moisture and atmospheric oxygen.
b) Thorough *mixing* of the relatively small amounts of catalyst/co-catalyst with the molten caprolactam.
c) Suitable *reaction zone length* depending on the kinetics of the reaction at the particular catalyst/co-catalyst combination.
d) Suitable *screw configuration* and *temperature control* to convey efficiently the very low viscosity fluid at the early stages of the reaction, as well as the much higher viscosity melt prior to discharge.
e) Minimal *degradation* of the polymer melt as a result of excessively high shear rates, high temperatures, or long residence time.
f) Provision for *devolatilizing* the unreacted monomer and *deactivating* the catalyst.
g) High *conversion* to sufficiently high *MW* polymer at reasonable *throughput* rates.

2.3.2.2 Meeting the Process Requirements

Given the complexities resulting from the differences in reactivities and mixing characteristics of the broad variety of the existing catalyst/co-catalyst combinations, studies in non-extrusion equipment alone cannot provide sufficient information to predict the course of the reaction under dynamic conditions. Recent process analysis studies in 20 to 35 mm diameter twin screw extruders have attempted to correlate operating parameters and screw geometry with product quality. In the work at IKV, Aachen, (Bartilla et al., 1986; Berghaus and Michaeli, 1990; Menges and Bartilla, 1987) and the studies of Van Buskirk and Akkapeddi (1988), the reaction was carried out in co-rotating fully intermeshing extruders (W & P , ZSK 30 and Leistritz, LSM 30.34, respectively). In the work reported by Welding Engineers, Inc. (Tucker and Nichols, 1987a and 1987b; Nichols, 1986), nylon 6 was produced in a counter-rotating non-intermeshing (CRNI) 20 mm extruder. The use of different catalyst/co-catalyst combinations does not permit direct comparison between the above studies. The following discussion, however, attempts to present some common approaches that met the requirements of continuous processes.

Co-rotating Intermeshing Extruders. The extrusion work at IKV was based on the preliminary batch reaction data that were discussed earlier. In applying these data to extrusion, *accurate metering* of the dry ingredients under inert atmosphere was ensured either through separate gravimetric feeders for the two reactants in powder form, or by pumping two molten caprolactam streams containing catalyst and co-catalyst, respectively. The screw configuration (see Chapter 4, Fig. 7) had different conveying elements for feeding, melting, reaction and discharging; thorough *mixing* of the reactants was accomplished through a set of kneading blocks just prior to the reaction zone. The *reaction zone length* estimated from the previous batch studies was more accurately determined experimentally and could be controlled by the barrel temperature and the overall temperature profile (see Chapter 4, Fig. 8). With this particular screw configuration, the MW of the product could be adjusted through controlled degradation in a second set of kneading blocks located after the reaction zone. Product characteristics were found to depend strongly on process conditions with typical MW of 80,000 produced at maximum throughput rates of 10 kg/hr. As predicted from the earlier batch studies, the MW was found to decrease at high screw speeds and temperatures, presumably as a result of *degradation* of the unstabilized melt. Residence times longer than 200 seconds were also shown to result in lower MW (Berghaus and Michaeli, 1990). The residual monomer content increased with increasing screw speed and could be reduced to less than 2% through *devolatilization* (with or without stripping agent). High residual monomer contents were found to decrease T_g and tensile strength, although impact strength increased slightly as a result of plasticization.

Based on the ampoule reactivity data discussed earlier, Van Buskirk and Akkapeddi (1988), used a short *reaction zone length* corresponding to an estimated residence time of 15-20 seconds for a given set of reactant concentrations and barrel temperature. Melt *mixing* of the preblended reactants fed under nitrogen was accomplished through kneading blocks. Conditions were such as to polymerize and build viscosity as early as possible in order to improve *conveying*. Improved conveying and, as a result, higher throughput rates were also obtained through the addition of a high viscosity polymer carrier to form *in situ* a compatibilized polymer blend. Details on process conditions, and conversion and properties of the extrudate are listed in Section 4.2.3b.

Counter-rotating non-intermeshing extruders. Kinetic data on the reaction system employed in the work by Welding Engineers were not available in the relevant publications; it is of interest to note that, by contrast to the IKV data, the polymerization of nylon 6 in a 20 mm CRNI with screw configured only with conveying elements, produced at 11 kg/hr a product whose MW was independent of screw speed, up to 525 rpm. The uniformity in MW was attributed to the continuous distributive mixing, a characteristic of the CRNI screw geometry used; apparently, distributive mixing promoted a *uniform reaction* at lower shear, without polymer degradation. *Accurate metering* under inert atmosphere was accomplished through piston pumps connected to two argon-purged tanks that contained caprolactam plus catalyst, and caprolactam plus co-catalyst, respectively. Conversion, MW and residual monomer content were controlled through the control of residence time, catalyst deactivation and devolatilization.

2.4 Free Radical Polymerization of Acrylic Monomers

The bulk free radical polymerization of acrylic monomers in batch reactors may present problems resulting from poor heat transfer, inadequate mixing, rapid viscosity increase, inadequate control of MWD, gel formation, etc. For some applications, high mixing efficiency equipment such as twin screw extruders can be used in processes requiring large batch type reactors; through proper removal of the polymerization exotherm, (as high as 600 kJ/kg), it is possible to obtain high yields, often in shorter times. Examples of such applications include hot melt adhesives (Kotnour et al., 1986), recyclable items from carboxylic copolymers (Belz, 1985; Deibig and Belz, 1985) and various impact modified thermoplastics listed in Section 4.2.3c.

2.4.1 Reaction Characteristics

2.4.1.1 Chemistry

The polymerization or copolymerization of acrylic monomers initiated by a free radical source proceeds through the classical initiation, propagation, chain transfer and termination through recombination or disproportionation steps. The reaction can be generally represented as follows:

```
nCH2=CR  +  mCH2=CR   ---->
     |           |
     R'          R'

-vvv-[CH2-CR]------[CH2-CR]-vvv-
          | n           | m
          R'            R''
```

where R = hydrogen or methyl, and R′, R″ = carboxyl, ester, nitrile, amide groups.

2.4.1.2 Materials/Concentrations

Suitable monomers are acrylic and methacrylic esters, acrylic acid and methacrylic acid, acrylonitrile and methacrylonitrile, substituted acrylamides, etc. Initiators are selected from azo compounds and peroxides such as diacyl, dialkyl, peroxyesters, hydroperoxides, and peroxydicarbonates. The initiator choice is based on half-life at processing temperatures and solubility in the monomers. Concentrations may be as low as 0.1 % or as high as 1% by wt. Occasionally, chain transfer agents such as mercaptans are added for MW control.

2.4.1.3 Kinetics – Fundamental Studies

In the work by Dey and Biesenberger (1987), an estimate of the time required to complete the reaction in an extruder as a function of type and concentration of initiator was obtained from isothermal differential scanning calorimetry (DSC) scans. Programmed DSC at various rates was used to simulate the temperature profile that was used subsequently in the non-isothermal polymerization of methyl methacrylate (MMA) in a counter-rotating Leistritz 34 mm twin screw extruder. A tubular prereactor containing monomer plus initiator was used to feed the extruder with a higher viscosity mixture for easier conveying. The temperature of the prereactor and the initiator concentration were determined from the DSC experiments.

A new instrument, called the rheocalorimeter, measures simultaneously the heat released and viscosity of a polymerizing solution as a function of time, and can provide useful information for the accurate modelling and control of the reactive extrusion process (Rosendale and Biesenberger, 1990). For example, the residence time in a prereactor required to reach a certain viscosity at a given low conversion can be estimated from data such as shown in Figs. 6 and 7. Fig. 6 shows the heat generation vs. time data and Fig. 7 shows the corresponding increase in viscosity of a comonomer system containing dissolved initiator at a reaction temperature of 85 °C (Grenci et al., 1989). Conversion may be calculated by integrating the heat curve and relating to the viscosity by eliminating the time coordinate.

Fundamental studies on RTD in a 53 mm co-rotating twin screw extruder with a suitable temperature profile (60 °C to 20 °C from entrance to exit) were conducted by using corn syrup to simulate the viscosity differences of the polymerizing medium (Nangeroni et al., 1985). The screw configuration consisted of varying pitch elements including a left handed seal at the devolatiliza-

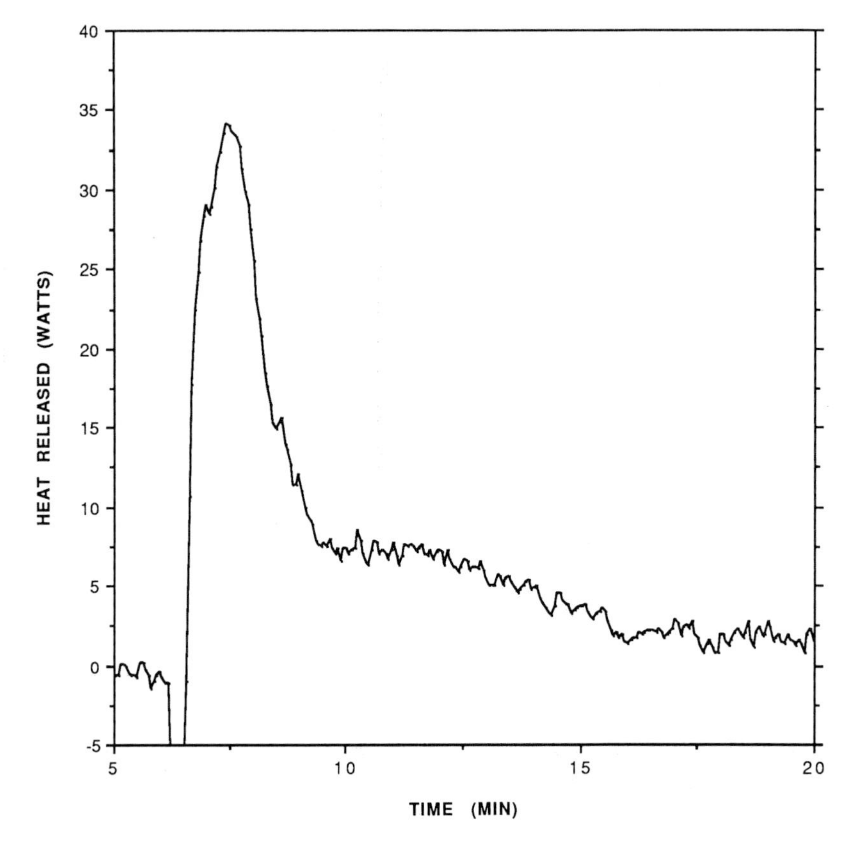

Figure 2.6 Heat generation vs. time of a polymerizing acrylic comonomer mixture containing dissolved initiator in the rheocalorimeter

tion zone. For this geometry, a narrow RTD was obtained at the lower screw speed (120 rpm). The results of the study indicated that geometries for reactions with strong residence time dependence could be designed to produce homogeneous products with narrow molecular weight distribution. The increase in polydispersity of PMMA with increasing RTD in a counter-rotating intermeshing 34 mm Leistritz twin screw extruder was demonstrated by Stuber and Tirrell (1985). In their experiments, MMA containing 0.25-1% predissolved initiator **C** (see Table 2) was introduced under pressure in the feed zone. Temperatures were 140-160 °C, well above the b.p. of the monomer and the T_g of the formed polymer. The RTD measured through UV absorption of an injected dye was found to be sharper at both higher operating temperatures and initiator concentrations (faster reaction leading to higher viscosity near the entrance and, as a result, reduced leakage backflow).

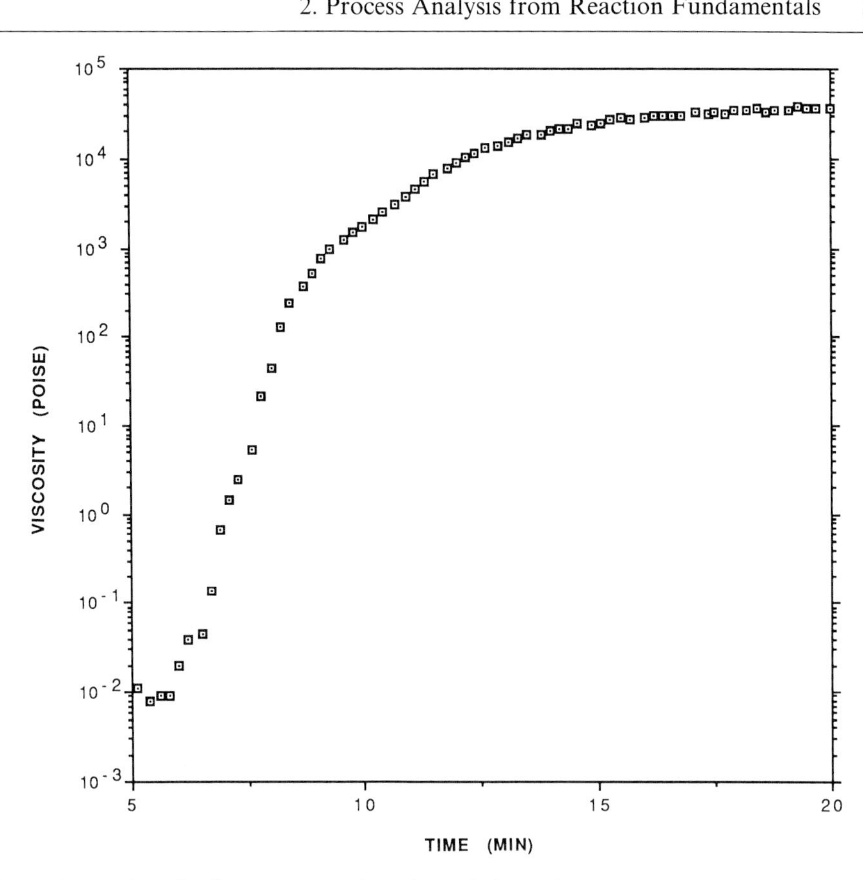

Figure 2.7 Viscosity increase as a function of time of a polymerizing acrylic comonomer mixture containing dissolved initiator in the rheocalorimeter

2.4.2 Reactive Extrusion Process Analysis

2.4.2.1 Defining the Process Requirements

The requirements of the bulk free radical polymerization in extruders may be summarized as follows:

a) *Inert atmosphere* to prevent oxygen inhibition
b) *Auxiliary equipment* for, i) preblending monomers with initiators, ii) advance the reaction if required, and, iii) feeding the liquid mixture, usually under pressure.
c) *Machine, screw geometry and temperature control* equipment able to handle the exothermicity of the reaction and the large viscosity differences from entrance to exit, and provide adequate control over residence time and RTD.
d) Ability to *add downstream* the comononer in order to enrich the mixture with the more reactive monomer, as well as other additives including chain transfer agents.

e) *Removal* of the unreacted monomer.
f) High *conversion* at reasonable residence time and throughput rates to products with sufficiently high *MW*, narrow *MWD*, and low gel content.

2.4.2.2 Meeting the Process Requirements – Industrial Applications

Literature examples from industrial practice are specific with respect to the particular monomer/initiator combination which controls the reactivity of the system. Although direct comparison is not always possible, common approaches that meet the requirements of a continuous bulk polymerization process can always be found. In the following discussion, reference is made to four comonomer systems having an acrylate ester as the major component:

System #1: Wielgolinski and Nangeroni (1983).
Hydroxylpropyl methacrylate/2-ethylhexyl methacrylate (90/10 by wt.) plus 0.3% t-butyl perbenzoate - 30 mm co-rotating, self-wiping twin screw extruder.

System #2: Kotnour et al. (1986).
Isooctyl acrylate/acrylic acid (from 90/10 to 95/5 by wt.) plus initiator mixture (methyltricaprylyl ammonium persulfate, azobisisobutyronitrile, 2,5-dimethyl-2,5-di(t-butyl-peroxy) hexane) - 34 or 67 mm counter-rotating twin screw extruder.

System #3: Matsuoka et al. (1984).
Methyl methacrylate/ethyl acrylate (50/50 by wt.) plus azobisisobutyronitrile - 50 mm single screw extruder.

System #4: Belz (1985) , Deibig and Belz (1985).
Ethyl acrylate/methacrylic acid (82/18 by wt.) plus dilauryl peroxide plus tetradecylmercaptan - 60 mm co-rotating twin screw extruder.

In all the above systems the monomers were nitrogen purged and the reactions were carried out in an essentially *oxygen-free* atmosphere. Monomers were usually *premixed* with initiators in a separate tank and pumped into the extruder throat. The viscosity of the feed was usually very low (less than 1 cp), although in some cases (System #2) a more viscous feed consisted of a prereacted polymer syrup or an *in situ* partially reacted monomer produced by means of a heated static mixer.

The particular chemistry of each system determined to a large extent the selection of extrusion *conditions* and *screw geometry*. In System #1, zones 1-3 were kept at 125 °C and zones 4-12 at 195 °C with a screw consisting mostly of short pitch conveying elements giving a residence time of about 10 min. In System #3, barrel temperatures and initiators were such as to promote early polymerization in the first half of the extruder (a 100-fold viscosity increase) in order to build viscosity and improve conveying. In System #2, efficiency was improved through a) the proper use of an initiator mixture containing components with different half-lives, and b) through increasing barrel temperature profile (e. g. from 60 to 180 °C) . For this system, the low temperature initiator yielded a rapid reaction to 10-15% conversion in the first barrel zone; the second initiator brought the reaction to 90% conversion in the second zone, whereas the high temperature initiator increased further the conversion at the third zone that was held at higher temperature. In the same System #2, the residence time ranging from 1 to 20 min was controlled through screw geometry (mostly conveying elements), temperature, speed and through the optional addition of a tubular extension at the end of the extruder. In System #4, control of the *exothermic reaction* was achieved by adjusting the ratio of two jacket heating fluids keeping a constant temperature profile (from 20 to 160 to 140 °C) along 10 barrel zones with very small tolerances.

The *downstream* addition of monomer, additional initiator, chain transfer agents etc., was, in principle, possible in all the above four systems. Unreacted monomer removal through *venting* was accomplished either in the extruder reactor, or in a second compounding extruder connected in series as described in System #4. *Conversions* were always fairly high (e.g. 92-99%) and products had sufficiently high *MW*. In System #3, the resulting *MWD* was as narrow as 7.3, whereas in System #2 the lower viscosity monomer feed gave a narrower *MWD* than the more viscous prereacted syrup (13-17 unimodal vs. 27 bimodal, respectively). In this last system, gel content could be minimized through the proper screw/barrel clearance, control of residence time, and the use of water cooled barrel sections, three component initiator mixtures, or chain transfer agents.

2.5 Concluding Remarks

In this chapter an attempt was made to provide a basis for a methodology that could be used to define the requirements of reactive extrusion processes from basic studies in laboratory scale equipment. It should be emphasized that the select case studies of this chapter are hardly representative of the broad spectrum of reactions that are possible in extruders. Space limitations do not allow to discuss important contributions to the design of functionalization and reactive compounding extrusion processes that have appeared in the recent literature. Examples include the fundamentals of the transesterification of EVA reported by Bouilloux et al. (1986), and the publications of Bourland et al. (1989), Scott and Macosko (1988), and Dagli et al. (1990), on the amine/anhydride and amine/acid reactions of functionalized polymers. Basic studies of this type clearly demonstrate the usefulness of specialized instruments and imaginative new experimental techniques to the design of reactive extrusion processes.

Chapter 3

Reactive Extrusion in the Preparation of Carboxyl-Containing Polymers and Their Utilization as Compatibilizing Agents

By Norman G. Gaylord, The Charles A. Dana Research Institute for Scientists, Emeriti, Drew University, Madison, NJ 07940

3.1 Introduction

The chemical modification of polymers is carried out to produce:
- New Copolymers
- Functionalized Copolymers
- Polymers with Controlled Crosslinking
- Compatibilized Polymer Blends
- Reinforced Structures

as well as to modify the processing or rheological characteristics and mechanical properties by modifying the molecular weight distribution of polymers.

The extruder is an ideal reactor for polymer modification in that it serves as a pressure vessel equipped for intensive mixing, shear, control of temperature, control of residence time, venting of by-product and unreacted monomer as well as the transport of molten polymer through the various sections of the extruder, each serving as a mini-reactor, and removal therefrom.

The present chapter deals with:

(a) The preparation of carboxyl-containing polymers by reactive extrusion, and

(b) The utilization of carboxyl-containing polymers as compatibilizing agents in the preparation of polyblends and alloys.

3.2 Preparation of Carboxyl-Containing Polymers

3.2.1 Carboxylation of Unsaturated Polymers

The reaction of an unsaturated polymer with maleic anhydride (MAH) proceeds through the "ene" reaction wherein a succinic anhydride moiety is appended to the polymer chain and the double bond shifts.

Unsaturated high density polyethylene (HDPE) has been prepared by cracking high molecular weight polymer in an extruder at 425-600 °C to yield a polyunsaturated PE wax having a molecular weight of 1,000-5,000. When the latter is mixed with MAH at 180-250 °C, the PE is carboxylated.

PE wax + MAH ⟶ PE wax–(succinic anhydride)

The PE-g-MAH may be blended with PE and used for hot melt or emulsion coatings or primers having adhesion to substrates due to the presence of carboxyl groups (Schaufelberger, 1963).

More recently, this reaction has been utilized in the preparation of an EPDM-MAH adduct, by the reaction of ethylene-propylene-diene terpolymer (EPDM) with MAH in a screw extruder at 260-300 °C (Caywood, 1975).

The carboxylated elastomer has been used in the preparation of rubber-toughened nylon, as discussed hereinafter.

3.2.2 Carboxylation of Saturated Polymers

3.2.2.1 Reaction with Acrylic Acid

The reaction of a polymerizable monomer with a saturated polymer in the presence of a radical catalyst generally results in the formation of a graft copolymer wherein the original polymer contains an appended branch of polymer derived from the newly polymerized monomer.

The reaction proceeds by hydrogen abstraction from the saturated polymer by a radical (cat·) resulting from the thermal decomposition of the catalyst:

$$P{-}C(R)(H){-}\sim \xrightarrow{\text{cat}\cdot} P{-}\dot{C}(R){-}\sim$$

Currently, a radical from the catalyst generally initiates homopolymerization of the monomer:

$$M \xrightarrow{\text{cat}\cdot} M\cdot \xrightarrow{M} M{-}M\cdot \xrightarrow{x\,M} M{-}M_x{-}M\cdot$$

$$P{-}\dot{C}(R){-}\sim \xrightarrow{M} P{-}C(R)(M\cdot){-}\sim \xrightarrow{x\,M} P{-}C(R)(M{-}M_x{-}M\cdot){-}\sim$$

The graft copolymerization of acrylic acid (AA) onto HDPE and polypropylene (PP), carried out in the melt in the presence of a radical catalyst in an extruder, yields a graft copolymer containing poly(acrylic acid) branches, accompanied by poly(acrylic acid) homopolymer. In a similar manner, the graft copolymerization of AA onto elastomeric ethylene-propylene copolymer (EPR) yields a poly(acrylic acid) grafted elastomer (Steinkamp and Grail, 1975 and 1976).

$$P{-}C(R)(H){-}\sim \; + \; CH_2{=}CH(COOH) \longrightarrow P{-}C(R){-}\sim \,\text{with branch}\, \left[CH_2CH(COOH){-}\right] \; + \; PAA$$

The resultant products, i.e. HDPE-g-AA, PP-g-AA and EP-g-AA, are carboxyl-containing polymers which cannot be made by copolymerization of AA with either ethylene, propylene or an ethylene - propylene mixture, because the polar carboxylic acid monomer reacts with the metal-containing catalysts used in the preparation of these hydrocarbon polymers and prevents polymerization of the respective monomers.

3.2.2.2 Reaction with Maleic Anhydride

The reaction of a molten saturated polymer with maleic anhydride in the presence of a radical catalyst, under the appropriate conditions, yields a polymer containing appended individual succinic anhydride and maleic anhydride units, without accompanying MAH homopolymer but accompanied by side reactions, including crosslinking and/or degradation not encountered in the reactions with AA.

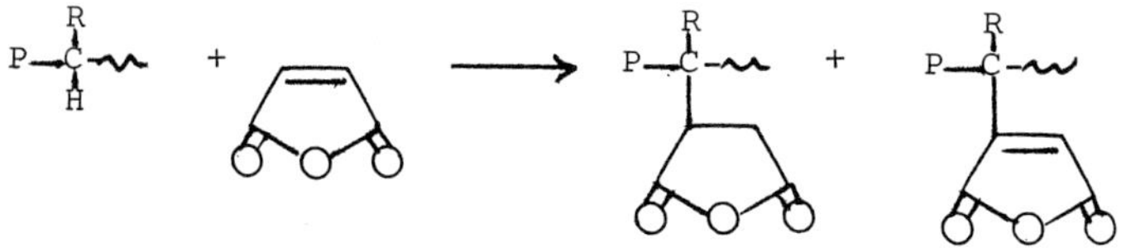

In contrast to AA, MAH is not a polymerizable monomer, in the sense that it undergoes radical initiated homopolymerization under the conditions normally effective with vinyl monomers such as methyl methacrylate, styrene, acrylic acid, etc. Thus, the mechanism shown previously for conventional graft copolymerization is not applicable to MAH.

MAH becomes appended to a saturated polymer in the presence of a radical catalyst under the conditions which promote the homopolymerization of MAH in the absence of the polymer. Therefore, it is necessary to understand the nature and mechanism of MAH homopolymerization in order to obtain insight into the mechanism of the grafting reaction.

MAH undergoes homopolymerization upon exposure to gamma radiation, ultraviolet radiation in the presence of a photosensitizer and shock waves. Radical catalysts are not effective under conventional conditions but are effective when used in high concentrations, when added to the monomer intermittently or continuously rather than in one shot, or when used at temperatures where the catalyst has a short half-life, i.e. $t_{1/2} < 60$ min (Gaylord, 1975).

Under the conditions indicated, monomeric MAH is converted to an excited dimer, which is the actual polymerizable species, rather than the monomer:

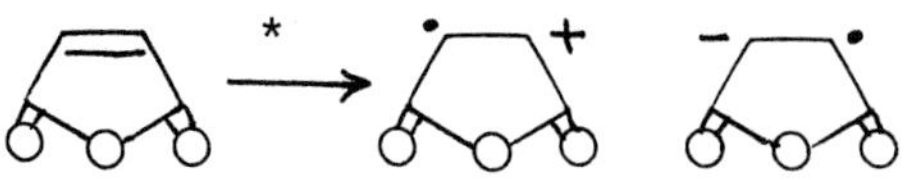

Evidence for the participation of the excited species in the homopolymerization is provided when the bulk polymerization in the presence of a peroxide catalyst is conducted in the dark in the presence of a photosensitizer. When the triplet energy of the sensitizer is 68 kcal/mole, the triplet energy of MAH, or higher, the yield and molecular weight of the MAH homopolymer is increased. When the triplet energy of the sensitizer is below 68 kcal/mole, quenching of the excited MAH decreases the homopolymer yield and molecular weight (Gaylord and Maiti, 1973).

The participation of cationic intermediates in the radical catalyzed homopolymerization of MAH is indicated by the failure to effect polymerization of MAH in the presence of redox systems containing amines, e.g. diisopropylpercarbonate-dimethylaniline and benzoyl peroxide-p-toluidine.

Dimethylformamide (DMF) does not inhibit the radical polymerization of vinyl monomers such as acrylic esters. Therefore, it does not interfere with the radicals generated upon thermal decomposition of a radical precursor or the propagating polymer radicals which are present during the polymerization. Nevertheless, the presence of small amounts of DMF inhibits the polymerization of MAH under the conditions which are normally effective, e.g. with a catalyst which has a short half-life at the reaction temperature (Gaylord and Koo, 1981).

DMF is an inhibitor for cationic polymerization since it is capable of donating electrons to the cationic intermediates and terminating propagation. The inhibition of MAH polymerization in the presence of DMF presumably follows the same route, i.e. electron donation to the cationic species in the excited dimer and/or to a cationic propagating chain end:

$$+ \; \text{DMF} \longrightarrow \; + \; \overset{+}{\underset{\cdot}{}}\text{DMF}$$

The DMF cation-radical recovers an electron from the ion-radical in the excited dimer to regenerate DMF and MAH:

$$+ \; \overset{+}{\underset{\cdot}{}}\text{DMF} \longrightarrow \; + \; \text{DMF}$$

The decomposition of dicumyl peroxide (DCP) in molten LDPE at 180 °C in an extruder results in partial crosslinking of the LDPE. When MAH is present, the amount of crosslinked LDPE is increased and the soluble LDPE contains appended MAH. Although the presence of benzoyl peroxide (BP) fails to produce crosslinked LDPE at 180 °C, crosslinking occurs when MAH is present and the soluble fraction contains appended MAH.

Crosslinking occurs in the peroxide-induced reaction of MAH with ethylene homopolymers, e.g. LDPE and HDPE, as well as copolymers with 1-butene and 1-octene, i.e. linear low density PE (LLDPF), and polar monomcrs, i.e. ethylene-vinyl acetate, ethylene-methyl acrylate and ethylene-acrylic acid copolymers.

Table 3.1 Reaction of MAH with LDPE at 180 °C

				LDPE-g-MAH *Xylene Solubility*		
					Soluble	
Peroxide	wt-%*	MAH wt-%*	DMF wt-%*	Insoluble %	%	MAH wt-%
DCP	2.0	0	0	51	42	0
	2.0	10	0	59	36	6.6
	2.0	10	1.5	0	100	2.8
	2.0	0	1.5	55	39	0
BP	2.0	0	0	0	100	0
	2.0	10	0	35	55	5.7
	2.0	10	1.5	0	100	1.8

* Based on LDPE

It is apparent that an intermediate in the homopolymerization of MAH is responsible for the crosslinking which accompanies the grafting of MAH onto ethylene-containing polymers. When DMF, which inhibits MAH polymerization, is added in admixture with the peroxide and MAH, in several portions, to molten LDPE, the amount of appended MAH decreases but crosslinked polymer is avoided. In the absence of MAH, the presence of DMF has no effect on the crosslinking of the LDPE (Table 1).

It has been proposed (Gaylord and Mehta, 1988) that the additional crosslinking of the ethylene copolymer which occurs during reaction with MAH, but not with acrylic acid, is due to the generation of additional radical sites on the polymer as a result of hydrogen abstraction by the excited MAH. These additional sites can couple and/or add excited MAH.

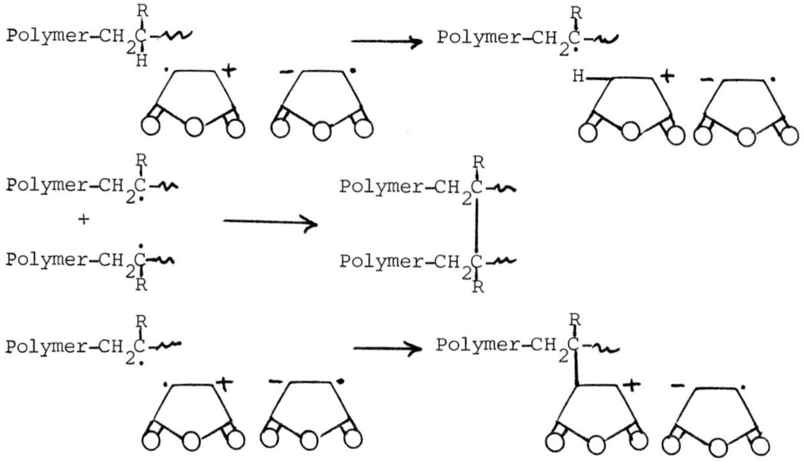

Electron donation by DMF converts the MAH cation in the excited dimer or the appended MAH cation to a radical which undergoes disproportionation.

Polymer-CH_2C(R)– DMF → Polymer-CH_2C(R)–

Polymer-CH_2C(R)– + Polymer-CH_2C(R)– →

Polymer-CH_2C(R)– H + Polymer-CH_2C(R)–

Thus, DMF prevents crosslinking by reducing the concentration of excited MAH while the residual excited MAH becomes appended before reaction with the DMF (Gaylord and Mehta, 1982).

The reaction of molten isotactic polypropylene (iPP) with a radical catalyst such as DCP or di-t-butyl peroxide (DBP) results in degradation, i.e. a decrease in the molecular weight and an increase in the melt flow, rather than crosslinking (Steinkamp and Grail, 1975 and 1976; Ide et al., 1968a, 1968b). This is due to the disproportionation of polymer radicals in lieu of coupling. The presence of MAH promotes further degradation due to the increased generation of polymer radicals. When DMF is present in the MAH-peroxide mixture which is added intermittently to the molten iPP, the degradation of the iPP is minimized, as shown by the higher intrinsic viscosity, [η], and the resultant iPP contains appended MAH units (Table 2) (Gaylord and Mishra, 1983).

Since excited species are quenched by the ground state species from which they originate, monomeric MAH quenches excited MAH. Therefore, in order to increase the efficiency and the amount of MAH which appends to the polymer, it is necessary to add the MAH to the molten polymer in several steps or intermittently.

The crosslinking and/or degradation which accompanies the molten polymer-MAH reaction occurs with all peroxides when they are used under the conditions necessary to promote MAH grafting on the polymer. However, these reactions are prevented by the presence of low or high molecular weight compounds which contain nitrogen, phosphorous or sulfur atoms and inhibit

The MAH-peroxide-polymer reaction in the melt in the extruder permits the preparation of carboxyl-containing polymer pellets, ready for fabrication, or of shaped objects directly.

MAH + peroxide → polymer → polymer–MAH → pellets, fiber, film, laminate

The presence of fillers during the MAH-peroxide-polymer reaction results in coupling and compatibilization to yield a composite with improved tensile strength and impact resistance, in addition to the increased modulus to be expected from the presence of the filler (Gaylord, 1976, 1978; Gaylord et al., 1980).

$\sim CH_2C(R)(H)CH_2\sim$ + MAH + HO–Si(–O–) $\xrightarrow{R^*}$ $\sim CH_2C(R)CH_2\sim$ (grafted succinic unit: HC—CH_2, O=C–O–Si– and C=O–OH) and/or hydrogen-bonded structure (H····O=C, O···H—O, –O–Si–, C=O–OH)

MAH is used in the preparation of carboxyl-containing high impact polystyrene (HIPS). The MAH is added to the reaction mixture in which styrene (S) is undergoing polymerization in the presence of polybutadiene (PBd).

PBd + S → PBd-g-S (HIPS); MAH → PBd-g-S-MAH (HIPS-MAH)

3.2.2.3 Reaction with Styrene-Maleic Anhydride

Styrene and maleic anhydride form a complex which undergoes spontaneous polymerization, in the absence of a catalyst, at 80 °C. When an equimolar S-MAH monomer mixture, maintained at 40-70 °C, is added to a molten polymer undergoing deformation at a temperature above 120 °C, a branch of alternating

S-co-MAH copolymer is appended to the substrate polymer. The higher the temperature, the shorter the branch length. The major deficiency of the process is the tendency of the highly reactive monomer mixture to undergo premature polymerization in the piping and auxiliary equipment before impinging upon the molten polymer in the extruder (Gaylord, 1973).

3.2.2.4 Reaction with Diels-Alder Adducts of Maleic Anhydride

The reaction of molten HDPE with the Diels-Alder adduct of MAH and cyclopentadiene or methylcyclopentadiene in the presence of a radical catalyst at 200-320 °C, results in the appendage of the carboxyl-containing adduct onto the polyolefin (Wu et al., 1975).

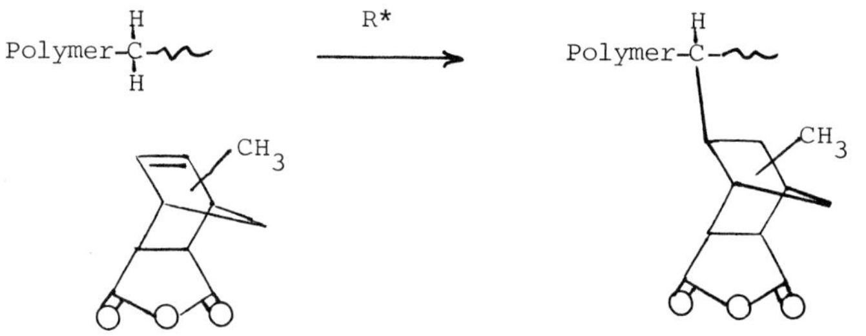

Due to the limited polymerizability of the adduct, the pendant moiety is probably present as a single unit rather than as a homopolymer branch. When blended with an ethylene-vinyl acetate (EVA) copolymer, the resultant blends are used as coextrudable adhesive layers in multilayer laminates (Shida et al., 1978).

3.3 Polyblends Containing Carboxylated Polymers

The compatibilization of two polymers, A and B, which are normally incompatible, e.g. polypropylene (PP) and nylon 6, may be enhanced by the presence of a compatibilizing agent. The latter is a block or graft copolymer having segments which are independently compatible with one of the polymers in the otherwise incompatible blend. In the blending process in an extruder in the melt, the compatibilizing agent becomes located at the interface of the incompatible polymers and a multiphase system is established in which particles of the polymer present in the lower concentration (A) are dispersed through the matrix of the polymer which is present in the higher concentration (B). The dispersed polymer may be present as < 5 micron particles which are in contact with the compatible segment of the compatibilizing agent. The other segment of the compatibilizing agent is in contact with the matrix polymer.

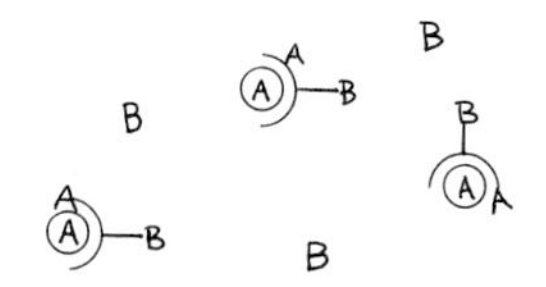

In lieu of mixing incompatible polymers and a compatibilizing block or graft copolymer, the latter may be generated *in situ* by the presence of reactive functionality. The carboxylated polymers are particularly useful for this purpose.

The blending of PE and nylon 6, under shear, was initially undertaken in an attempt to improve the impact resistance of nylon 6. Subsequently, it was found (Armstrong, 1968b) that the presence of an ethylene-methacrylic acid copolymer (EMAA) improved the compatibility. This was due to the interaction between the carboxyl group on the PE and the amine end group on the nylon or an amide group in the nylon chain.

PE
PE ———————————— x ———————————— nylon 6
nylon 6

The *in situ* formed compatibilizing agent may involve covalent bond, i.e. amide group formation or hydrogen bonding.

PE–C(=O)–NH-nylon   PE–C(=O······H–N)–O–H····O=C-nylon

A further improvement in PE-nylon polyblends involved the use of a partially neutralized ethylene-methacrylic acid copolymer, commonly referred to as an ionomer. The ionomer-nylon polyblends, having the characteristics of "toughened nylon", are fabricated by molding into shaped objects. In addition, laminates prepared by coextrusion of the individual polymers and possessing good interfacial adhesion are used in multilayer barrier packaging (Murch, 1974).

In an effort to improve the impact resistance of nylon, various elastomeric polymers containing carboxyl groups have been blended in the melt with nylon 6 and nylon 6,6. "Acrylic modifiers" for nylon 6, for use in blow molding applications, included random copolymers of butyl acrylate, butylene glycol dimethacrylate, methyl methacrylate and acrylic acid, as well as graft copolymers of methyl methacrylate and methacrylic acid onto poly-(butadiene-co-styrene) (Owens and Clovis, 1972).

$$\text{P}\sim(\text{COOH}) + \text{nylon–NH}_2 \xrightarrow{225^\circ\text{C}} \text{P}\sim(\text{C=O–NH–nylon})$$

Polymers such as copolymers of ethylene with vinyl acetate (VA), butyl acrylate (BA) and ethyl acrylate (EA) containing MAH or a precursor thereof, e.g. a diester or the diacid, which is converted to the anhydride at the elevated temperatures of melt blending, have greater interaction with the nylon end-group (Hammer and Sinclair, 1976 and 1977).

E-VA-CH—CH~ (O=C, C=O; R-O, O-R)
E-BA-CH—CH~ (O=C, C=O; HO, OH)
E-EA-CH—CH~ (O=C, C=O; O)
nylon-6, $225^{\circ}C$, 1-10 min →
~CH—CH~ (O=C, C=O; HO, NH; nylon) → ~CH—CH~ (O=C, C=O; N; nylon)

EPDM —nylon-6→ EPDM (N-nylon-6)

The initially formed amic acid is converted to the heat-stable succinimide grouping during melt processing. The polyblend:

EPDM
EPDM-MAH-nylon 6
nylon 6

has significantly greater impact strength than the previously described impact-modified nylons.

Considerably higher impact strengths are obtained from the EPDM-MAH reaction with nylon 6,6. Since nylon 6,6 has two amine end groups, a single chain can react with and crosslink the EPDM. The resultant toughened or rubber-modified nylon 6,6 containing about 20% EPDM has an Izod notched impact strength of 25 ft-lb/in. as compared with 1 ft-lb/in. for the unmodified nylon. The "super-tough" nylon 6,6, trademarked Zytel ST by the Du Pont Co., contains:

EPDM
EPDM-MAH-nylon-MAH-EPDM
nylon 6,6

The particle size of the dispersed rubber particles is about 1 micron or less and its tensile modulus is less than 20,000 psi. The incorporation of glass fibers or a mineral filler greatly increases the modulus of the polyblend while still retaining an impact strength higher than that of the unmodified nylon 6,6 (Epstein, 1979a).

A toughened thermoplastic polyester, i.e. polycarbonate, polyethylene terephthalate (PET) or polybutylene terephthalate (PBT) is prepared in a similar manner using a carboxyl-containing elastomeric copolymer or EPDM-MAH (Epstein, 1979b).

Blending PP-g-MAH with nylon 6, in the melt at 230 °C, results in the generation of a compatibilizing agent containing segments of PP and segments of nylon 6.

PP + nylon-6 —230°C→ PP N-nylon-6

The presence of PP without carboxyl groups and excess nylon yields a compatible polyblend whose properties are derived from both polymers, e.g. mechanical properties from nylon and water resistance from PP (Ide et al., 1972 and 1974).

PP
PP-g-MAH-nylon 6
nylon 6

The PP-nylon 6 blend may also contain glass fibers whose dispersion in the polyblend is increased by the presence and probable reaction with the PP-g-MAH.

ABS imparts impact resistance to blends with nylon (Grabowski, 1964b). Improved impact resistance results from the addition of a carboxyl-containing polymer, e.g. poly(styrene-co-acrylic acid) or poly(styrene-co-acrylic acid-co-MAH), presumably due to improved compatibilization resulting from hydrogen-bonding between the carboxyl group and the amide and/or the a-methylene group of the acrylonitrile (Brandstetter et al., 1982a).

An oil-resistant thermoplastic elastomer is prepared by melt-blending PP containing about 10% PP-g-MAH and nitrile rubber (NBR) containing 0.2-3% amine-terminated liquid NBR. The resultant polyblend, trademarked Geolast by the Monsanto Co., is compounded with a phenolic curative and stannous chloride to vulcanize the NBR (Coran and Patel, 1983b).

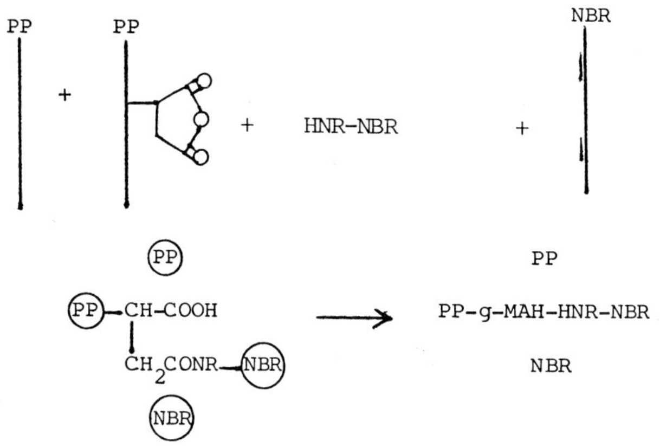

PP + PP (grafted anhydride) + HNR-NBR + NBR ⟶

(PP)–CH–COOH ⟶ PP

CH$_2$CONR–(NBR) PP-g-MAH-HNR-NBR

(NBR) NBR

An aromatic polycarbonate forms a compatible blend with an ABS resin (Grabowski, 1964a). The properties of the blend are improved by the addition of poly (styrene-co-acrylic acid-co-MAH) (Brandstetter, 1982b), or poly (styrene-co-methyl methacrylate-co-MAH) (Jones and Mendelson, 1986).

Hydrogen-bonding interactions are involved in the compatibilization of polyvinyl chloride (PVC) with polyolefins using a polyolefin grafted with diethyl maleate as compatibilizing agent (Benedetti et al., 1985 and 1986).

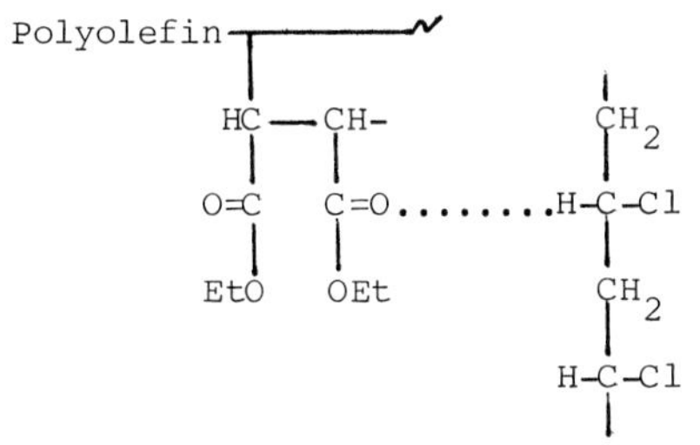

Polyolefin

HC — CH- CH$_2$

O=C C=O.......H-C-Cl

EtO OEt CH$_2$

H-C-Cl

Similarly, a styrene-maleic anhydride terpolymer is compatible with PVC through hydrogen bonding and increases the thermal resistance of the latter (Anon., 1986).

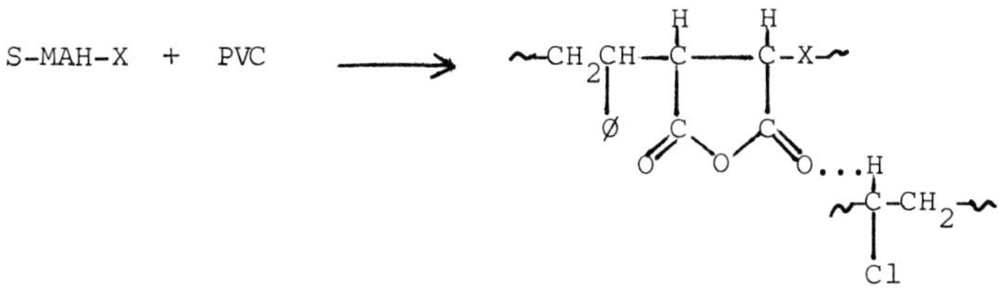

S-MAH-X + PVC ⟶ ~CH$_2$CH–C(H)–C(H)-X~ (Ø; anhydride ring C(=O)–O–C(=O)...H–C(Cl)–CH$_2$~)

The presence of a hydrogenated triblock S-Bd-S copolymer having an S-EB-S i.e. styrene-(ethylene-alt-butene)-styrene structure, in a blend of a thermoplastic polyester with a dissimilar resin such as a polyamide, polyolefin, polyurethane or nitrile barrier resin, results in an interlocking network structure with increased heat distortion temperature and impact strength (Gergen and Davison, 1978).

The properties of polyblends containing polar thermoplastic polymers and an S-Bd-S triblock copolymer are greatly improved by the presence of as little as 2% MAH grafted on the latter. The MAH is grafted on a partially unsaturated triblock copolymer by the "ene" reaction and on the hydrogenated saturated triblock copolymer in the presence of a radical catalyst. The MAH-modified triblock copolymers are alloyed with polyamides, polyesters, polyurethanes, polyacetals, polysulfones, polycarbonates, polyphenylene ethers and sulfides, nitrile polymers, ionomers and vinyl alcohol-ethylene copolymers and vinyl acetate copolymers (Shiraki et al., 1986, 1987a and 1987b).

Interactions between the carbonyl groups on PET and the acid groups on MAH-grafted S-EB-S hydrogenated triblock copolymer compatibilizes a blend of PET and HDPE with significant improvement in impact strength. Interpenetrating PET and S-EB-S phases are formed (Chen and Shiah, 1989).

An impact resistant blend of polyphenylene ether (PPE) and PS containing S-EB-S triblock copolymer is compatibilized with nylon by the presence of a sulfonated PS ionomer.

PPE
PS
P(S-EB-S)
$PS\text{-}SO_3^-\ Zn^{++}$
H_2N-nylon 6

A blend of PPE and PS is compatibilized with a sulfonated EPDM ionomer by the presence of a sulfonated polystyrene (Golba and Seeger, 1987).

PPE
PS
$PS\text{-}SO_3^-\ Zn^{++}$
$EPDM\text{-}SO_3^-\ Zn^{++}$

A reactive polystyrene (RPS) containing 1% oxazoline functionality is capable of reacting with polymers containing carboxyl, anhydride, epoxy, amine, phenolic hydroxyl, isocyanate and other functionality (Dow, 1985).

PS–(oxazoline) + HO–C(=O)–~ → PS–C(=O)–$NHCH_2CH_2$–C(=O)–~

This reaction is utilized in the preparation of polyblends (Anon, 1985b), as well as in the reactive coextrusion of multilayer film or sheet (Anon, 1985a).

Anhydride-containing polymers which have been reacted with the oxazoline-containing polystyrene include EPDM-MAH, PP-g-MAH, HDPE-g-MAH adduct and MAH-containing styrene copolymers. Acid-containing polymers which have been alloyed with RPS include EAA, acrylic rubber-AA, EMAA ionomer, EVA-MAH and carboxyl-containing nitrile rubber.

3.4 Recapitulation

The process of reactive processing permits the preparation of functionalized polymers, including copolymers containing carboxyl groups which cannot be made directly by the polymerization process, and opens up significant possibilities for the creation of new specialty and engineering plastics. Carboxyl-containing polymers are of particular interest since they may serve as compatibilizing agents acting through covalent, ionic or hydrogen bonding. In addition to imparting specific properties to polyblends and alloys, e.g. increased toughness, impact strength and/or heat distortion temperature, carboxyl-containing compatibilizing agents are useful in scrap reclamation.

Part II

Review of Reactive Extrusion Processes

Chapter 4

Reactive Extrusion: A Survey of Chemical Reactions of Monomers and Polymers during Extrusion Processing

By S. Bruce Brown, Polymer Chemistry and Materials Laboratory, General Electric Research and Development Center, Schenectady, NY 12301

4.1 General Information

4.1.1 Introduction

Reactive extrusion refers to the deliberate performance of chemical reactions during continuous extrusion of polymers and/or polymerizable monomers. The reactants must be in a physical form suitable for extrusion processing. Reactions have been performed on molten polymers, on liquified monomers, or on polymers dissolved or suspended in or plasticized by solvent. Although sometimes seen in the literature as "reactive compounding" or "reactive processing", the term "reactive extrusion" is preferred in this review because it distinguishes the performance of chemical reactions in a continuous extrusion process with short residence time from chemical reactions run in long residence time batch mixers or kneader reactors (which are types of reactive compounding), and from Reaction Injection Molding (which is a type of reactive processing).

There has been intense recent activity in the field of reactive extrusion, mostly in industrial laboratories and at extruder companies rather than in academic laboratories. The majority of greater than 700 patents and approximately 80 published papers on the subject of reactive extrusion have appeared within the last 20 years. The purpose of this article is to organize the published examples of reactive extrusion into specific categories and to review within each category general examples which best illustrate the use of an extruder as a chemical reactor for the particular chemical processes of that category. Although the chemical literature has been searched through 1989, it is virtually impossible to retrieve all published works referring to reactive extrusion because the majority of articles are patents which sometimes do not refer to reactive extrusion (or "reactive compounding" or "reactive processing") either in the title or abstract. With few exceptions the examples included in this review are from patents and journal articles which have been evaluated from the full text rather than from abstracts. As a result, primarily patents filed in the United States, Great Britain, or as European Patent Applications are included. Only a small number of selected examples of Reactive Extrusion published as abstracts from conference proceedings are included, again because of lack of detail. Despite the large amount of excellent work being done in Japan, only those patents or publications by Japanese workers which have appeared in English language or in United States, British, or European Patent Application form are included, primarily because it is often difficult to determine if a reactive extrusion process was actually employed from examining a Japanese patent abstract alone. French, Dutch, Belgian, Canadian, East and West German patents, and patents of other countries are also only covered if they have appeared in an equivalent form in the United States, Britain, or as a European or World Patent Application.

Even when the full text of a patent dealing with reactive extrusion has been available there are often distressingly few details given, and, as is often the case in dealing with patent literature, one doesn't always know from the published details whether the process actually works. The author would greatly appreciate having any errors or additional examples of Reactive Extrusion processes being brought to his attention.

Selected aspects of reactive extrusion have been treated previously in short reviews by Mack and Herter (1976), Wielgolinski and Nangeroni (1983), Kowalski (1985a), Eise (1986), Frund (1986), Ratzsch (1987), Lambla (1988), and Tzoganakis (1989). A logical organization of the categories of chemical reaction performed by reactive extrusion was first presented by Brown and Orlando (1988) and is utilized in the present work. In this review examples of reactive extrusion are limited to polymerization or chemical reactions of polymers in single screw (SSE) or twin screw extruders (TSE) in continuous processes. Processes in oscillatory kneaders, roll-mills, or "extruder-molders" have been excluded. Emphasis is on continuous processes which can be practiced commercially or which can serve as models for commercial processes. Batch processing in Brabender mixing equipment can frequently be scaled to continuous extrusion and many examples of reactions in Brabender equipment are given, particularly when the reaction times are less than 15 min. Batch mixing processes with reaction times longer than 30 min are not emphasized.

To be considered an example of reactive extrusion a chemical reaction must take place as a result of mixing in an extruder. Hence, the production of chemically bonded laminates is not included since the actual chemical reaction takes place outside the extruder without requiring mixing. Production of foams and the use of blowing agents are not considered since these processes seldom involve actual chemical reaction on the polymer itself. Chemical reactions which take place during mixing in injection molders are also omitted.

4.1.2 Types of Reactions Performed by Reactive Extrusion

The types of chemical reactions which have been performed by reactive extrusion may be conveniently divided into six categories as described in Table 1.

Table 4.1 Types of Chemical Reactions Performed by Reactive Extrusion

Type	Description
Bulk Polymerization	Preparation of high molecular weight polymer from monomer or low molecular weight prepolymer, or from mixture of monomers or monomer and prepolymer.
Graft Reaction	Formation of grafted polymer or copolymer from reaction of polymer and monomer.
Interchain Copolymer Formation	Reaction of two or more polymers to form random, graft, or block copolymer either through ionic or covalent bonds.
Coupling/Crosslinking Reactions	Reaction of polymer with polyfunctional coupling or branching agent to build molecular weight by chain extension or branching, *or* reaction of polymer with condensing agent to build molecular weight by chain extension, *or* reaction of polymer with crosslinking agent to build melt viscosity by crosslinking.
Controlled Degradation	Controlled molecular weight degradation of high molecular weight polymer (controlled rheology), *or* controlled degradation to monomer.
Functionalization/ Functional Group Modification	Introduction of functional groups into polymer backbone, endgroup, *or* sidechain, *or* modification of existing functional groups.

Representative examples from each category of reaction will be presented with emphasis whenever possible on what particular features of the extruder were critical for the success of the reaction. There is some unavoidable overlap in the categories of the above classification scheme. The following decisions have been made regarding choice of category for placement of specific examples. These decisions were often difficult because of the sometimes vague nature of the patent literature.

Bulk Polymerization vs. Polymer Coupling Reaction = = >

Chain extension of low molecular weight oligomers or prepolymers by reaction with coupling agents are included under Bulk Polymerization when the oligomers were usually < 10,000 molecular weight and had few, if any, practical applications other than to serve as precursor for high molecular weight polymer. When oligomers had > 10,000 molecular weight and/or had obvious commercial applications in their own right, then use of coupling agents for chain extension is included under Coupling Reactions. Similarly, when the coupling agent was not "normally" a component of the polymerization reaction, then the example is included under Coupling Reactions.

Graft Reaction vs. Functionalization Reaction = = >

The section on Graft Reactions also includes functionalization of polymers by graft reaction with unsaturated "monomers" (such as vinyl trialkoxysilanes and maleic anhydride) which may have little or no tendency to homopolymerize. Other types of polymer functionalization are included under Functionalization / Functional Group Modification.

4.2 Bulk Polymerization

4.2.1 Introduction

In bulk polymerization a monomer or a mixture of monomers is converted to high molecular weight polymer with little or no solvent dilution. Extruder reactors have been designed which handle pure monomer as feed or which take low viscosity prepolymer from a CSTR for final polymerization to high viscosity. Bulk polymerization in an extruder has been considered in general review articles by Mack (1972), and Mack and Herter (1976), and in more specialized articles on engineering aspects by Meyuhas et al. (1973), Siadat et al. (1979), Chella and Ottino (1981), and Lindt (1983).

During the course of the reaction a separate polymer phase is formed which is often but not always soluble in the monomer phase. In some examples of bulk polymerization a second polymer is predissolved in the monomer feed solution so that an interpenetrating polymer network or a two phase system forms in the extruder. In all cases the viscosity of the reaction mixture may increase sharply as the reaction proceeds. The change in viscosity is often from less than 50 Pa s to greater than 1,000 Pa s over the length of the extruder. Heat transfer becomes more difficult with increasing viscosity. Extruder reactors for bulk polymerization have been designed to convey simultaneously in different barrel segments both starting material and product with a large difference in viscosity, and to control within narrow limits the temperature gradient in the reaction mixture arising from the exothermic heat of polymerization. Provision is also made for the removal of unreacted monomer or volatile by-products by devolatilization of the product at reduced pressure before exiting the extruder. In this manner adequate control of the degree of polymerization may be achieved and consistently uniform product may be obtained.

For maximum rate of reaction and most economical operation bulk polymerization is carried out at the highest possible temperature. Heat transfer not only through the barrel walls but also via the screw itself through circulation of a heat transfer fluid in the screw interior has been used for temperature control. SSEs with screw diameter of 1.2 m, screw length of 15.2 m, and production rates of 900 kg/hr have been described. More efficient heat transfer and shorter extruder residence time is possible using TSEs with intermeshing, self-wiping screws. The high rate of material surface renewal results in increased contact with heat transfer surface in the extruder, in which case temperature tolerance of ± 5 °C may often be achieved. Intermeshing screws also provide improved mixing and higher rate of reaction. When applicable, TSEs allow for higher output of final product than SSEs. Two types of bulk polymerization have been performed in extruder reactors: condensation polymerization and addition polymerization.

4.2.2 Condensation Polymerization

Condensation polymers can arise through a repeated condensation process of two distinct monomers to give high molecular weight polymer and a low molecular weight by-product such as water or a low boiling alcohol. To ensure high conversion to product, the reaction equilibrium must be optimized by efficient removal of low molecular weight by-product. Extruder reactors for condensation polymerization typically provide for vacuum venting at one or more barrel segments to remove volatile by-product. When two monomers are condensed, exact stoichiometry matching is critical for production of high molecular weight polymer because condensation polymers are formed by a step-growth mechanism. Since it is difficult to feed solids into an extruder feed throat with > 1 wt % accuracy, the reactants are often fed to the extruder in the form of a melt, or liquid reactants are substituted when possible. Stoichiometry problems are avoided when a self-condensing monomer is employed. Such monomers also release a small volatile side product requiring a vacuum vented extruder reactor. An example is use of an alpha, omega aminocarboxylic acid to form polyamide with water as side product.

4.2.2a Polyetherimides

Takekoshi and Kochanowski (1974, 1977), Banucci and Mellinger (1978), and Schmidt and Lovgren (1983, 1984) have described condensation of bisphenol A dianhydride with different aromatic diamines to give polyetherimides using an extruder reactor.

polyether imide

The basic reactor design shown in Fig. 1 (Banucci and Mellinger, 1978) consisted of a TSE with intermeshing screws and barrel length such that the average material residence time was 4.5 min. In Fig. 1 a solid mixture of

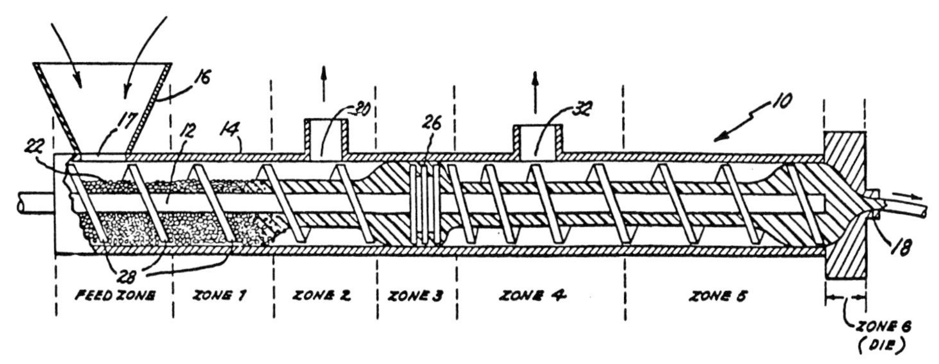

Figure 4.1 Extruder reactor for synthesis of polyetherimide (Banucci and Mellinger, 1978)

bisphenol A dianhydride, a diamine such as m-phenylene diamine, and phthalic anhydride chain stopper was fed to the extruder throat where the materials melted in zone 2 and began polymerizing. The temperatures across the extruder were typically 45 °C in zone 1, 250 °C in zone 2, and 320 °C in the remaining zones. Zone 2 of the reactor was only partially filled so that surface renewal allowed removal of some of the water of reaction through the first vent (20) maintained at atmospheric pressure. At zone 4 there was a vacuum vent (32) for more complete removal of water. Fig. 2 shows the screw geometry in zone

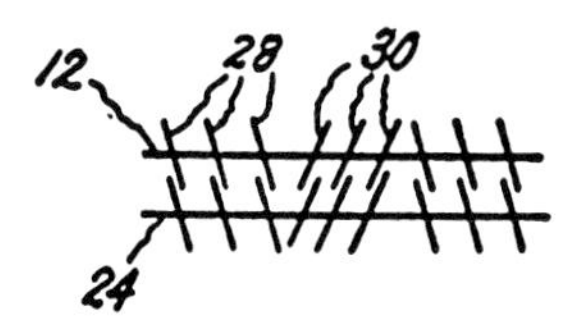

Figure 4.2 Diagram of screw elements in zone 3 to promote melt sealing (Banucci and Mellinger, 1978)

3 which promoted melt sealing to prevent back mixing of material between zones 2 and 4. The screw elements here consisted of a right handed (forward conveying; 28) followed by a left handed (backward conveying; 30) screw element to retard the flow of material between zones.

Lo and Schlich (1986) achieved higher throughput and improved product properties by performing the melt polymerization in two stages. For example

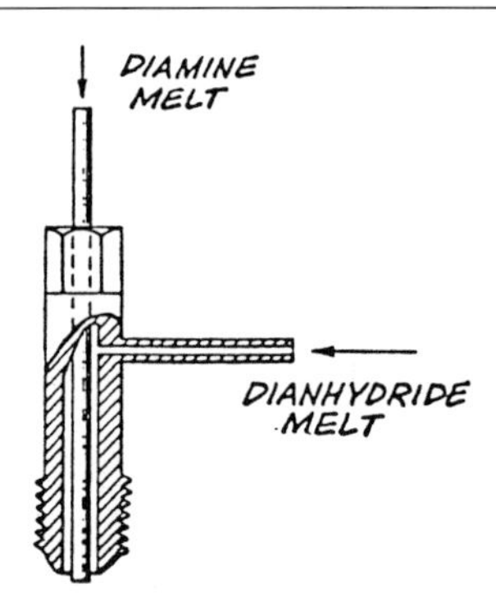

Figure 4.3 Concentric feed pipe for feeding molten reactants to extruder (Schmidt et al., 1984)

in the first extrusion a low molecular weight prepolymer with .23-.31 dl/g IV was formed through polymerization of a m-phenylene diamine with 4.71% molar excess of bisphenol A dianhydride using either a 1 inch or a 2 inch SSE with efficient water removal. The prepolymer was reextruded with an amount of diamine sufficient to achieve stoichiometric balance to give product with IV .47-.49 dl/g.

Schmidt et al. (1984) and Schmidt and Lovgren (1985) achieved improved stoichiometry control in a single extrusion step by feeding the dianhydride and the diamine to the extruder as separate melt streams via a concentric tube feed inlet shown in Fig. 3. Product inhomogeneity was prevented by keeping the two molten reactants separate until they could be mixed properly in the extruder.

4.2.2b Polyesters

Polyesters have been synthesized in extruder reactors from low molecular weight prepolymer.

$$HO-C_6H_4-C(CH_3)_2-C_6H_4-OH + C_6H_4(CO_2Ph)_2 \xrightarrow{\text{KOPh catalyst}}$$

iso:tere = 75:25

$$\left[O-C_6H_4-C(CH_3)_2-C_6H_4-O-\overset{O}{\overset{\|}{C}}-C_6H_4-\overset{O}{\overset{\|}{C}}\right]_n + 2\ PhOH$$

polyarylate

Kosanovich and Salee (1983, 1984a,b) took condensation prepolymer of bisphenol A and 75:25 isophthalate : terephthalate diphenyl esters from a CSTR and completed the polymerization in a 5 stage TSE equipped with five vacuum vents to remove phenol byproduct. Operating at 125 rpm and 320-340 °C, the extruder produced polyester of IV .50-.57 dl/g starting from IV .40 dl/g.

Gouinlock et al. (1968) have demonstrated the use of an extruder reactor to advance the molecular weight of low molecular weight copolyester derived from bisphenol A, neopentyl glycol, and terephthaloyl chloride. A prepolymer with 0.16 dl/g IV was fed to a Welding Engineers Point-Eight TSE with six barrel sections fitted with one large vent and two small vents adjacent to one another and placed either before or after the large vent as shown in Fig. 4.

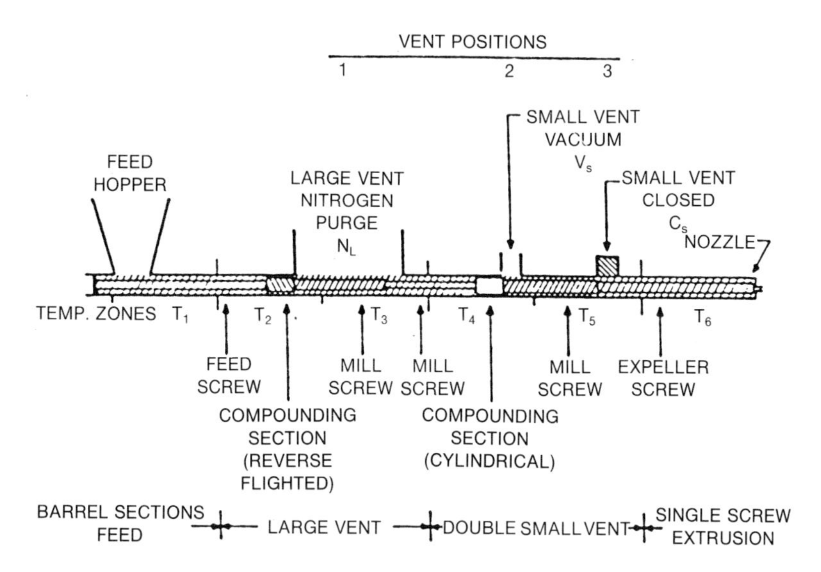

Figure 4.4 Extruder reactor for synthesis of polyester (Gouinlock et al.,1968). Reproduced by permission of John Wiley and Sons.

Each vent was either closed, provided with nitrogen purge, or subjected to vacuum (2 mm Hg). Material residence time varied from 15 min at 100 rpm to 30 min at 50 rpm. Final product IV at 300°-325 °C varied from 0.53-0.76 dl/g depending upon the vent configuration, screw speed, and whether or not triphenyl phosphine stabilizer (1.25%) was added. Compared to solution synthesized polyester, melt synthesized product showed an increase in melt viscosity due to chain branching, probably due to Friedel-Crafts acylation of bisphenol A aromatic rings by terephthaloyl chloride promoted by hydrogen chloride at the elevated extruder temperature.

4.2.2c Melamine-Formaldehyde Resin

H_2N, N, NH_2 / N, N / NH_2 + xCH_2O ⟶ R_2N, N, $N(R)-CH_2OH$ / N, N / NR_2

R = H, CH_2OH

curing ⟶ $-\!\!(N(R)$–triazine(N, N, N; NR_2)–$N(R)-CH_2\!)_n-$ ⟶ network polymer + xH_2O

Melamine-formaldehyde prepolymer has been synthesized by Streetman (1985) in an SSE reactor at 130 °C. A typical solid feed stock of 2:1 paraformaldehyde : melamine gave 95-100% conversion from 3 min extruder residence time. Typical prepolymer melt viscosity was 250 Pa s. Final curing of the extrudate was carried out in a heated reaction chamber where water by-product was removed.

4.2.2d Cyanoacrylate Monomer

CO_2CH_3 / CH_2 / CN + CH_2O $\xrightarrow{\text{base}}$ $-\!\!(CH_2-C(CO_2CH_3)(CN))_n-$ + H_2O

$\xrightarrow[\Delta]{}$ $CH_2=C(CO_2CH_3)(CN)$

cyanoacrylate monomer

In a reactive extrusion process described by Waniczek and Bartl (1984) to produce cyanoacrylate monomer, cyanoacetic ester was first condensed with formaldehyde to give a low molecular weight polymer in a TSE with L/D 40. For example, a liquid mixture of methyl cyanoacetate (or its ethyl analog) and paraformaldehyde was fed to a 5 zone extruder at 120 °C and combined with a 50% solution of piperidine in ethyleneglycol dimethylether introduced into zone 3 of the extruder. Zone 4 was vacuum vented to remove water by-product. The polymer extrudate with average MW 600-1,000 was fed along with hydroquinone antioxidant to a second 32 mm TSE where the temperature was raised in successive barrel segments to 300 °C to crack the polymer to methyl alpha-cyanoacrylate monomer. Through a separate feed port a heat transfer fluid such as tricresyl phosphate was added to the reaction mixture. The monomer was removed through a vacuum vent and the black viscous extrudate was discarded.

4.2.3 Addition Polymerization

Although addition polymer formation does not result in a low molecular weight byproduct, the synthesis of addition polymers by bulk polymerization in an extruder is often done with vacuum venting to remove unreacted monomer. Also, because of the high heat of polymerization associated with some types of addition polymerization such as free radical reactions, it is sometimes advantageous to add a volatile inert material to the monomer reactant. In this way the polymerizing reaction mixture may be cooled by volatilization of the inert material, which may be removed through vacuum venting at an appropriate extruder barrel segment.

4.2.3a Polyurethanes

The synthesis of polyurethanes and polyurethane-ureas in extruder reactors has been reported by a number of investigators. Since these reactions proceed by step-growth polymerization, stoichiometry matching among the reactants is important for production of high molecular weight polymer and often the reactants are fed to the extruder as melts or liquids.

$$HO{+}(CH_2)_4{-}O{-}\overset{O}{\overset{\|}{C}}{-}(CH_2)_4{-}\overset{O}{\overset{\|}{C}}{-}O{+}_n(CH_2)_4{-}OH$$

typical low mol. wt. polyester
OH terminated

+

$$HO{-}(CH_2)_4{-}OH$$

typical
diol chain extender

+

$$OCN{-}C_6H_4{-}CH_2{-}C_6H_4{-}NCO$$

typical
polyisocyanate

↓

$$+O{-}\overset{O}{\overset{\|}{C}}{-}\overset{H}{N}{-}C_6H_4{-}CH_2{-}C_6H_4{-}\overset{H}{N}{-}\overset{O}{\overset{\|}{C}}{-}O{-}(CH_2)_4{+}\ \text{etc.}$$

representative
polyurethane

Early work by Frye and coworkers (1966) described polyurethane synthesis by premixing 100 parts of a polyester of adipic acid and ethylene glycol (MW = 2,000; hydroxyl number = 56) with 9.4 parts 1,4-butanediol and 40 parts 4,4′-diphenylmethane diisocyanate at 50°-100 °C for 30-45 sec followed by addition of the liquid product to a 0.75 inch TSE at 14 g/min. The reaction mixture was conveyed through a 15 inch zone at 177 °C to the extruder die at 150 °C. The product was reextruded under the same conditions to give material with tensile strength of 6,000 psi and 500% tensile elongation.

Rausch and McClellan (1972) synthesized polyurethane using a corotating TSE consisting of a feed section at 93 °C, mixing section at 103°-143 °C, extruder section at 177°-188 °C, and a die at 216 °C with 40 orifices. The screw elements in the mixing section consisted of eliptical paddles with tips designed to provide shearing action against the inner wall of the barrel and against the corotating shaft. The extruder had an internal capacity of 11 lbs polyurethane mixture and average residence time of 150 sec. The extruder was fed with a mixture of polytetramethylene ether glycol (500 parts), 1,4-butanediol (67.5 parts), and stabilizers and lubricants from one feed tank held at 60 °C, and with 4,4′-methylenebis(phenyl isocyanate) from a separate feed tank at 60 °C. The feed rates were such that the NCO : OH ratio was maintained at 1.05 : 1. A 50% solution of stannous octoate catalyst in dioctyl phthalate was injected into the glycol feed line before it entered the extruder at a rate of 0.05 wt % of the total reaction mixture. The product had tensile strength of 5,580 psi and 340% tensile elongation.

In a process described by Ullrich et al. (1976), a 53 mm TSE was used as a reactor. A mixture of 9 parts butane-1,4-diol chain extender and 91 parts low molecular weight polyester (hydroxyl number = 51.7) from adipic acid and butane-1,4-diol, and 35 parts 4,4′-diisocyanatodiphenyl-methane were pumped separately as melts into the extruder entry zone at 90-120 °C. The middle section of the extruder was held at 180-260 °C and the last section at 100-180 °C. Operating at 70-300 rpm, the extruder produced polyurethane sheet through a sheet die at 30-100 kg/hr throughput and 2.5-0.8 min residence time. The specific net energy requirement was .18-.54 MJ/kg of material passed through the machine. A key to the success of this process was the presence of at least two or at most three kneading zones each 240 mm in length (at 7, 8, and 9) preceeded by short metering segments shown in Fig. 5. Kneading screw ele-

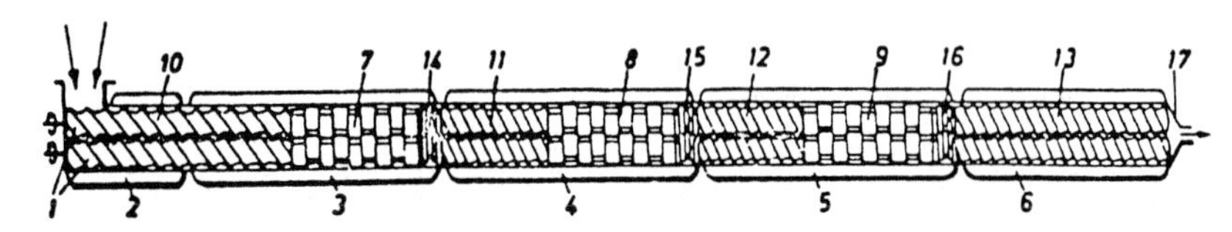

Figure 4.5 Extruder reactor design for polyurethane synthesis (Ullrich et al., 1976)

ments were combined in series in each of these kneading zones to provide sufficient mixing and shearing of the reactant mixture to prevent the formation of gel inhomogeneities in the extruded sheet. Suitable intermeshing kneading elements are shown in Fig. 6. A variety of other polyurethane and polyurethane-urea precursors and coreactants were described in this same patent.

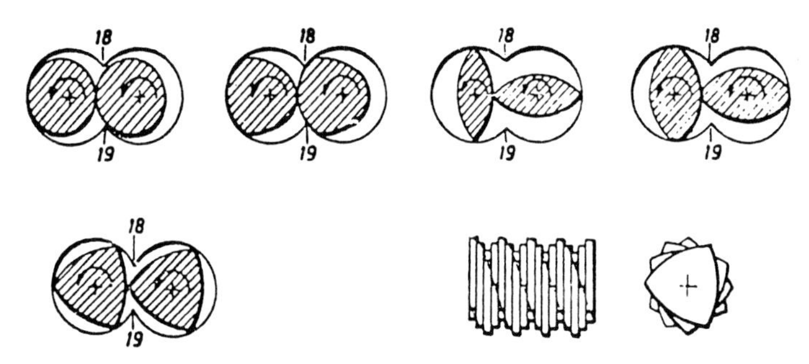

Figure 4.6 Kneading screw elements suitable for intensive mixing in polyurethane synthesis (Ullrich et al., 1976)

Other patents have covered use of mixtures of diol chain extenders, e.g. an equimolar mixture of 1,4-butanediol and ethylene glycol, in polyurethane synthesis in an extruder (Quiring et al., 1981a) addition of a thermoplastic polyurethane at the extruder feed hopper followed by injection of diisocyanate plus diol chain extender at a later barrel segment (Goyert et al., 1981) addition at the extruder feed hopper of an ABS graft copolymer prepared from 14.4% styrene, 5.6% acrylonitrile, and 80% butadiene (optionally with glass filler) along with the polyurethane reaction ingredients (Goyert et al., 1982, 1985) and a two-stage process for polyurethane synthesis in which low molecular weight polyester, butanediol, and diisocyanate were intensively mixed and allowed to solidify on a steel band conveyer before feeding to a 200 mm SSE at 160-210 °C for completion of polymerization (Zeitler et al., 1986).

A related patent (Goyert et al., 1988) described polymerization in an extruder of hydroxy-terminated polyester diol, diol chain extender, and diisocyanate in the presence of a crosslinking agent free of hydroxyl groups such as trimethylol-propane trismethacrylate. The products are polyurethane elastomers which are crosslinkable with radiation.

Reischl (1986) combined toluene diisocyanate distillation residues (about 27% free isocyanate groups) with 34.6% diethylene glycol in a 53 mm corotating TSE at $\leq$ 165 °C with 1-2 min residence time. The polyurethane product could be molded into board with or without woodchip filler.

Quiring et al. (1981b) produced polyurethane-ureas by reaction of a diisocyanate with hydroxy-terminated polyester, diol chain extender and water. For example, 100 parts polybutanediol adipate (hydroxyl number = 52), 15 parts 1,4-butanediol, 1.2 parts ethylene-bis-stearylamide, and 0.4 parts 2,6-di-t-butyl-4-methylphenol were combined and 200 parts of this mixture at 70 °C were fed along with 100 parts 4,4′-diisocyanato- diphenylmethane at 50 °C to a 12 zone, 53 mm TSE (L/D = 41) with corotating, self-cleaning screws and kneading elements located in the fourth, seventh, eighth, and tenth zones. The first three zones were maintained at 190 °C. Water (15 parts) was injected into the reaction mixture at the fifth barrel segment at 230 °C after which the temperature was gradually decreased to 120 °C over the remainder of the extruder segments. The product had melt index of 40 g/10 min and was storage stable. A thermoplastic resin such as BPA polycarbonate or ABS could be introduced at the feed hopper along with the above-described ingredients.

Roberts (1977) reacted diisocyanate with water in a separate step to form isocyanate-terminated polyurea followed by reaction in an extruder with hydroxy-terminated polyester and diol chain extender to produce a polyurethane. For example, a mixture of 1,974 g toluenediisocyanate and 9 g water was reacted in a flask and the solid N,N′-di(isocyanato)- tolueneurea was ground to a powder. A mixture of 610 g of the urea derivative, 667 g polycaprolactone diol (MW ≈ 2,000), and 141 g diethylene glycol was fed to a vented SSE (L/D = 36) with temperature profile 100-180 °C. The downstream vent was operated at atmospheric pressure. The polyurethane product had tensile strength of 3,200 psi and tensile modulus of 23,800. Reextrusion of the material gave test parts with tensile strength of 4,200 psi and tensile modulus of 43,500, indicating that polymerization was not complete after one extrusion.

Low molecular weight PBT copolymers with terminal hydroxyl groups have been polymerized with diisocyanates to give polyurethanes in extruder reactors by Mizuno et al. (1978). For example, flame retarded polyurethanes have been prepared in 2 min residence time in 65 mm SSE at 250 °C by reacting a mixture of 5.4 parts diphenylmethane-4,4′- diisocyanate, 8 parts antimony oxide, and 100 parts PBT copolymer (MW 7,000; prepared from 100 parts dimethyl terephthalate, 56 parts butanediol, and 25 parts ethylene oxide adduct of tetrabromo bisphenol A). The isolated product had MW 25,000. The formulations could be glass filled and could contain an epoxy resin.

Related patents (Mizuno and Sugie, 1979; Mizuno and Adachi, 1981; Adachi and Mizuno,1981) describe polyurethane formation from low molecular weight PBT with terminal hydroxyl groups and an isocyanate prepolymer prepared from diphenylmethane-4,4′-diisocyanate and either bishydroxybutyl terephthalate or dihydroxyethyl terephthalate or tetrabromobisphenol A polycarbonate. Glass fibers or beads were present in the formulations.

Studies of poly(esterurethane) synthesis in a Brabender mixer have been published by Schollenberger et al. (1981, 1982). The polymer was prepared from diphenylmethane-4,4′- diisocyanate, poly(tetramethylene adipate) glycol, and 1,4-butanediol in a ratio of either 2:1:1 or 3:1:2. Polymerization variables such as temperature, stabilizer presence, and macroglycol acid number were investigated. The catalytic effect of stannous octoate and the quenching effect of added 1-propanol or 1-decanol were also examined.

A mathematical model describing flow of a polymerizing polyurethane mixture in an SSE channel has been developed by Hyun and Kim (1988) and compared to experimental data generated on a 19 mm Brabender extruder (L/D = 20) with constant channel depth of 2.5 mm. The reactants consisted of 4,4′-diphenylmethane diisocyanate, polycaprolactone diol, and 1,4-butanediol chain extender (equivalent wt. ratio 2:1:1). Dibutyltin dilaurate was added as catalyst.

An optimized process for synthesizing polyurethane in a TSE from polyester alcohol, MDI, and butanediol chain extender has been published by Rotermund (1987). It was claimed that almost ideal molecular weight distributions could be prepared with minimum content of low molecular weight compounds.

4.2.3b Polyamides

Na^+

n (caprolactam) —initiator→ $\left[C(=O)-(CH_2)_5-NH \right]_n$

Nylon 6

Polyamide synthesis in extruder reactors by ring-opening polymerization of lactams has been reported in a number of publications. In an early example Illing (1969) polymerized lactams such as caprolactam or dodecalactam in a TSE equipped with corotating, intermeshing screws. The reaction was initiated with 0.2-0.3 wt % sodium lactamate and an accelerator was also added. The components were either all preblended for throat feeding, or two feed streams consisting of lactam plus initiator and lactam plus accelerator could be fed simultaneously. Material was melted at 90-100 °C in an extruder zone equipped with kneading blocks and conveyed to a polymerization zone held at 140-160 °C. In this section the shear forces on the polyamide melt could be varied using an adjustable valve, which controlled the clearance between the screw elements and the outer sleeve of the barrel. By varying the gap width between 0.5-13 mm and simultaneously varying the screw speed, the average MW of the polymer could be controlled at a certain temperature in the shear zone. The final product passed through a melt seal and was expanded into devolatilization zone at 230 °C. Throughput rate was 450 lb/hr for an 83 mm TSE and 900 lb/hr for a 120 mm TSE. Average residence time of material was only 2-3 min.

backbone. Free radical initiators and, less commonly, air or ionizing radiation have been used to initiate the reaction. Extruder reactors for performing graft reactions may include intensive mixing sections and screw segments designed to expose the maximum surface area of polymer substrate to grafting agent. In some cases the monomer to be grafted is injected into a molten stream of the substrate polymer under conditions where a high surface area of the polymer is exposed for grafting. An initiator such as peroxide may be added at a separate barrel segment before or after monomer injection, or premixed with polymer substrate in the form of a concentrate. Unreacted monomer and other volatiles are removed by vacuum venting in a devolatilization zone prior to exiting through the die.

Depending on the type of monomer employed, monomer homopolymerization may compete with grafting. Efficient mixing of the monomer with polymer is essential to minimize homopolymer formation although in some examples given below homopolymer formation is actually desired in addition to grafting. With monomers prone to homopolymerization, the chain length of the resulting grafts is usually long enough so that the resulting product may be considered a true graft copolymer which has physical properties different from those of the substrate polymer. However, depending upon the individual reactivities and mole ratios of monomer and polymer, the initiator level, the processing temperature, and other factors, the graft chain length may be quite short or even consist of a single monomer unit. In this case the substrate polymer may have relatively unchanged mechanical properties but markedly different chemical properties. An example is functionalization of polyolefins through grafting with maleic anhydride. While the maleic anhydride graft segment may consist of only one unit, the polyolefin-g-maleic anhydride has adhesive properties and may form copolymers with other nucleophile-containing polymers through reaction at the anhydride site (see Section 4.4). Since the chain lengths of grafted blocks made by reactive extrusion have seldom been determined, no distinction is made in this section between true graft copolymers and simple polymer functionalization which results from grafting of a potentially polymerizable monomer. A small number of other examples of polymer functionalization through a grafting reaction *not involving a potentially functionalizable monomer* are given in Section 4.7.

4.3.1a Vinyl Silanes

Grafting of vinyl silanes to polyolefin substrates in the presence of peroxide is the most common example of a graft reaction performed in extruder reactors. Polyolefins grafted with vinyl silanes are readily crosslinked by moisture and the materials have a large commercial market as wire coating and pipe insulation. Reviews summarizing aspects of the various grafting and crosslinking processes have been published by Scott and Humphries (1973), Bloor (1981), Munteanu (1985), and Cartasegna (1986).

The overall process is shown in the scheme below.

$$\left(CH_2-\underset{}{\overset{R}{\overset{|}{C}}}H-CH_2 \right)_n \quad + \quad CH_2{=}CHSi(OCH_3)_3 \quad \xrightarrow{\text{peroxide initiator}}$$

some polymer
with graftable site

$$\left(CH_2-\underset{\underset{CH_2CH_2Si(OCH_3)_3}{|}}{\overset{\overset{R}{|}}{C}}-CH_2 \right)_n \quad \xrightarrow{H_2O\ \text{cure}}$$

$$R-\underset{\underset{\left(CH_2\right)_n}{|}}{\overset{\overset{\left(CH_2\right)}{|}}{C}}-CH_2CH_2\underset{\underset{OCH_3}{|}}{\overset{\overset{OCH_3}{|}}{Si}}-O-\underset{\underset{OCH_3}{|}}{\overset{\overset{OCH_3}{|}}{Si}}-CH_2CH_2-\underset{\underset{\left(CH_2\right)_n}{|}}{\overset{\overset{\left(CH_2\right)}{|}}{C}}-R \quad + \quad CH_3OH$$

crosslinked

Two-Step Processes/Vinyl Silane Grafting A two-step commercial process has been described which involves grafting vinyl silane to polyolefin in an SSE or TSE at 180-200 °C to give a grafted product with a storage life of several months. For crosslinking, the grafted polyolefin is mixed with a crosslinking catalyst which is often dibutyltin dilaurate in the form of a concentrate with ungrafted polyolefin. This mixture is passed through a second extruder or pair of extruders for mixing and shaping to give a product in the form of a wire or cable coating. The grafted polyolefin product is then crosslinked in the presence of moisture. Dow Corning has marketed technology for crosslinked polyolefin produced by the two-step process under the name Sioplas E.

In an early example (Scott, 1972), a mixture of 100 parts PE with 3 parts vinyl triethoxysilane and 0.12 parts DCP, was fed to a continuous kneader extruder at 180-184 °C, 31 rpm, and 1.5-2.5 min residence time at 13 lbs/hr output. Extraction of the extrudate showed that 88.5% of the silane was covalently bound. Before curing with moisture, the product was extruded with 5% of a concentrate containing 100 parts PE, 1 part dibutyltin dilaurate, and 0.12 parts DCP as curing catalyst. Both LDPE and HDPE were used in this process.

peroxyisopropyl)benzene, 2.6 parts VTMOS, and 0.1 parts triallyl cyanurate were mixed at 1,700 rpm until the temperature reached 95 °C, after which 650 rpm. The mixer was cooled to 70 °C and a mixture of 5 parts 1% dibutyltin dilaurate on PE was added. The dibutyltin dilaurate concentrate had been previously prepared in the same mixer under similar conditions. The combined formulation was extruded on an SSE at 160-230 °C with 2.5 min residence time and 25 rpm to effect grafting.

Hagger et al. (1988) have employed continuous monitoring of dynamic viscosity of silane grafted polymer for process control. For example, 100 parts PE, 1.5 parts VTMOS, 0.1 part DCP, 0.5 parts stabilizer, and 0.05 parts dibutyltin dilaurate were fed to a Maillefer 120 mm extruder (L/D = 30) having a feed zone with 8D length, a homogenizing zone of 6D length, a converging zone of 6D, and a metering zone of 10D. The first two zones were held at 130 °C and the last two at 230 °C. A gear pump withdrew grafted polymer at 1,000 mm^3/sec from a point one diameter from the extruder outlet and fed the material to a dynamic rheometer. The supply of peroxide fed to the extruder feed throat was adjusted to keep the product melt viscosity at 1.5-2.4 x 10^4 poise at 190 °C and 1 sec^{-1} shear rate.

Spielau et al. (1987) have grafted vinyl silane onto PP with MFI < 0.7 g/10 min at 190 °C in the presence of radical initiators with a half-life of 1 min at about 160°-240 °C. A certain amount of degradation of the PP was purposely intended so that extruder energy consumption would be decreased, and extrudate products with MFI 25-70 g/10 min at 230 °C were claimed. For example, 100 parts PP, 2 parts methacryloxypropyl- trimethoxy silane, 0.05 parts dibutyltin dilaurate, and 0.4 parts t-butylperoxy-(3,5,5-trimethyl)-hexanoate were mixed and extruded on an SSE at 220 °C to give product in the form of a band which showed 90% crosslinking after two days storage in hot water. VTMOS could also be used as silane component. BOP was among the other peroxides that could be used.

In processes in which the extruder was not actually used as the reactor, PE was melted in an extruder fitted with a cavity transfer mixer immediately before the die. VTMOS, DCP, and dibutyltin dilaurate were fed to the molten PE in the cavity transfer mixer where grafting took place (Gale and Sorio, 1985; Wheeler, 1986).

Other examples of one-step processes include addition of hydrated alumina to blend and curing through thermal treatment to promote release of water from the hydrated salt (Schleese et al., 1981), grafting of vinyl silane to EVA (Hochstrasser and Kertscher, 1982), production of film from LLDPE and VTMOS using masterbatch containing PE/silane/peroxide (Columbo et al., 1986), grafting process with static mixer at end of extruder (Kerschbaum and Konicka, 1987), grafting of vinylsilanes to SEBS for adhesive compositions (St. Clair and Chin, 1988), and grafting of VTMOS to mixture of PE, EVA, and nitrile rubber in the presence of DCP and dibutyltin dilaurate (Umpleby, 1986).

Silane Grafting by Ester Exchange An entirely different method for grafting silane functionality involves coextruding a polymer containing pendant ester groups with a carboxylic acid ester containing a silane functionality in the alcohol portion of the ester as depicted in the Scheme. Silane grafting takes place in the presence of a transesterification catalyst such as a tetraalkoxy titanate.

$$\left(CH_2-\underset{\underset{CO_2R}{|}}{CH}\right) + CH_3\overset{\overset{O}{\|}}{C}OCH_2CH_2Si(OCH_3)_3$$

$$\xrightarrow{Ti(OR')_4} \left(CH_2-\underset{\underset{CO_2CH_2CH_2Si(OCH_3)_3}{|}}{CH}\right) + CH_3\overset{\overset{O}{\|}}{C}OR$$

In one example (Keogh, 1981), EEA copolymer (18% acrylate) and 3.7 wt % 2-(triethoxysilyl)ethyl acetate ("acetooxyethyltrimethoxy silane") were combined in a Brabender mixer at 160 °C until homogeneous. Tetraisopropyl titanate (1 wt %) was added after which mixing was continued for about 15 min. The product was combined with 5 wt % EEA copolymer containing 1 wt % dibutyltin dilaurate and pressed into test plaques which were crosslinked by exposure to moisture. The material had only 20% wt loss on extraction with decalin, which indicated a high degree of crosslinking. For comparison, the same material made using either 1-(2-trimethoxysilylethyl)-3,4- epoxycyclohexane or N-3-(trimethylsilyl)propyl acetamide showed essentially no crosslinking. Other silanes such as methacryloxypropyl- trimethoxysilane could also be used successfully. Ethylene copolymers with 2-18% ethyl acrylate or with 10% butyl acrylate could be used.

A related patent (Keogh, 1982) described grafting to EEA copolymer using a polysiloxane derived from polymerization of acetooxyethyltrimethoxy silane with tetraisopropyl titanate in a separate reaction step. Subsequent mixing of 244 g EEA copolymer (16% acrylate) with 5.8 g of a mixture of 97% polysiloxane and 3% dibutyltin dilaurate followed by addition of 1.26 g tetraisopropyl titanate in a Brabender mixer at 160-170 °C for 30 min gave grafted copolymer which was molded into test plaques and cured with water. The degree of crosslinking was found to be the same as when acetooxyethyltrimethoxy silane itself was used for grafting, but the amount of volatiles released during grafting

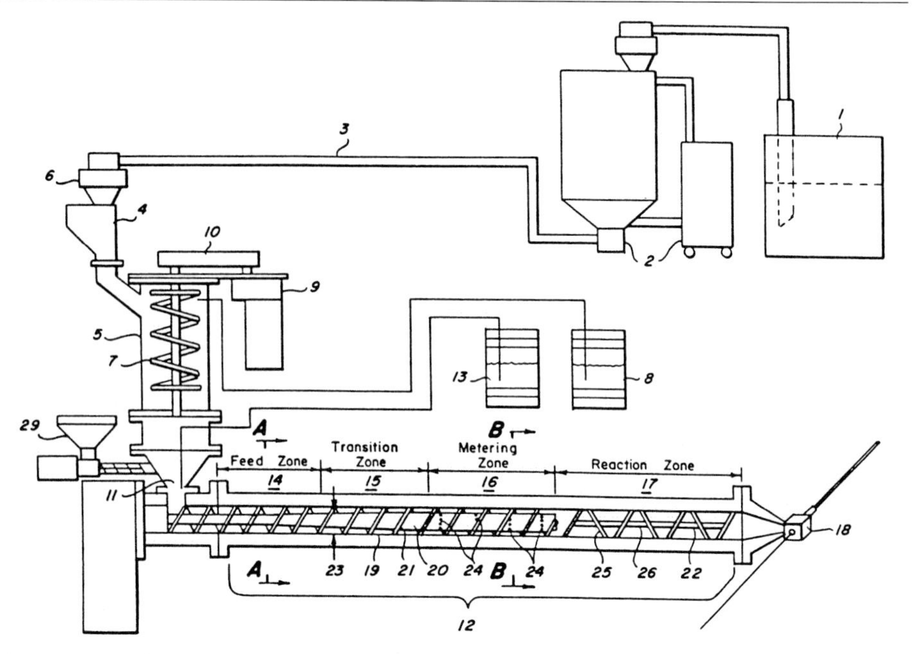

Figure 4.12 Extruder system for grafting silane by exchange reaction (Brown, 1986)

with the polysiloxane was significantly lower. Polysiloxanes end-capped through reaction with a monoester such as ethyl benzoate could also be used.

Related patents (Keogh et al., 1984; Brown, 1986) have described a process where the extruder was modified to insure intensive mixing and high grafting efficiency (Fig. 12). EEA copolymer was mixed with dibutyltin dilaurate, dried, and mixed in a helical mixer (7) with tetraisopropyl titanate (from reservoir 8) such that the titanate soaked completely into the pellets. The helical mixer fed directly to an extruder feed hopper where the copolymer pellets were contacted with an appropriate polysiloxane or monomer siloxane as described above pumped in from reservoir 13. The 2.5 inch extruder (L/D 30) had a hollow screw for circulation of coolant water for better temperature control, and an axially grooved barrel to improve friction between reactants and barrel, thus offsetting lubrication by grafted product. The extruder consisted of a feed zone (140-160 °C), a transition zone where most melting occurs (150-175 °C), a metering zone (160-190 °C) containing four sets of radial mixing pins where sufficient pressure was generated to force the material through a final reaction zone. This zone consisted of a dynamic or a static mixer (22) with increased channel depth for longer residence time of an increased volume of material allowing the reaction to go to completion at 160-190 °C. The product was coated directly onto wire followed by curing with moisture. Colorants could be added via an auger-type hopper at 29.

Polyolefins have also been grafted by reaction with a vinyl silane and radical initiator followed by treatment in a separate step with an organotitanate catalyst for facilitation of crosslinking (Keogh, 1985). For example, 100 parts PE, 18.1 parts vinyl tris(n-dodecyloxy)silane, and 17.6 parts of a masterbatch of PE containing 1.5% DCP and 0.1% hindered phenol stabilizer were mixed in a Brabender mixer at 185 °C for 7 min. A molded test plaque immersed in a water bath at 70 °C for 16 hrs showed a Monsanto Rheometer reading of 6 lbs-inch as a measure of crosslinking. Material which was remelted with 5 wt % tetramethyl titanate gave test plaques which showed 36 lbs-inch. For comparison, test plaques of a blend containing 0.03 wt % dibutyltin dilaurate showed a reading of 19 lbs-inch. EEA copolymer (17-19% acrylate) could be used in place of PE.

4.3.1b Acrylic Acid, Acrylic Esters, and Analogs

$$-\!\left(CH_2-\underset{}{\overset{R}{\overset{|}{C}}}H-CH_2\right)_n- \quad + \quad CH_2=\overset{R}{\overset{|}{C}}-\overset{O}{\overset{\|}{C}}-Z \longrightarrow$$

some polymer with graftable sites

R = H, alkyl
Z = NR_2', OR', OH

$$-\!\left(CH_2-\overset{R}{\overset{|}{\underset{\underset{CH_2\underset{\underset{R}{|}}{C}H-\underset{\underset{O}{\|}}{C}-Z}{|}}{C}}}-CH_2\right)_n-$$

Grafting to Polyolefins or Olefin Copolymers Grafting of AA and its analogs to polyolefins such as PE and PP has been disclosed by a number of workers. In an early example (Nowak and Jones, 1965), PE was fed to a 1.25 inch extruder at 24.5 g/min at 190 °C. At the midsection of the extruder a solution of AA containing 0.75 wt % t-butylperacetate and 0.04 wt % t-butyl catechol inhibitor was injected at 3 g/min. Material residence time was 2.5 min. Grafting level was assessed by measuring the change in tensile strength of molded test parts. Tensile strength increased with increasing amount of AA added to the extruder. Many other initiators were used including DCP, TBP, BOP, DTB, and cumene hydroperoxide. PP and EP were also grafted under these conditions in the presence of DCP at 175-200 °C. The initiator solution could also be throat-fed with polymer substrate. Methacrylic acid was also used.

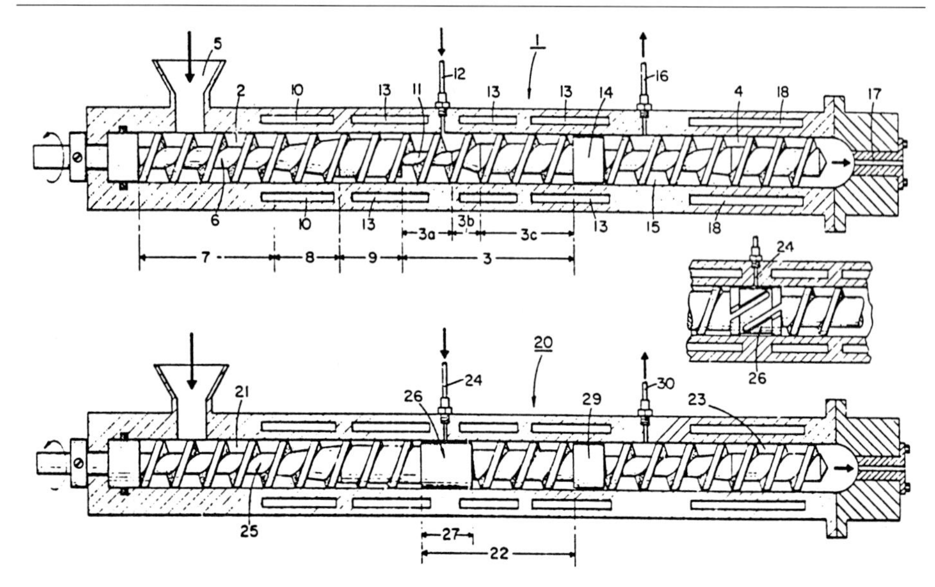

Figure 4.13 Extruder screw configurations for grafting acrylic acid onto polyolefins (Steinkamp and Grail, 1975)

In a related patent (Nowak, 1966), PE was treated with high energy ionizing radiation in a separate step before extrusion with AA under similar conditions. Grafting level was similar to that obtained using added peroxide as initiator.

Many of the fundamental principles connected with efficient grafting of immiscible reactants using an extruder reactor were disclosed in a series of patents from Exxon. An example of a 2 inch Egan SSE used for grafting AA to PP is shown in Fig. 13 (Steinkamp and Grail, 1975). The extruder reactor consisted of a feed zone at 204-288 °C, a reaction zone at 121-232 °C, and a metering zone at 177-232 °C. The figure shows three different screw configurations in the reaction zone for generating high surface area of molten polymer. In the top configuration the screw geometry was specially designed so that the polymer was melted and compressed in the feed zone, and then decompressed into a reaction zone with greater volume (11) where it presented high surface area for efficient reaction with a liquid mixture of AA and peroxide initiator which was injected into this zone under pressure at 12. In the middle configuration the reaction zone 26 has a screw root of large cross sectional area having a series of dead end channels cut into its perimeter, out of which polymer was forced in the form of a thin film presenting high surface area for reaction with monomer and initiator fed through line 24. In the bottom configuration the reaction zone 27 has a screw root 26 of very large cross sectional area with clearance between root and extruder wall of about 10-20 mils. At this point the polymer was present as a thin film for reaction with monomer and initiator fed through line 24. As shown in the top and bottom configuration the grafted product is again compressed at a screw element with large cross sectional area (14 or 29) and then decompressed into a final metering zone which contains a vacuum vent (16 or 30) for removal of excess monomer. The two compression zones acted as melt seals, which prevented loss of pressurized AA from the reaction zone.

In one example, PP was reacted at 152-204 °C at 160 rpm. A solution containing 15 g DCP per 100 g AA was injected into the reaction zone at a rate of 9.92 wt % to give 8.6 wt % grafted monomer in the final product with AA homopolymer comprising less than 30% of the grafted PP product. PP molecular weight degradation occurred simultaneously with grafting so that the product had a lower value for molecular weight distribution, improved melt flow rate, and reduced die swell compared to starting material. The grafted products were useful in compositions requiring good adhesion between grafted polymer and some substrate.

A related patent (Stenmark and Heinrich, 1975) discloses a specially designed screw geometry for a SSE which permits more intensive mixing of PP with AA and initiator for high grafting efficiency.

Grafting of AA to PP, PE, PS, or poly(4-methylpentene-1) under apparently similar conditions has been reported by Ide and Sasaki (1977) in connection with improving adhesion of the polymers for use in laminates or glass-reinforced blends.

PP, EP, or mixtures have been grafted with AA, n-butyl methacrylate, or lauryl methacrylate using DCP in a 60 mm extruder reactor at 180 °C (Anonymous, 1970). Polymer and 0.3-1.0% initiator were throat-fed while different levels of monomer were fed 2 min downstream. Total material residence time was 5 min. Depending on conditions, grafting efficiencies of 17-54% were obtained. The remainder of the monomer was present as homopolymer. Lowest grafting efficiency was observed when all ingredients were throat-fed.

Ogihara et al. (1977) have grafted EP copolymer with mixtures of a methacrylate trialkoxysilane and a second acrylate monomer to make adhesive compositions. For example, 100 parts EP copolymer (8 wt % ethylene) was mixed with 0.5 parts TBP initiator, 0.5-2.5 parts methacryloyloxypropyl trimethoxy silane, and 0.5-2.5 parts zinc acrylate and extruded at 220 °C and 5 min residence time to give grafted compositions which displayed better adhesion to aluminum plate than did an EP copolymer grafted without zinc acrylate. Calcium or aluminum acrylate, tris(allyloxy)-s-triazine, or allyl glycidyl ether could be used in place of zinc acrylate.

In a process in which homopolymerization of AA was said to be minimized (Clementini and Spagnoli, 1986), PP was functionalized with hydroperoxide groups before extrusion with AA. For example, 2 kg PP (IV = 2.28 dl/g) was mixed 1 hr at 110 °C with 6 g lauroyl peroxide and 1 g DBPH to give material with 0.07% hydroperoxide groups and IV = 1.5 dl/g. The product was mixed at 30 °C with 40 g AA and extruded on a 45 mm SSE at 200 °C and 30 sec residence time. The product had 1.92% grafted AA. As noted in Section 4.7.3 (Binsack et al.,1981), the hydroperoxidation step can also be carried out in an extruder before grafting.

Zeitler et al. (1976) have grafted EVA with AA in the presence of radical initiator. For example, 100 parts EVA premixed with 3 parts AA and 0.03 parts BOP was fed to a SSE at 140 °C and 12 rpm to give product with acid number of 45. EBA was grafted under similar conditions.

Waniczek et al. (1986) have grafted molten EVA by peroxygenation with air in a 32 mm TSE followed by treatment with methacrylic acid in a subsequent extruder zone. Peroxygenation produces material with 50-3,000 ppm active oxygen and also gives some oxidative degradation of the polymer. The level of monomer incorporated was inversely proportional to the melt index of the product. Degradation led to a final grafted product with more favorable melt index for application as a hot melt adhesive. Mixtures of methacrylic acid and either styrene or n-butyl acrylate were also grafted using this process.

Grafting of methacrylic acid to natural rubber by a mechanochemical process has been described by Shuttleworth and Watson (1974). For example, 150 g methacrylic acid was impregnated onto 600 g rubber sheet and then passed through a 1 inch SSE (L/D = 12) at 20 rpm. The extruder was efficiently cooled to maintain temperature of the extruded product at ≤ 70 °C, which is a convenient temperature for mechanochemical scission of rubber chains and radical generation. One to 10 passes of the material through the extruder were necessary for high conversion of monomer. The product was characterized by selective solvent extraction, Mooney viscosity, SEM, and IR spectroscopy. Yields varied from 53 to > 80% grafted monomer depending on the number of passes.

Polyolefins have been grafted with various monomers in extruder reactors using oxygen as radical initiator. For example (Binsack et al., 1981), high pressure PE was fed at 75 kg/hr to a 57 mm corotating TSE (L/D = 41) with self-cleaning screws at 100 rpm. The extruder was divided into an induction zone at 220-235 °C, an oxidation zone at 210 °C, and a polymerization zone at 210 °C. Air was introduced at 100-130 bar at 400 l/hr to the induction zone and also at 70-90 bar at 1,200 l/hr to the oxidation zone. A 20/80 mixture of AA/n-butyl acrylate was introduced at 15-18 bar at 11.3 kg/hr into the polymerization zone. Following reaction, the material was devolatilized in the extruder and had total residence time of 2.3 min. The grafted product contained 1.1% AA, 11% n-butyl acrylate and 300 ppm residual peroxide. In another example EP was grafted with n-butyl acrylate using a 20 mm TSE (L/D = 48) with counterrotating screws at 190 °C.

In a related patent (Korber et al., 1982), a similar process was used to graft monomers such as butyl acrylate, ethylhexyl acrylate, acrylate-AA mixtures, or acrylate-MA mixtures to high pressure PE or EVA copolymers (8-45% VA).

Michel (1984) pretreated PP with ozone to generate radical sites for grafting of unsaturated monomers such as GMA in the melt.

Togo et al. (1988) have grafted GMA onto PP in the presence of styrene monomer (see also below). For example, 3 kg PP was mixed with 90 g GMA, 300 g styrene, and 15 g DCP, and the mixture was fed to a TSE at 180-220 °C. The extracted product had 6.2 wt % bound styrene and GMA by IR. PPE was also grafted under these conditions at 300-320 °C but without added styrene monomer. The level of grafted epoxide was 1.3 wt % by IR.

Phadke (1987) has grafted EPDM with GMA in a 30 mm corotating TSE. For example, EPDM (66/34/8 ethylene/ propylene/ diene) was fed at 4.6 lb/hr to a feed zone at 70 °C. The remainder of the extruder was at 200 °C. A 10/1 (wt/wt) mixture of GMA and DBPH was fed downstream at 0.385 lb/hr. The product had a high level of grafted epoxide by IR and 17% gel by toluene extraction.

Taubitz et al. (1988a) have grafted ABS with GMA by reactive extrusion. For example, 98.5 parts ABS and 1.5 parts GMA were extruded on a W&P 53 mm TSE with 4 min residence time at 180 °C (melting zone), 200 °C (reaction zone with kneading screw elements), and 180 °C (devolatilization zone). The product was used without further characterization to prepare compatibilized blends with PS functionalized with carboxylic acid groups (see Section 4.4).

Gallucci and Going (1982) have published a study on grafting of GMA to LDPE, HDPE, or EP copolymer. For example, 19 g LDPE was mixed with 1.9 g GMA and 0.12 g DCP in a mixing bowl attached to a torque rheometer. After 5 min at 120 °C, the temperature was raised to 175 °C for an additional 5 min. No torque increase was observed and no grafting occurred without peroxide initiator. The polymer was analyzed by IR spectroscopy after purification by extraction. Grafting efficiency as a function of DCP/GMA ratio was determined. BOP was also used as initiator. PP and PS failed to graft under these conditions. PE-g-GMA underwent typical reactions of the epoxide group.

In an example (Andersen, 1984) where homopolymerization of monomer was desired, EPDM was starve-fed to the throat of a 30 mm TSE with 14 segments (L/D 42) operated at 250 rpm (Fig. 14). The feed throat was held at 215 °C. Methyl methacrylate and t-butyl peroxide were fed under pressure through a liquid feed port at the third barrel segment. Monomer was prevented from escaping this segment by the presence of dynamic melt seals consisting of reverse screw flight elements (44) before and after the third segment. The temperature was gradually increased from 135 °C in segment 3 (23) to 210 °C in segment 12 (32). A vacuum vent removed unreacted monomer at segment 13 (33). About 35% of monomer grafted while 65% was homopolymerized to PMMA. The final product mixture was blended with additional PMMA to give 20-23% total rubber. Physical properties of this material were superior to those for a comparable blend without copolymer since the copolymer acted as a compatibilizer for EPDM-PMMA (see Section 4.4). Polybutadiene rubber

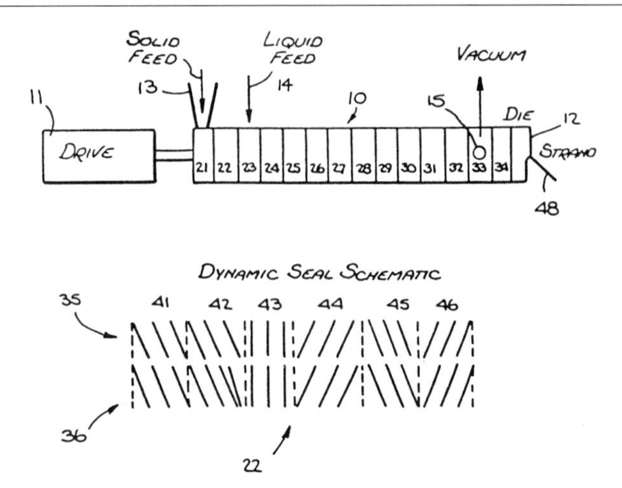

Figure 4.14 Extruder configuration with dynamic melt seals for grafting acrylate ester (Andersen, 1984)

was also grafted using this procedure. Other extruders which could be used in this process were described.

Masked AA derivatives which can undergo reverse Diels-Alder reaction to generate AA (or related monomer) on heating to 120-300 °C have been used by Tabor et al. (1988) as grafting agents for polyolefins. For example, 40 g ethylene-octene-1 copolymer was treated in a mixing head at 50 rpm and 220 °C with .0082 moles 5-norbornene-2-carboxylic acid and 2.8 mg L130, and mixing was continued for 6 min at 200 rpm. The product was purified by precipitation from xylene to give material with 0.27 wt % grafted AA, which was virtually identical to the grafting level achieved using AA itself. HDPE was also grafted using this procedure. Other masked monomers used included the Diels-Alder adducts of cyclopentadiene or furan with maleic anhydride. The actual grafting agent in these examples was the monomer derived from Diels-Alder adduct decomposition and not the starting Diels-Alder adduct itself since the IR spectra of polymers reacted with either pure monomer or Diels-Alder adduct were essentially identical.

An example of an extruder reactor for grafting EP rubber with methyl methacrylate is shown in Fig. 15 (Staas, 1981). The rubber was fed to the feed throat of a counterrotating, tangential TSE with 5 zones as indicated, each zone separated by nonflighted screw elements. Methyl methacrylate was fed through injection tube 15 to zone 2 (12) followed by addition of L130 initiator in toluene solution through injection tube 13 to zone 3 (20). Reactants were conveyed to a reaction zone 22 held at 175 °C followed by a devolatilization zone with vacuum venting at 14 to produce 140 g graft polymer per 100 g polymer feed. Other substrates used in this process included LDPE, EVA, EEA,

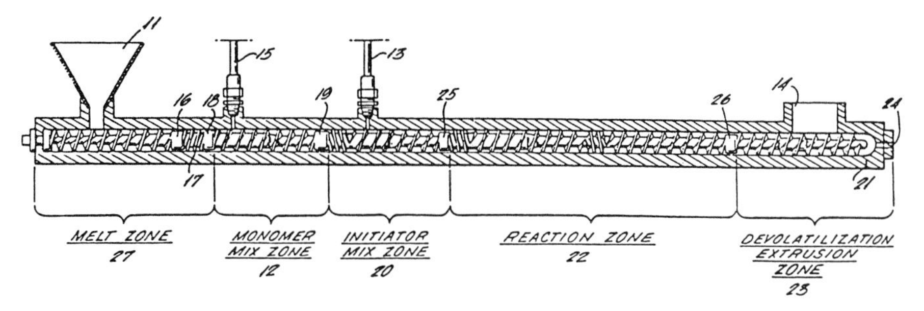

Figure 4.15 Extruder configuration for grafting methyl methacrylate (Staas, 1981)

and EPDM grafted with hydroxyethyl methacrylate, vinyl acetate mixtures, or maleic anhydride mixtures.

Chiang and Yang (1988) have published a study on grafting of AA to PP in an unspecified extruder in the presence of BOP. AA and BOP were dispersed onto PP as an acetone solution and the acetone was evaporated before melt mixing at 130-200 °C and 15 rpm. Grafting efficiency and assessment of AA homopolymer formation were determined by selective solvent extraction and IR spectroscopy. PP-g-AA was used in preparing mica-filled composites.

Simmons and Baker (1989) have published a study on grafting dimethylamino ethyl methacrylate (DMAEMA) to LLDPE in a Rheomix mixer in the presence of a radical initiator such as L130. In a typical reaction, polymer was mixed with 10% DMAEMA and 1% L130 at 160 °C and 100 rpm for 15 min. Product was analyzed by solvent extraction, DSC, melt index, and IR and NMR spectroscopy. Grafting level, depending on temperature, and initiator and monomer concentrations, was from 0.5 to 2.8 wt %. No grafting occurred in the presence of AIBN. LLDPE-g-DMAEMA formed compatibilized blends with styrene-MA copolymer supposedly through strong polar attractive interaction (Simmons and Baker, 1990).

Grafting to Polyphenylene Ethers Johnson et al. (1989) have grafted PPE with acrylate derivatives in extruder reactors. For example, PPE was mixed with 0.75 mole ratio phenyl acrylate and extruded on a TSE at 320 °C to give material with 0.3 mole ratio incorporated monomer. p-Methoxy-, p-fluoro-, and p-cumylphenyl acrylate were also used as monomers. The grafted products were used to produce PPE-nylon copolymer during lactam polymerization.

PPE has also been grafted with N-methacryloylcaprolactam (NMCL). For example (Taubitz et al., 1987a), a mixture of 97 parts PPE and 3 parts NMCL was extruded on a 53 mm TSE at 270-280 °C with devolatilization and 3 min residence time to give product with 20% toluene insolubles and 1 wt % bound NMCL. PPE was also grafted with NMCL in the presence of PS and DCP.

A related patent (Taubitz et al., 1987b) described grafting of 2 wt % methacrylamide to PPE under similar conditions to give product with 1.2 wt % methacrylamide incorporation.

4.3.1c Styrene, Styrene Analogs, Styrene-Acrylonitrile

Jones and Nowak (1965) have grafted styrene to PE in a reactive extrusion process. For example, styrene monomer in admixture with dicumyl peroxide and t-butyl cathechol inhibitor was fed under pressure to a molten stream of polyethylene in a 1.25 inch extruder with 2.5 min residence time at 190 °C. After styrene homopolymer was removed by extraction, copolymer containing 4-15% PS graft was obtained. Styrene derivatives such as vinyl toluene isomers and dichlorostyrene, or mixtures of styrene with acrylonitrile were also used.

Toyama et al. (1978) have grafted PPE with styrene by reactive extrusion. For example, 700 g PPE and 300 g PS were mixed with a solution of 30 g DBPH in 200 g styrene monomer and melt-kneaded on a counterrotating TSE at 250 °C with residence time about 4 min. Although the extruder was provided with a 30 mm vent, virtually no styrene volatilized and the extruded product contained 52% grafted PS. A mixture of styrene and acrylonitrile was also grafted using this process.

In related patents (Kasahara et al., 1982; Fukuda and Kasahara, 1983; Kosaka et al., 1984) 50 parts PPE, 10-15 parts styrene, and 0.6-1.5 parts di-t-butyl peroxide were blended in a Henschel mixer and extruded on a 40 mm TSE at 260-280 °C to give graft copolymer.

Izawa et al. (1979) have grafted styrene and substituted styrenes to mixtures of PPE with EMMA copolymer. For example, 1000 g PPE, 300 g EMMA (18% acrylate), 700 g p-methylstyrene, and 20 g di-t-butyl peroxide were mixed and extruded on a 40 mm SSE at 60 rpm and 250 °C maximum temperature to give graft copolymer with improved mechanical properties and increased resistance to weathering and oil. Styrene and styrene/acrylonitrile were also grafted using this process. PS or styrene-butadiene block copolymer were used in place of or in addition to EMMA.

Taubitz et al. (1989a) have grafted styrene to an olefin-functionalized PPE. For example, 500 parts PS, 200 parts styrene monomer, and 30 parts L130 were combined with 700 parts of a mixture of 0.5 parts PS and 9.35 parts PPE, which had been grafted with maleic acid mono-2-acryloxyethyl ester in a prior extrusion step (see below). The mixture was extruded on a TSE at 250 °C with 2 min residence time to give material with 60 wt % PS grafted to PPE compared to 12 wt % for a similar blend containing PPE first grafted with maleic acid monoethyl ester and 11 wt % for a similar blend containing unfunctionalized PPE.

Binsack et al. (1981) have grafted styrene/acrylonitrile mixtures to EP in a 20 mm counterrotating TSE with air as radical initiator using the procedure described in the preceding section under the same reference. For example, EP was fed at 0.9 kg/hr to the extruder where it was mixed with air in both an induction zone and an oxidation zone followed by mixing with 72/28 styrene/acrylonitrile in the polymerization zone and finally devolatilization. The grafted product contained 11% styrene, 4% acrylonitrile, and 600 ppm residual peroxide.

Andersen (1984) has grafted styrene or styrene/acrylonitrile mixtures to EPDM in the presence of di-t-butyl peroxide using the procedure described in the preceding section under the same reference. For example, 1 part EPDM was fed to a 53 mm corotating TSE (L/D = 36) with 12 segments with temperature profile 200-230 °C. Styrene monomer (0.5 parts) mixed with 0.00175 parts di-t-butyl peroxide was introduced at the third extruder segment at 12 lb/hr. After devolatilization, the isolated product contained 72% rubber and 28% grafted and ungrafted PS.

Vroomans (1988) has grafted a mixture of styrene and MA onto PE by reactive extrusion. For example, LDPE was fed to a W&P TSE at 150-170 °C at 48-50 g/min. A mixture of 2:2:1 styrene / MA / acetone containing 0.025-0.100 wt % t-butyl benzoylperoxide was introduced at 5-7 ml/min into the molten polymer stream to give a mixture with 4-6 wt % MA. The final product contained 3.7-6.1 wt % bound MA and could be used as an impact modifier for nylon or PET. An SSE followed by three static mixers in series could also be used for grafting.

Togo et al. (1988) have grafted a mixture of styrene and hydroxyethyl acrylate or methacrylate to PP. For example, 3 kg PP, 90 g 2-hydroxyethyl acrylate, 300 g styrene, and 15 g DCP was extruded on a TSE at 180-220 °C to give material with 6.5 wt % grafted styrene and 1.9 wt % grafted 2-hydroxyethyl acrylate. Use of 120 g hydroxyethyl acrylate instead of 90 g in the same process gave material with 6.2 wt % grafted styrene and 3.5 wt % grafted 2-hydroxyethyl acrylate. GMA could be used in place of hydroxyethyl acrylate. The material was used to prepare compatibilized blends with functionalized PPE as described in Section 4.4.

4.3.1d Maleic Anhydride, Fumaric Acid, and Related Compounds

One of the most common subjects of reactive extrusion patents and refereed publications is grafting of maleic anhydride and its analogs such as fumaric acid, itaconic acid, citraconic acid, citraconic anhydride, or alkenyl succinic anhydrides. In this section examples are given primarily of published references which cover only grafting of MA or analogous compounds to polymers. A number of published references which cover grafting with MA followed by reactive extrusion of the MA-functionalized polymer with a second polymer to form a graft copolymer are also included under Interchain Copolymer Forma-

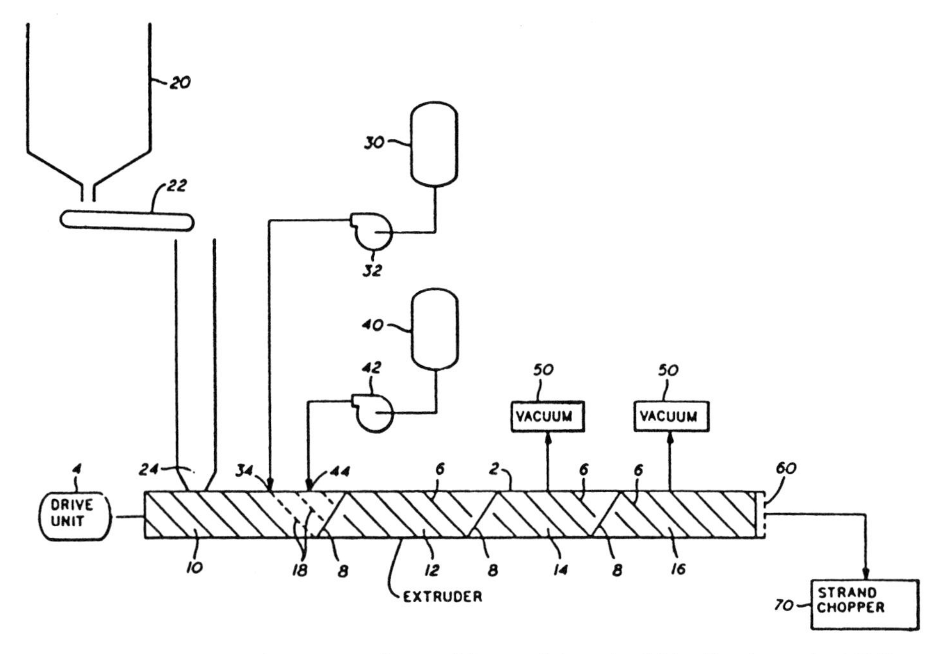

Figure 4.16 Extruder reactor for grafting maleic anhydride (Strait et al., 1988)

FA has been grafted to EPDM using a 28 mm TSE at 200 rpm and 260 °C in the presence of dicumyl peroxide supported on clay (Waggoner, 1987). The extruder contained an intensive mixing zone containing a kneading element followed by at least one screw element with reverse pitch compared to the normal conveying elements where grafting took place in < 30 sec. This section was followed by a reduced pressure zone where unreacted FA was removed within 20 sec residence time. EPDM extruded with 3 wt % FA and 1 wt % initiator had 1.07 wt % incorporation of monomer into the final product by IR determination.

Ethylene-propylene-1,4-hexadiene copolymer (61.4 wt %, 32%, 6.6%) has been grafted with MA by coextrusion on a 53 mm TSE at 12 rpm (Caywood, 1975, 1977). The extruder consisted of a short feed section, four reaction sections at 300 °C, a vacuum port section at 260 °C and 25 inches Hg pressure, and a cooling section and die at 150 °C. The screw geometry was arranged to generate 100-200 psi pressure in sections 1-4 and no pressure in the devolatilization section. MA was fed as a molten stream to the first reaction zone at 4.8% of the polymer weight. The extruded product had 2.23 wt % MA incorporation by IR and 2.19 wt % by titration. The values changed only slightly after purification. Ethylene-propylene-1,4-hexadiene-2,5-norbornadiene copolymer was also grafted with MA using this process.

Gaylord (1985), and Gaylord and Mehta (1988) have described grafting of MA to a variety of polymers in the presence of free radical initiators and additives which suppress both homopolymerization of MA and radical degradation and crosslinking of the substrate polymer. Reactions were performed in Brabender mixing equipment at 60 rpm by adding in portions a mixture containing 2-20 wt % MA, 0.25-2 wt % initiator, and 2-20 mole % (based on MA) additive to the molten polymer at 120-180 °C (depending upon the polymer). Total residence time was around 10 min. The isolated products were fractionated in refluxing xylene (or THF in the case of PVC) to determine % insolubles. The soluble portion was analyzed for MA content by titration or by oxygen analysis. Substrate polymers grafted with MA under these conditions were LDPE, LLDPE, HDPE, PP, EVA, ethylene/methyl acrylate copolymer, polybutene-1, polyisobutylene, EP, EPDM, poly(ethyl acrylate), and PVC. Radical initiators employed were dicumyl peroxide, benzoyl peroxide, t-butyl peroctoate, t-butyl perbenzoate, and azobisisobutyronitrile. Additives which prevented side reactions were nitrogen, phosphorus, and sulfur containing compounds such as N,N-dimethylformamide, N,N-dimethylacetamide, caprolactam, diphenyl or triphenyl phosphite, triphenyl phosphate, dimethyl methylphosphonate, triphenylphosphine oxide, p-tolyl disulfide, and dimethylsulfoxide. In one example LDPE reacted with 10 wt % MA and 1 wt % dicumyl peroxide had 76% insolubles in the absence of additive but 0% insolubles in the presence of either 10 mole % caprolactam, triethyl phosphite, or dimethylsulfoxide with 3.9, 3.1, and 6.7 wt % incorporation of MA, respectively. The effect of dimethylformamide on grafting of MA to LDPE in the presence of DCP or BOP was the subject of a separate published study (Gaylord and Mehta, 1982; Gaylord et al., 1983).

Gaylord and Mishra (1983) have also described efficient grafting of MA to PP without significant PP degradation under similar melt conditions in the presence of dimethylformamide or dimethylacetamide.

A related study has appeared describing similar effects of stearamide on the grafting of MA to EP rubber in the presence of DCP or t-butyl perbenzoate (Gaylord et al., 1987).

MA grafting to styrene-butadiene or styrene-isoprene copolymer has been reported in a 20 mm SSE (L/D 22) at 30 rpm and 200-220 °C with 180 sec residence time (Saito et al., 1981a). Radical inhibitors such as hindered phenols or phenothiazine were added. For example, extrusion of styrene/butadiene/styrene (15/70/15) block copolymer with 2% MA and 0.5% 2,6-di-t-butyl-4-methylphenol gave material with 0.12 wt % toluene insoluble portion and 1.06 wt % MA incorporation by titration representing 53% conversion of MA. Grafting with increased residence time (e.g. 700 sec) or in the absence of radical inhibitors or in the presence of DCP gave product with 2-58% toluene insolubles.

Gergen et al. (1986) have grafted block copolymers of styrene/ethylene-butylene/styrene (SEBS) with MA in a 30 mm corotating TSE at 350 rpm in the presence of DBPH at 150 °C (feed zone) to 260 °C (die). The products were fractionated in THF to determine gel content, and bound MA in the soluble fraction was measured by titration. In one example SEBS reacted with 5 wt % MA and 0.5 wt % initiator had 2% gel and 4.6 wt % bound MA after fractionation. Control experiments showed that no grafting took place without radical initiator and that the site of attachment of MA was at the EB midblock portion of the copolymer.

Grafting of FA or N-phenylmaleimide to styrene-butadiene copolymer has been claimed by Taubitz et al. (1988b) in an extrusion process in the absence of radical initiator. For example, 99.2% styrene-butadiene copolymer and 0.8% FA were extruded on a 53 mm TSE with melting zone at 180 °C, reaction zone at 105 °C, and devolatilization zone at 180 °C. The product had 0.7 wt % FA incorporation and showed improved mechanical properties in molded test parts compared to a control in which MA was grafted in the presence of dibenzoyl peroxide. According to a related patent (Taubitz et al., 1988c), grafting 98 parts PS alone with 2 parts FA at 230-260 °C gave only 0.1 wt % FA incorporation.

Wong (1989) has grafted PP with a mixture of MA and styrene monomer. For example, PP coated with 1.1 wt % MA was fed at 30 kg/hr to a 53 mm TSE with temperature profile: zone 1 170°, zone 2 188°, zone 3 230°, zone 4 230°, and die 230 °C. Styrene (1.4 wt %) was injected into zone 2. The reaction mixture was devolatilized before exiting. The product had 0.8 wt % bound MA and MFI 17 dg/min. No grafting was obtained in the presence of antioxidant. PP extruded under the same conditions without MA or styrene had MFI 20 dg/min. For comparison, PP grafted under these conditions with MA using 500 ppm L130 as radical initiator and no styrene gave product with 0.15 wt % bound MA and MFI 258 indicating degradation of PP. MA and styrene could also be introduced separately but simultaneously to the extruder through concentrically mounted tubes, but attempts to add MA as a mixture with styrene either resulted in undesired copolymerization or precipitation of MA from solution. LDPE was also grafted with MA using this process. The products exhibited good adhesion to aluminum.

Shyu and Woodhead (1988) have grafted PP with either MA or a mixture of 1:1 MA : styrene monomer in the presence of peroxide initiator and a catalyst. For example, 36 parts PP was melted in a Brabender mixer at 180 °C / 60 rpm under nitrogen after which a mixture of MA (5.4 parts), a peroxide such as DCP (0.53 parts), and a catalyst such as N,N-diisopropylethanolamine (0.32 parts) was added in four portions over 8 min. Total material residence time was 10 min. The isolated product had 2.28 wt % MA grafted but polymer degradation was observed. PP was also grafted using 3.7 parts 1:1 styrene/MA mixture along with t-butylperoxy pivalate (0.24 parts) as initiator and N,N-diisopropylethanolamine (0.10 parts). The process was run for about 8 min residence time at 90 °C, which is below the melting point of PP. The product had 1.95 wt % bound MA and showed greatly reduced degradation of PP.

In a study of PP-g-MA adhesive properties (Bratawidjaja et al., 1989), PP was reacted with 0.4-1.6 wt % MA and 0.4-1.6 wt % BOP at 200 °C in an "extrusion molder" with 3 cm diameter and 100 cm length equipped with pelletizer. After precipitation of the product from xylene, bound MA content was determined by IR spectroscopy to be between 0.1 and 0.55 mol. % in direct proportion to the amount of MA and BOP added.

Wu et al. (1975) have grafted HDPE with olefinic anhydrides or carboxylic acids such as tetrahydrophthalic anhydride (THPA). For example, a mixture of 150 lbs HDPE and 15 lbs THPA was extruded on a corotating TSE at 50 lbs/hr and 300 rpm. The extruder consisted of five heating zones and die at 200 °C (zone 1), 270 °C, 320 °C, 270 °C, 230 °C, and 180 °C (die). A mixture of 0.3 pph t-butyl hydroperoxide and 3.1 pph o-dichlorobenzene was fed to zone 2 and the product was devolatilized at zone 4. The product had 0.5 wt % incorporated THPA and showed 150% tensile elongation in molded parts containing 40 wt. % titanium dioxide filler. A control blend of unfunctionalized HDPE and titanium dioxide had < 10% elongation. Other grafting agents included nadic anhydride and methyl nadic anhydride. Other polymeric substrates were LDPE and EPDM.

In a related patent (Krebaum et al., 1975), polyolefins were grafted with mixtures of monomers containing olefinic anhydrides. For example, a mixture of 150 lbs HDPE and 15 lbs MA was extruded on a corotating TSE at 50 lbs/hr and 300 rpm. The extruder configuration and temperature was the same as in the preceding example. A mixture of 5 pph dibutyl maleate (DBM) and 0.5 pph TBHP was added at zone 2. The product showed 110% tensile elongation in molded parts containing 40 wt % titanium dioxide filler. A control blend of titanium dioxide and HDPE extruded with MA but without DBM had < 12% elongation. Other grafting agents included nadic anhydride/DBM and methyl nadic anhydride/DBM. Other polymeric substrates were LDPE, ethylene-hexene-1 copolymer, and EPDM.

Grafting to various polyolefins using maleic anhydride masked as its Diels-Alder adduct with either cyclopentadiene or furan was described in the preceding section (Tabor et al., 1988).

Techniques have been reported by Yamamoto et al. (1979) for the prevention of irritating fumes and mold plate-out due the presence of unreacted anhydride monomer. For example, maleic anhydride or nadic anhydride (1 wt % of polymer) have been grafted to PP by premixing with polymer and peroxide initiator and adding to the feed throat of a 50 mm extruder (L/D 32) at 190°-200 °C and 110 rpm. Glycidoxypropyl trimethoxysilane was added as a liquid under pressure at a distance L/D 16 from the feed throat inlet. A vent was present at L/D 24. Typically 91-96% of remaining uncombined anhydride reacted with silane, compared with only 20-28% of bound anhydride. The product was claimed to have improved odor, improved peel strength to aluminum, and improved miscibility with glass fibers. Variations included use of

PE and EP copolymer (7.2% ethylene) as polymer substrates, use of cis-4-cyclohexene-1,2- dicarboxylic anhydride as graft monomer, and use of different epoxy- and aminosilanes as anhydride traps.

Grafting of MA or Its Analogs to Polyphenylene Ethers Polyphenylene ethers (PPE) such as poly(2,6-dimethyl-1,4-phenylene ether) have been grafted with maleic anhydride and its analogs in extruders. The majority of patents mentioning MA grafting to PPE have as their primary purpose and claim the subsequent reactive extrusion of the PPE-g-MA with some nucleophile-containing polymer such as nylon to form a copolymer. These references are covered under Interchain Copolymer Formation.

Taubitz et al. (1988c) have grafted mixtures of PPE and polystyrene with FA in the absence of radical initiators. For example, 94 parts PPE, 13.5 parts PS, and 2.5 parts FA were fed to a 53 mm TSE and were melted in the first zone using kneading elements at 255 °C, then reacted in the second zone with kneading at 265 °C, and finally devolatilized at 255 °C under vacuum. Residence time was about 2.5 min. The product had 4 wt % insolubles in toluene and contained 1.6 wt % bound FA. PPE-g-FA synthesized in this manner was also used to prepare compatibilized blends of PPE with nylon 6 or 6/6 (Taubitz et al., 1989b).

Extrusion of PPE with N-phenylmaleimide (NPM) with or without PS resulted in functionalized PPE (Taubitz et al., 1987c). For example, extrusion of 97.1 parts PPE and 2.9 parts NPM on a 53 mm TSE at 280 °C with devolatilization gave product with 1.0 wt % maleimide content by IR spectroscopy. The reaction also succeeded in the presence of DCP radical initiator.

Reaction of 99 parts PPE with 1 part 2-ethylhexyl monomaleate under similar extrusion conditions at 270°-280 °C gave product with 0.8 wt % incorporation of half ester by IR spectroscopy (Taubitz et al., 1989c).

In a related patent (Taubitz et al., 1989a) 9.35 kg PPE mixed with 0.5 kg PS was grafted with 0.15 kg maleic acid mono-2-acryloxyethyl ester in a TSE operated at 270 °C with devolatilization zone at 280 °C. Residence time was 1.8 min. The product was reextruded with PS and styrene monomer to give PPE grafted with high levels of PS (see above). Grafting to PPE under similar conditions using substituted MA derivatives such as dimethoxy maleic anhydride or dimethoxy maleimide has also been claimed (Becker and Rohr, 1988).

Akkappeddi et al. (1988) have grafted PPE with MA or FA in the presence of N-bromosuccinimide (NBS). For example, a mixture of 98.9/1.0/0.1 PPE / MA / NBS was fed to a 1 inch SSE (L/D = 24) with temperature profile 260-300 °C. The extracted product had 0.6 wt % bound MA by titration. Other monomers grafted included cinnamic acid, 2-isopropenyl-2-oxazoline, bis(2-hydroxyethyl)fumaramide, and morpholinylmaleamic acid.

Summary of Additional Grafting Reactions of MA or Its Analogs Other examples of grafting of MA or its analogs to polymers in extrusion processes are summarized in the table below.

Table 4.2 Other Examples of Grafting Reactions with MA or Its Analogs

Grafting Agent	Substrate Polymer	Initiators	Reference
MA	HDPE	DCP	Tabor and Allen, 1987
MA[a]	HDPE	(DCP)	Inoue et al., 1987
MA	UHPE[b]	DCP (optional)	Motooka and Mantoku, 1986
MA	EP	TBP	Fukui et al., 1985
MA	EP	TBPL[c]	Nishio et al., 1988
NA[d]	EP	DTB	Okamoto et al., 1989
MA	SEBS	DBPH	Nakazima and Izawa, 1988; Shiraki et al., 1986
MA	H2-PB[e]	–	Hergenrother et al., 1984, 1985
MA, FA, IA	PP	BOP	Ide and Sasaki, 1977
MA, IA	PP+EP+SAN-g-EPDM	L130	Kawada et al., 1988
MA	PP+EP or EPDM	–	Perron and Bourbonais, 1988a
MA[f]	PP, PPE	DCP	Togo et al., 1988
phenyl fumarate	PPE	–	Johnson et al., 1989
MA	PPE	DTB	Nakazima and Izawa, 1988
FA	PPE+SEBS	–	van der Meer and Yates, 1987
MA, IA	EBA, EVA, LDPE+EBA	DTB	Bergström et al., 1988
FA	EBA, EVA	DTB	Bergström and Palmgren, 1987
MA	EVA, EEA, EBA	L130	Wong, 1988

(Continued on page 126)

[a] patent discloses a solution purification scheme for removing unreacted MA from grafted polymer
[b] ultrahigh molecular weight PE (IV = 8.20 dl/g)
[c] t-butylperoxy laurate
[d] nadic anhydride (endo-bicyclo[2.2.1]-5-heptene-2,3-dicarboxylic anhydride)
[e] polybutadiene hydrogenated to residual unsaturation level of 8-10%
[f] grafting to PP was performed in the presence of styrene monomer

Interchain copolymer formation by either covalent or ionic *crosslinking* is also possible in extruder reactors. In covalent crosslinking, pendant functionalities of one polymer react with pendant functionalities of a second polymer to form a covalent bond. In ionic crosslinking acidic groups such as carboxylic acids bound to one polymer may mediate interchain copolymer formation by protonation of basic groups on a second polymer leading to chain association. However, in the majority of examples of copolymer formation by ionic crosslinking, ionic groups such as carboxylic, sulfonic, or phosphonic acid are present on both polymers and are at least partially neutralized by a metal cation, usually divalent, which may form a bridge between the two polymers during extrusion.

Ionic crosslinks are usually thermally reversible which may limit the usefulness of blends containing them in certain commercial applications. Thermal reversibility in the melt, however, is necessary to overcome self-crosslinking of neutralized ionic groups within each homopolymer before interchain crosslinking can occur. Often a high degree of plasticization of the ion-containing polymer melt is required so that high processing temperatures which might lead to polymer decomposition need not be used. In some cases polymers containing masked ionomers, chemical groups which form ionic species during the actual extrusion process, have been used to form copolymers during reactive extrusion. Use of masked ionomers may result in lower energy during extrusion processing since ionic self-association need not be overcome before interchain copolymer formation can occur.

4.4.2 Compatibilization of Immiscible Polymer Blends

Numerous general reviews discussing compatibilization of immiscible polymer blends have been published (see for example Bucknall, 1977; Fox and Allen, 1985; Menges, 1989; Paul and Newman, 1978; Paul et al., 1988; Rudin, 1980; Solc, 1981; Sperling, 1987; Teyssie et al., 1988). Interchain copolymer formation by reactive extrusion is particularly useful for compatibilization of immiscible polymer blends so that a product may be obtained with combinations of desirable properties arising from both polymers. The most common examples involve compatibilization of thermoplastics with immiscible, rubbery impact modifiers, and compatibilization of amorphous polymers with crystalline polymers for improved solvent resistance. Compatibilization in this sense refers to "operational compatibility" as defined by Gaylord (1989) in which the polymer blend exhibits useful technological properties over the lifetime of a molded part.

The majority of known polymers are immiscible with one another and, upon coextrusion, one polymer will form a dispersed phase in a continuous matrix of the other polymer. Which polymer forms the continuous phase depends upon the relative amounts and viscosities of the polymers with the more viscous polymer generally forming the dispersed phase. Usually the dispersed phase will agglomerate into large domains when the blend is subjected to further thermal processing such as molding or heat aging. Phase agglomeration in molded parts results in delamination, brittleness and/or poor surface appearance. In a compatibilized blend of immiscible polymers the dispersed phase is stabilized against agglomeration by interphase adhesion and/or lowered interfacial tension between the two phases. Interphase adhesion may be achieved by several mechanisms but a common and efficient method is through the presence of a block or graft copolymer of the two homopolymers which, when present at the interface of the two immiscible phases, acts as an emulsifying agent and lowers interfacial tension. Often as little as 0.5-2.0 wt % copolymer is sufficient to achieve phase stabilization but more frequently 10-20 wt % is necessary to obtain optimum physical properties of the blend. Block or graft copolymers are most economically formed by reactive extrusion to form covalent or, less commonly, ionic bonds during the extrusion mixing step to disperse one polymer in the other. The fundamental requirements for compatibilization by covalent (or ionic) bond formation by reactive extrusion are as follows:

==> Sufficient mixing to achieve desired morphology of one polymer phase well-dispersed in second polymer phase
==> Presence of reactive functionality (e.g. nucleophilic functionality on one polymer and electrophilic functionality on second polymer) for covalent bond formation, or ionic (or masked ionomeric) sites on both polymers, or acidic sites on one polymer and basic sites on the second polymer
==> Functionalities are of suitable reactivity to react across melt phase boundary
==> Reaction must take place within time constraints of extruder
==> Covalent (or ionic) bond formed is stable to subsequent processing conditions

Due to the above requirements, TSEs rather than SSEs are often more efficient for interchain copolymer formation because of their ability to generate a large interfacial area for chemical reaction across the phase boundary.

The majority of commercially available polymers of interest for forming copolymers have nucleophilic end groups such as carboxylic acid, amine, and hydroxyl. These nucleophilic end groups can form covalent bonds with suitable electrophilic functionality leading to copolymer formation when a suitable electrophilic group is attached to a second polymer. As will be seen in the examples for Interchain Copolymer Formation of Types 2, 3 and 4 , the kinds of electrophilic functionality suitable for reaction to form covalent bonds across a phase interface within the time constraints of an extruder are most often *cyclic anhydride, epoxide, oxazoline, isocyanate, and carbodiimide*. These same electrophilic functionalities are useful for forming covalent bonds in a homogeneous phase leading to coupling or branching of homopolymers which have nucleophilic end groups as shown in examples included in the Section 4.5 on Coupling/Crosslinking Reactions. The same electrophilic functionalities are also useful for capping carboxylic acid end groups of polyesters as shown in Section 4.7. The scheme below illustrates specific reactions of polymer amine and carboxylic acid terminal groups (representative polymer B = P^B) with the most important polymer-bound electrophilic groups mentioned above (representative polymer A = P^A).

Cyclic Anhydride

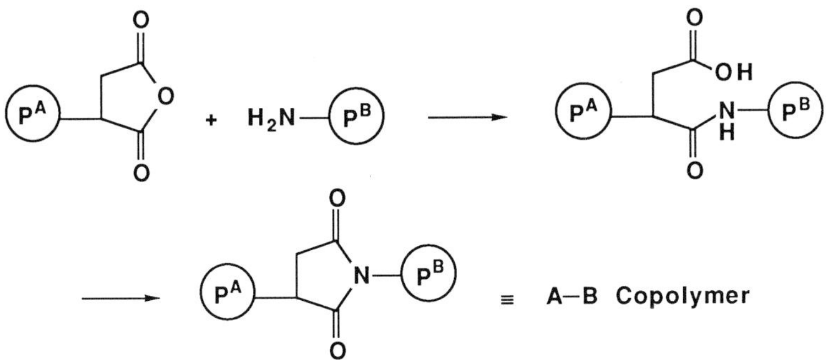

P^A + H_2N–P^B ⟶ ⟶ ≡ A–B Copolymer

Epoxide

P^A + H_2N–P^B ⟶

P^A + HO_2C–P^B ⟶ [P^A + ^{-}OC–P^B] ⟶

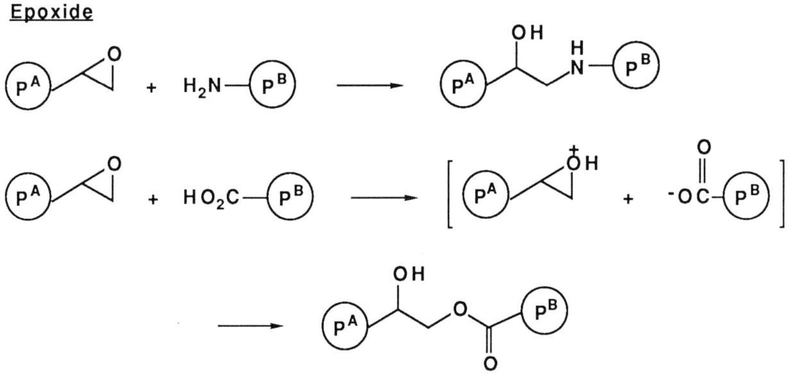

Oxazoline

P^A + H_2N–P^B ⟶

P^A + HO_2C–P^B ⟶ [P^A + ^{-}O–C–P^B] ⟶

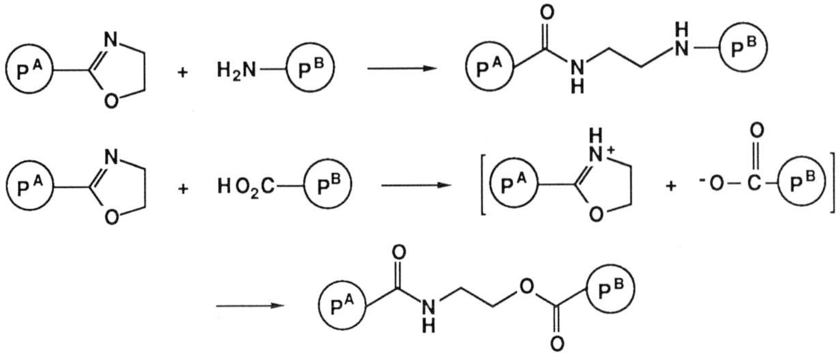

Under the appropriate conditions the oxazoline and epoxide rings may react with both strong and weak nucleophiles, e.g. with polymers containing amine, mercaptan, hydroxy, epoxy, anhydride, and, most particularly, carboxylic acid groups. In some cases reaction may be catalyzed by a Lewis acid or quaternary ammonium salt. Under extruder processing conditions cyclic anhydrides react irreversibly only with relatively strong nucleophiles such as amine, thiol, or in certain cases, hydroxyl or epoxide.

Carbodiimide

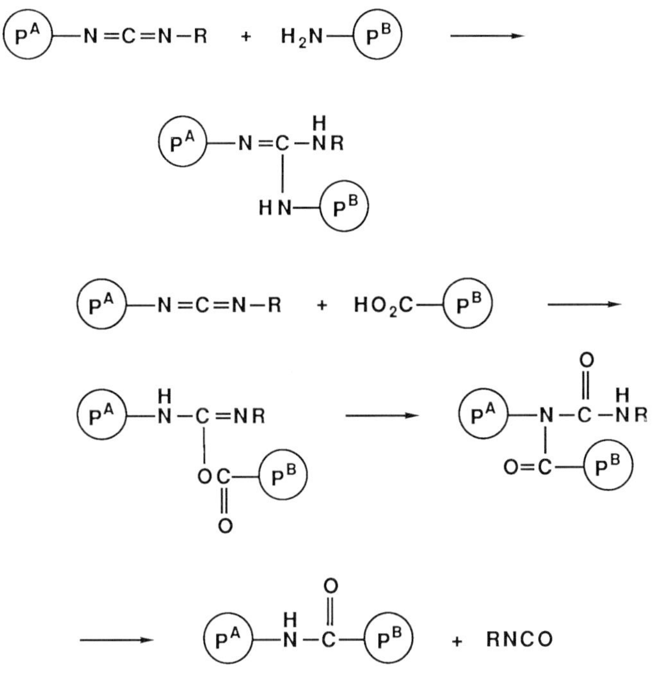

$$P^A\text{—N=C=N—R} + H_2N\text{—}P^B \longrightarrow P^A\text{—N=C(—NHR)—HN—}P^B$$

$$P^A\text{—N=C=N—R} + HO_2C\text{—}P^B \longrightarrow$$

$$P^A\text{—NH—C(=NR)—OC(=O)—}P^B \longrightarrow P^A\text{—N(—C(=O)—}P^B\text{)—C(=O)—NHR}$$

$$\longrightarrow P^A\text{—NH—C(=O)—}P^B + RNCO$$

Isocyanate

$$P^A\text{—NCO} + H_2N\text{—}P^B \longrightarrow P^A\text{—NHC(=O)NH—}P^B$$

$$P^A\text{—NCO} + HO_2C\text{—}P^B \longrightarrow P^A\text{—NHC(=O)—O—C(=O)—}P^B$$

$$\longrightarrow P^A\text{—NHC(=O)—}P^B + CO_2$$

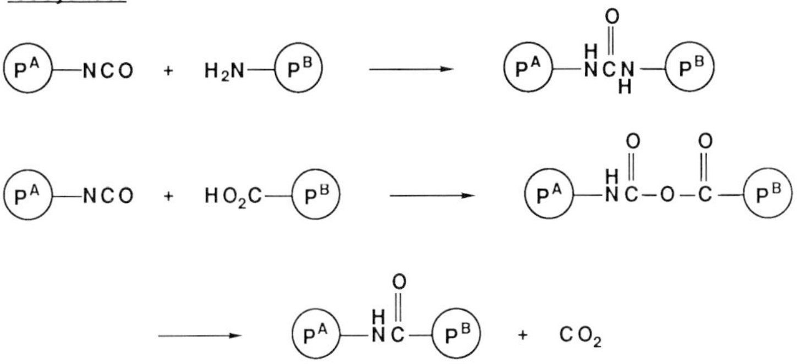

The reactions of carbodiimides and isocyanates with amines or carboxylic acids are highly dependent upon the substituents bound to these functional groups. The reaction products depicted in the above scheme are only meant to be representative of routes which can lead to copolymer formation.

For copolymer formation by reaction Types 2,3,4, and 5, the degree of compatibilization as reflected in the blend physical properties is directly dependent upon the concentrations of functionality on each polymer, and this is specifically illustrated in some of the examples which follow. In the majority of examples the interchain copolymer which presumably forms during reactive extrusion was not characterized. In a small number of cases there may be little or no copolymer formed at all, but when a judgment was required on copolymer formation, it was felt best to include even questionable examples with the goal of tabulating as wide a variety of examples as possible.

4.4.3 Interchain Copolymer Formation (Type 1): Random and/or Block Copolymers by Chain Cleavage/Recombination

Casale and Porter (1975, 1978) have published reviews on mechanochemical generation of radicals by chain scission of addition polymers during extrusion under high shear, and recombination of these radicals to form block and graft copolymers. Among the copolymers formed by this process were PE/PIB; nitrile rubber/PVC; PP/polyamide; and PP/lignin.

Transesterification reactions of polyesters during extrusion are common and are often associated with deterioration in certain polymer blend properties because of the broadening in molecular weight distribution. Hence, most experimental studies have sought to prevent transesterification rather than promote it. A brief review covering transreactions in polyester and in polycarbonate binary blends has appeared (Porter et al., 1989).

A general review of "Interchange Reactions Involving Condensation Polymers" has been published by Kotliar (1981) which describes a number of different interchange reactions in the melt between polyesters, polyamides, and polyester + polyamide. In polymer blends these reactions may be considered to involve chain cleavage followed by recombination of free end groups to give an equilibrium molecular weight distribution of random and/or block copolymer.

Polyester + Polyester The ester interchange reaction between bisphenol A polycarbonate and a polyarylate derived from bisphenol A and iso/terephthalic acid has been studied by Golovoy et al. (1987). Blends between 15/85 to 85/15 parts polycarbonate: polyarylate were prepared on a 30 mm W&P TSE at 300 rpm and temperature profile 260°, 290°, 280°, 260°, and 300 °C (die). A 50:50 blend was also prepared using a Haake 1.9 cm SSE at 285°. The materials were characterized by DSC which showed two Tg values for blends compounded on the SSE (or solution cast) but only one Tg for blends prepared on the TSE. Blends prepared on the SSE phase separated during annealing, while those prepared on the TSE did not. Molded parts from blends prepared on the TSE showed tensile and impact values below those values predicted by the rule of mixtures due to a combination of molecular weight degradation and transesterification during processing.

Transesterification has also been studied in polyarylate/ polycarbonate blends prepared by extrusion in a Brabender apparatus at 30-40 rpm and 250°-300 °C (Mondragon and Nazabal, 1985, 1986).

Transesterification in 50/20/30 blends of polyarylate (as above), BPA polycarbonate, and PET was reported by Cheung et al. (1989) for blends prepared at various temperatures on a Haake SSE. The products were characterized by DSC and IR spectroscopy. Transesterification was prevented up to 280 °C by addition of a phosphite stabilizer, up to 300 °C by addition of both of phosphite and carbodiimide, and up to 325 °C by addition of phosphite, carbodiimide, and hindered phenolic stabilizers.

In a related study (Golovoy et al., 1989), it was shown using ^{31}P NMR that in situ hydrolysis of a specific phosphite stabilizer to a phosphonate was necessary for effective stabilization of a polyarylate/BPA polycarbonate/PET blend against transesterification during melt processing. The phosphite employed was bis(2,4-di-t-butylphenyl)pentaerythritol diphosphite (Ultranox 624).

A detailed study of transesterification in blends of polyarylate with either PET, BPA polycarbonate or a copolymer of BPA with epichlorohydrin has appeared (Robeson, 1985). Blends were prepared by extrusion at 265-270 °C with a maximum of 3 min residence time, and compression molded at 260-350 °C with short and long cycle times. Only those blends molded at higher temperature and longer cycle time showed evidence of transesterification. A 50:50 blend of polyarylate with PET extruded on a Killion SSE (L/D = 30) showed significant transesterification only above 320 °C extruder set temperature (292 °C melt temperature).

Transesterification of polyarylate with polycarbonate occurs during devolatilization of the mixed polymer solutions in an extruder (Freitag et al., 1985). For example, a polyester composed of bisphenol A and 1:1 iso/terephthalic acid with p-t-butyl phenol as chain terminator was dissolved in 1:1 dichloromethane/ chlorobenzene to form a 10% solution and mixed with a similar 10% solution of a polycarbonate composed of bisphenol A, phosgene, and p-t-butyl phenol chain terminator. The relative amounts of the two solutions varied between 80:20 and 10:90. The combined solution was concentrated to 45% solids and fed to a devolatilizing TSE with 50 sec residence time at 220-280 °C and 20-60

PPE end-capped with anhydride functionality has been coextruded with nylon to give PPE-nylon copolymer which acts as compatibilizer for the PPE-nylon blend (Aycock and Ting, 1986, 1987). For example, PPE-OH end groups were capped in solution with trimellitic anhydride acid chloride. The level of anhydride functionalization was measured by appearance of carbonyl bands and diminution of hydroxyl bands in the IR spectrum. A mixture of 24.5 parts anhydride-functionalized PPE, 24.5 parts unfunctionalized PPE, 41 parts nylon 6/6, and 10 parts of a styrene/hydrogenated butadiene/styrene copolymer as impact modifier was extruded on a corotating 28 mm TSE at 290 °C. Molded test parts showed higher impact and tensile strength than test parts for blends containing only unfunctionalized PPE or PPE grafted with MA. Solubility measurements in formic acid and toluene showed that blends containing PPE end-capped with anhydride functionality had high levels of PPE-nylon copolymer.

In a related patent (Sivavec, 1989) PPE capped with anhydride functionality was prepared by coextrusion with various 4-substituted esters of trimellitic anhydride. For example, 412 g PPE was extruded with 8% 4-(o-carbophenoxyphenyl)trimellitic anhydride and 0.5% triphenyl phosphite on a TSE at 170-300 °C and 300 rpm. The extrudate contained PPE with 40% carboxylation and 55% capping of hydroxy end groups. Extrusion of 49 parts of capped PPE with 41 parts nylon 6/6 and 10 parts SEBS under similar conditions gave a blend with Izod impact strength 247 J/m.

Aharoni and Largman (1983a), and Aharoni et al. (1984a) prepared block copolymers of nylon with carboxylic acid-terminated polymers such as PET or PBT by extrusion in the presence of tertiary phosphite esters. For example, a mixture of 80 g nylon 6, 20 g PET, and 0.6 g tributyl phosphite was extruded on a 1 inch Wayne extruder (L/D = 25) with mixing screw at a temperature greater than the melting points of both polymers, at least 150 °C and preferably > 200 °C. Extruder residence time was preferably > 2 min. The extrudate had reduced viscosity of 2.66 compared to 1.68 for a control blend made without phosphite and 1.60 for 100% nylon 6. Copolymer formation between nylon 6 and nylon 6/6, nylon 11, or nylon 12 was also reported. Triphenyl phosphite was used in certain examples. The process was also claimed to be suitable for making graft copolymers of nylon with EAA copolymers. The mechanism is discussed in Section 4.5. Note that Khanna et al. (1983) above disclosed random copolymers of nylon 6 with nylon 6/6 made under similar conditions but apparently at higher temperature.

Polycarbonate-nylon compatibilized blends have been prepared by reactive extrusion by Hathaway and Pyles (1988, 1989). For example, a hydroxy-terminated bisphenol A polycarbonate was end-capped with anhydride groups through reaction with trimellitic anhydride acid chloride. A mixture of 40 parts functionalized polycarbonate, 40 parts nylon 6, and 20 parts of an acrylate-styrene core/shell impact modifier was extruded on a counter-rotating TSE at

277 °C. Molded test parts showed notched Izod impact strength of 6.5 ft-lb/in. and 192% tensile elongation. The presence of PC-nylon copolymer was determined by selective extraction. An amorphous polyamide could also be used in these blends.

4.4.5 Interchain Copolymer Formation (Type 3): Graft Copolymers

4.4.5a Nylon/Polyolefin Blends

Many commercial nylon products are two-phase blends which are toughened by the presence of a dispersed polyolefin-containing phase with lower modulus than the nylon matrix. Extrusion of nylon with anhydride- or acid-functionalized polyolefin may form at least some nylon-polyolefin copolymer which acts as compatibilizing agent leading to stabilized nylon-polyolefin blends with excellent physical properties. The success of this process depends upon sufficient concentrations of amine terminal groups on nylon and anhydride or acid groups on polyolefin to give adequate levels of copolymer under the mixing and residence time limitations of the extruder. Nylon copolymers have also been formed through reactive extrusion with polymers containing other types of functionality such as epoxy- or benzyl halide groups.

Nylon/Modified Polyethylene or Ethylene Copolymer, or Modified Polypropylene A large number of studies have been published on nylon/polyolefin blends compatibilized by copolymer formation during reactive extrusion. For convenience this subsection is divided into examples from refereed papers followed by examples from patents.

Early work by Ide and Hasegawa (1974) described nylon 6/PP blends compatibilized by formation of nylon-PP copolymer using PP-g-MA. The copolymer formed during "extrusion molding" of mixtures of the materials at 230 °C. Evidence for copolymer was obtained by selective solvent extraction and DSC. Analysis of residual amine groups on nylon showed decreasing level with increasing MA level on PP. The blends contained a much more finely dispersed phase of nylon than did control blends made without functionalized PP and containing no copolymer. Physical properties such as tensile strength and elongation also improved markedly compared to blends containing no copolymer.

Blends of nylon 6/6 with PE-g-MA have been described by Hobbs et al. (1983). The blends were prepared on a W&P TSE at 270 °C. Impact failure was discussed in terms of fracture morphology, composition, and particle adhesion and dispersion. Impact strength increased with amount of PE-g-MA (0.4% MA) added to the blend up to 14 ft-lb/in. at 40 wt % loading. A control blend containing PE-g-MA prereacted with n-octyl amine showed relatively poor impact strength.

Studies have appeared on blends of nylon 6 with EP-g-MA (2.9 wt % MA) with and without added unfunctionalized EP copolymer (Cimmino et al., 1984, 1986; Martuscelli et al., 1984; Greco et al., 1987; D'Orazio et al., 1987). In one example, blends were prepared in a Haake Rheocord mixer at 260 °C with 20 min mixing time. The blends showed improved physical properties with increasing proportion of EP-g-MA at 20 wt % rubber loading. Blend morphology showed decreased size of dispersed EP phase with the presence of EP-g-MA and evidence for adhesion between dispersed phase and nylon matrix.

Blends of nylon 6 with ethylene/methacrylic acid copolymer have been prepared by McKnight et al. (1985) by mixing for 20 min in a Brabender mixer at 250 °C. The domain size of the dispersed EMAA phase decreased regularly with increasing MAA content of the copolymer while tensile properties of the blend increased regularly. The presence of copolymer was shown by selective extraction experiments and IR analysis. It was concluded that both copolymer formation and hydrogen bonding between the nylon and EMAA phases could affect ultimate blend properties.

Graft copolymer formation between nylon 6 and EAA copolymer has been studied by Braun and Illing (1987) in blends made on a TSE at ≥ 270 °C. Blend compositions were 90/10 and 80/20 nylon/EAA. Products were characterized by IR and NMR spectroscopy, DTA, GPC, selective solvent extraction, determination of residual amine and carboxylic acid groups, and morphology. Based on carboxylic acid groups remaining in the product, about 25-35 polyamide chains were grafted to each EAA chain.

Interfacial and rheological effects on the formation of dispersed rubber phase were studied by Wu (1987) in blends of nylon with an EP rubber containing $<$ 1 wt % of an unspecified carboxyl group. Blends of 15% rubber and 85% nylon were prepared on a corotating 28 mm TSE with temperature profile 280° (zone 1), 280°, 250°, 230°, 230°, and 280 °C (die). Dispersed particle size was related to the interfacial tension between nylon and rubber. The interfacial tension was related to the polarities of the materials and whether or not copolymer was present, and was much lower for blends containing carboxy-functionalized rubber than for control blends containing unfunctionalized rubber.

The effect of morphology on permeability properties has been studied in nylon-polyethylene blends compatibilized through the addition of PE functionalized with unspecified anhydride or carboxyl groups (Subramanian and Mehra, 1987).

More recently, Curto et al. (1990) have described blends of nylon 6 with oxidized PE (cf. Armstrong, 1968a, below). For example, LDPE was photooxidized to give material with 0.07-0.28 mol/l carbonyl groups and 11-25% gel content depending upon time of exposure (24-72 hrs). Blends of 25 wt % oxidized PE and 75 wt % nylon 6 (55 meq/kg amine groups) were prepared in a Brabender mixer at 240 °C and 150 rpm with 15 min residence time. Blends containing photooxidized PE had higher tensile strength and elongation than control blends containing unoxidized PE with no carbonyl groups. SEM photomicrographs also showed markedly better phase dispersion in compatibilized blends compared to a control blend.

Nylon 6 blends with EPDM-g-MA (0.4 wt % MA) have been studied by Borggreve et al. (1988a,b). The blends were prepared in a 40 mm SSE fitted with a 10 cm cavity transfer mixing head. Impact strength was measured as a function of temperature, volume % rubber, and wt average particle diameter. Impact strength at low temperature increased with decreasing particle diameter and increasing amount of rubber. A correlation was found between the ductile-brittle transition temperature and the interparticle distance. The effect on final blend properties of varying the level of MA grafted to the rubber was also investigated (Borggreve and Gaymans, 1989a). MA was grafted in the presence of BBPD. The presence of nylon-rubber copolymer was inferred from selective solvent extraction tests, nitrogen analysis, and IR spectroscopy.

In related studies Borggreve and Gaymans (1989b) prepared 90/10 blends of nylon 6 with either EPDM-g-MA, EP-g-MA, or PE-g-MA as described above and measured impact strength as a function of temperature. Rubber particle size and type of impact modifier were found to have a strong effect on mechanical properties. Borggreve and Gaymans (1989c) have also published the results of dilatometry tests for nylon 6 blends containing various MA-grafted rubbers.

Nylon 6/6 blends with EP-g-MA (0.7 wt % MA) have been studied by Ban et al. (1988, 1989). This blend along with a control blend containing unfunctionalized EP rubber was prepared on a 30 mm non-intermeshing TSE with a compounding screw configuration at 280 °C. Rubber domains in the blend with EP-g-MA contained typically 40 vol % occluded nylon while domains in the control blend contained < 1%.

A study of nylon 6/6 blends with EPDM-g-MA has been published by Crespy et al. (1986). The effect on blend properties of varying the level of MA from 0-10% of rubber phase was studied. Blends of 96 parts nylon : 4 parts MA-grafted rubber were prepared by extrusion on a W&P TSE at six different temperature profiles from 204-347 °C. Extruded materials were characterized by IR spectroscopy, DSC, and SEM which showed evidence for copolymer formation between the two phases. Maximum mechanical properties of molded test parts were observed at 6% MA grafted to rubber.

An early Du Pont patent (Anonymous, 1965) described blends of nylon 6/6 with ethylene-methacrylic acid copolymer (5.7 mol % acid) prepared on an SSE at 280 °C having a polyolefin dispersed phase in the form of particles less than 5 microns in diameter. The blends had improved barrier properties in blow molded parts.

In a related patent (Kohan et al., 1972), blends were disclosed of ethylene-methacrylic acid copolymer (9.8-16.4 wt % acid) with a nylon having at least 60% amine end groups prepared on an SSE at 280 °C. In one example a 1:1 blend containing nylon 6/6 with 30 meq/kg less amine than carboxylic acid end groups gave lower impact strength than the same blend made using nylon 6/6 having 33 meq/kg more amine than carboxylic acid end groups.

Blends of nylon 6/6 were also prepared by Murch (1974) with ethylene-methacrylic acid copolymer (11-12 wt % acid) which was 66% neutralized with zinc ions. Extrusion on a W&P 83 mm TSE at 280 °C and 300 lb/hr of a mixture of 100 parts nylon and 25 parts copolymer gave a blend whose molded test parts showed higher tensile strength than a similar blend containing unneutralized copolymer. Blends were also extruded on a 2 inch Egan SSE at 280 °C.

Coextrusion of nylon with oxidized, linear polyethylene gave blends with improved barrier properties (Armstrong, 1968a). For example, a mixture of 47.5% nylon 6, 47.5% polyethylene, and 5% oxidized polyethylene (acid number = 20; melt index = 0.1; density = 0.96) was extruded on a 1 inch extruder at 222-226 °C and 11 rpm, resulting in a throughput of about 800 g/hr. The extrudate product was substantially more resistant to dispersed phase agglomeration upon heat aging than was a similar sample made without oxidized PE. Blow molded bottles made from the blend with oxidized PE had higher drop impact resistance than bottles made from a blend without oxidized PE. Similar results were obtained when EAA copolymer (3 mol. % acid) was substituted for all of the oxidized PE.

In a related patent (Mesrobian et al., 1968) a blend of 50 parts nylon 6 with 40 parts PE and 10 parts ethylene-methacrylic acid copolymer (4 mol % acid; 38% neutralized with Na ions) was prepared by double-pass extrusion on a laboratory extruder at 229 °C. Bottles blow molded from this blend had improved barrier and impact properties compared to bottles made from blends without copolymer.

Swiger and Juliano (1979) and Swiger and Mango (1977) grafted LDPE (42 parts) with 0.53 parts MA in the presence of 0.042 parts DCP in the melt in a batch reactor at 110°-165 °C for 12 min. The resulting LDPE-g-MA contained 0.012 mol bound MA /100 g LDPE by IR and oxygen analysis. A blend of 10.5 parts LDPE-g-MA with 24.5 parts nylon 6/6 was prepared by extrusion on a 28 mm W&P ZDS-K TSE at 271 °C and 200 rpm. Molded test parts showed markedly higher impact strength at both 25 °C and -18 °C than control blends containing LDPE free of MA. Diethyl maleate was also efficiently grafted to PE under these conditions but maleic acid required 195 °C and use of di-t-butyl peroxide for grafting. Use of dibenzoyl peroxide in place of DCP was much less effective for grafting MA. Nylon 6 could also be used in these blends.

Vroomans (1988) grafted LDPE in an extruder with a mixture of styrene and MA and used the product as an impact modifier for nylon. For example, an 80/20 blend of nylon 6 and LDPE containing 1.8-3.0 wt % covalently bound MA was extruded on a W&P TSE at 240 °C. Molded test parts showed 105-118 KJ/m^2 Izod impact strength at 23 °C. Nylon 6/6 and 4/6 could also be impact modified using similar graft copolymers.

Hammer and Sinclair (1976, 1977) have described preparation of novel thermoplastic elastomers by blending low molecular weight (< 5,000) amine-terminated polyamides with copolymers containing amine-reactive sites such as anhydrides or vicinal carboxylic acid groups. Blends were prepared using a TSE, Brabender mixing equipment, or a roll mill typically at 210°-230 °C and 10-20 min residence time. Amine-reactive copolymers included ethylene/methyl acrylate/monoethyl maleate, ethylene/ vinyl acetate/maleic anhydride, ethyl acrylate/monoethyl fumarate, ethyl acrylate/maleic anhydride, PE-g-maleic anhydride, and EPDM-g-maleic anhydride. Polyamide oligomers included polycaprolactam, polylaurolactam, and caprolactam/laurolactam copolymers. Physical properties of molded test parts were reported.

In a Du Pont patent with numerous examples (Epstein, 1979a), nylon blends with high impact strength were prepared by coextrusion with a variety of functionalized ethylene copolymers bearing sites which may adhere to the nylon matrix and having tensile modulus at least 10 and preferably 20 times less than that of the nylon. Blends were extruded on a W&P 28 mm TSE at a temperature 5°-100 °C above the melting point of the nylon matrix, usually around 310 °C. The extruder was equipped with a vacuum vent operated at 25-30 inch vacuum pressure for efficient devolatilization. All of the ethylene copolymers contained either an anhydride, carboxylic acid, or epoxide group. Examples of functionalized copolymers included ethylene/methyl acrylate/ monoethyl maleate, ethylene/vinyl acetate/maleic anhydride, ethylene/vinyl acetate/GMA, EPDM-g-MA, EPDM-g-monoethyl maleate, and EPDM-g-FA. In certain blends with high impact strength free carboxylic acid groups on the copolymer were at least partially neutralized with a zinc salt. Nylons which were used included crystalline types such as different molecular weight grades of nylon 6/6, and amorphous types such as poly(trimethyl hexamethylene terephthalamide).

In a related patent (Richardson, 1981), blends of nylon 6 with FA-grafted ethylene / propylene / 1,4-hexadiene / norbornadiene copolymer were extruded at 270 °C on a 28 mm TSE to give test parts with improved tensile strength and mold shrinkage.

Epstein and Pagilagan (1983) prepared toughened blends of amorphous and semicrystalline polyamides by extrusion with a functionalized polyolefin. For example, 54 parts nylon 6/6, 27 parts amorphous polyamide (derived from iso/terephthalic acid and hexamethylenediamine + bis[p-aminocyclohexyl]methane), 9 parts EPDM, and 10 parts EPDM-g-fumaric acid (1.8 wt % FA) were extruded on a 28 mm TSE with a vacuum port in the same manner as described in the preceding example. Molded test parts showed higher notched Izod impact strength than toughened blends containing only one of the two polyamides.

Blends of nylon 6, 6/6, or a mixture of the two with PE or EVA (8-45% VA) grafted with monomers such as butyl acrylate, ethylhexyl acrylate, acrylate-AA mixtures, or acrylate-MA mixtures were formed by extrusion on a W&P 53 mm TSE at 265-290 °C (Korber et al., 1982). Molded test parts showed good homogeneity and impact strength.

Nylon 6 blends with MA-functionalized ethylene-butene-1 copolymer have been reported by Ohmura et al. (1982). For example, an 80:20 blend of nylon 6 with ethylene-butene-1 copolymer (14% butene; 20% crystallinity) functionalized with 0.35% MA was extruded at 250 °C at 7 kg/hr on a 40 mm extruder equipped with vacuum vent. Molding compositions with superior mechanical properties were obtained after addition of a urea compound to aid in mold release. Nylon 6/6 was also blended in this manner.

Blends of nylon 6, 6/6 copolymer with FA-functionalized ethylene- butene-1 copolymer have been described by Sawden (1988). For example, a masterbatch extrudate containing 83.3 wt % nylon 6, 6/6 copolymer (80:20; mp $\approx$ 225 °C) and 16.7 wt % of 1.0 wt % FA-grafted ethylene-butene-1 copolymer was diluted to produce a mixture of 81 wt % ethylene-butene-1 copolymer, 15.83 wt % nylon 6, 6/6, and 3.17 wt % of 1.0 wt % FA grafted ethylene-butene-1 copolymer which was extruded at 232 °C into plastic pipe which had a better combination of physical properties than similar pipe made without functionalized ethylene-butene-1 copolymer.

Roura (1982, 1984) prepared nylon blends with improved low temperature impact resistance by combining a mixture of nylon 6/6 with nylon 6 and an anhydride or carboxylic acid-functionalized polyolefin. For example, a mixture of 41 parts nylon 6/6, 40 parts nylon 6, 9 parts EPDM (72% ethylene, 24% propylene, 4% 1,4-hexadiene), and 10 parts EPDM-g-FA (1.5-2.0% FA) was extruded on a vented TSE at 310-320 °C. Molded test parts showed notched Izod impact strength of 12.1 ft-lb/in. at -40 °C compared to 4.5 ft-lb/in. for the same blend made without nylon 6 and 3.6 ft-lb/in. for the same blend made without nylon 6/6.

Blends of EPDM-g-MA with an elastomeric copolyamide synthesized from hexamethylenediamine, a bis(fatty acid), and caprolactam have also been prepared by reactive extrusion (Giroud-Abel and Goletto, 1984).

Polyamide blends with EPDM-g-FA had improved impact strength when an ionomer resin was included in the formulation (Dunphy, 1989). For example, 84 parts of a polyamide prepared from a mixture of iso/terephthalic acid (65:35) and hexamethylene diamine was mixed with 8.4 parts EPDM grafted with 1.5-2.0 wt % FA, and 7.6 parts ionomer resin containing 80% ethylene, 10% isobutyl acrylate, and 10% methacrylic acid (72% neutralized with zinc cation). The mixture was extruded with vacuum venting at 235 rpm and 321 °C on a 53 mm W&P TSE (L/D = 35) with mixing elements 750 mm and 1390 mm from the feed end of the screw. The extrudate was molded into test parts which showed impact strength of 19.4 ft-lb/in. at 0 °C compared to 15.0 ft-lb/in. for a blend containing 3.8 parts unfunctionalized EPDM and 3.8 parts EPDM-g-FA along with 8.4 parts ionomer resin, and 16.3 ft-lb/in. for a blend with 25 parts EPDM-g-FA and no ionomer resin.

Nylon blends with low molecular weight MA-grafted EP rubber have been reported by Olivier (1986a). For example, 1:1 blends of nylon 6 and MA-grafted EP (55% ethylene) were passed three times through a 1 inch SSE (L/D = 23) at 215-232 °C at 10-20 g/min, and then diluted with nylon to the desired composition before molding. Blends with 15% rubber loading containing EP-g-MA with Mooney viscosity (1 + 4 at 125 °C) 64 showed higher impact strength than the same blends with EP-g-MA with Mooney viscosity 98 at the same MA loading (1.74-1.88%).

In a related patent (Phadke, 1988a) nylon-EP copolymer blends with acceptable properties were formed by extruding nylon 6 with a masterbatch of nylon 6, EP, and MA. For example, a masterbatch containing 80 parts EP copolymer (60% ethylene), 20 parts nylon 6, 1.69 parts MA in acetone (10 parts in 6.6 parts), and 1 part DBPH was extruded three times on a 1 inch SSE (L/D = 20) at 218 °C. A blend of 1 part masterbatch and 2.8 parts nylon 6 gave molded test parts with notched Izod impact strength 18 ft-lb/in. at both -20° and 25 °C.

Blends of EP-g-MA with a copolyamide of (e.g.) 60% polyhexamethylene-adipamide and 40% polyhexamethylene-terephthalamide have also been described by Neilinger et al. (1988).

Nylon-polyarylate blends with improved physical properties were prepared by Okamoto et al. (1989) by blending with both an anhydride- and an epoxide-functionalized olefin copolymer. For example, a mixture of 35% nylon 6, 35% polyarylate derived from BPA and 1:1 iso/terephthalic acid, 15% EP copolymer grafted with endo-bicyclo-[2.2.1]-5-heptene-2,3-dicarboxylic anhydride, and 15% of an ethylene/vinyl acetate/GMA terpolymer was extruded on a TSE at 270 °C. Molded test parts had Izod impact strength of 84.6 kg-cm/cm. Control blends with only one of the two functionalized olefin copolymers had decreased impact strength and surface abnormalities in the molded parts. Ethylene/AA/MA terpolymer could also be used in these blends.

Blends of nylon 6/6 with PP-g-MA have been disclosed by Davis (1975). For example, PP-g-MA was prepared by extrusion of 100 parts PP with 4 parts MA and 1 part DCP at 270 °C with 1.5 min residence time to give product with about 1.5% grafted and 1.5% free MA. This material was reextruded with 90-95% PP at 200 °C. The product was extruded with 75-90% nylon 6/6 at 285 °C, 60 rpm, and 3 min residence time. Molded test parts showed slight improvement in tensile elongation compared to controls with no PP-g-MA.

Polypropylene-polyamide blends have been prepared by Mashita et al. (1988) by extrusion in combination with an epoxide-functionalized copolymer. For example, a mixture of 10 parts PP-g-MA (0.11% MA), 85 parts nylon 6/6, and 5 parts EVA-GMA copolymer (10% GMA) was kneaded and extruded on a 65 mm vented extruder at 280 °C. Molded test parts showed impact strength markedly higher than that of control blends with either unfunctionalized PP or without epoxide-containing polymer. Ethylene/GMA (6 parts GMA) copolymer could be used as epoxide-containing polymer. Nylon 6 was also used in these blends.

Blends of nylon with a mixture of PP-g-MA and EP-g-MA have been disclosed by Perron and Bourbonais (1988a). For example, a mixture of 3:2 PP and EP copolymer with 1 wt % MA and 0.045 wt % DBPH (as 45% solid on inert filler) was extruded on an 83 mm compounding TSE at 150 rpm and temperature profile 177° (feed), 193° (transition), 232° (metering), and 249 °C (die). Material was fed at 200 lb/hr and had residence time of about 3 min. A blend of 1 part grafted polyolefins with 1.8 parts nylon 6/6 was extruded using the same extruder conditions except that the temperature profile was 204° (feed), 246° (transition), 271° (metering), and 271 °C (die). Molded test parts had improved resistance to moisture absorption compared to blends without MA-grafted polyolefins.

Nylon-PP blends containing nadic anhydride-grafted polyolefins have been described by Fujita et al. (1987). For example, a mixture of 30 parts nylon 6, 50 parts PP, and 20 parts of a 30:70 mixture of PP and EP copolymer which had been previously extruded together with 0.2% nadic anhydride was extruded on a 45 mm vented TSE (L/D = 30) at 250 °C. Molded test parts showed impact strength of 33 kg-cm/cm and 40% tensile elongation compared to 25 kg-cm/cm and 20% tensile elongation for the same blend made using PP and EP copolymer which had been individually extruded with nadic anhydride and then recombined in the blend. In other examples, PP, nylon, and MA were preextruded and mixed with more nylon and PP.

Blends of nylon with MA-grafted EEA copolymer and a crosslinking agent have been disclosed by Ohmae et al. (1988). The crosslinking agent could be reactive with either nylon or EEA-g-MA, and was believed to mediate intramolecular crosslinking under the processing conditions. For example, 40 parts nylon 6, 60 parts EEA-g-MA (27% EA; 3.1% MA), and 0.5-1.9 parts hexamethylenediamine carbamate were melt-kneaded on a 30 mm vented SSE at 280 °C. Molded test parts showed excellent low temperature impact strength and a better combination of physical properties than was obtained in blends without either crosslinking agent or MA-grafted EEA. Nylon 6/6 could be used in place of nylon 6. Other crosslinking agents included PE-g-GMA. An earlier, related patent described similar nylon/EEA-g-MA blends prepared by extrusion in the absence of crosslinking agent (Mashita et al., 1987).

Compatibilized blends of nylon with polycarbonate were formed by Perron and Bourbonais (1988b) through coextrusion with MA-grafted polyolefin and a polyesteramide derived from diphenylmethane diisocyanate. For example, 30 parts nylon 6, 55 parts bisphenol A polycarbonate, 5 parts polyesteramide, and 10 parts of a mixture of 3 parts PP-g-MA and 2 parts EP-g-MA were extruded on a W&P ZSK 83 TSE operated at 1,000 psi with 3 min residence time and a production rate of 200 lb/hr at 250 °C. Molded test parts exhibited notched Izod impact strength of 22 ft-lb/in.

Blends of nylon with PPE were formed through coextrusion with a combination of EP-g-MA and EVA-g-GMA by Nishio et al. (1988). For example, a mixture of 50 parts nylon 6/6, 50 parts PPE, 11.8 parts EP-g-MA (0.7 wt % MA), and 5.9 parts EVA-g-GMA (5 wt % VA; 10 wt % GMA) was extruded on a twin screw kneader. Molded test parts showed a good combination of flow and impact strength. Nylon 6 could be used in place of nylon 6/6.

Blends of nylon with PPE were formed through coextrusion with a combination of EP copolymer, MA, a peroxide initiator and PS-g-GMA (Mawatari et al., 1989). For example, a mixture of 27 parts PPE, 2 parts PS-g-GMA, 8 parts EP copolymer, 0.08% MA, and 0.04% DBPH initiator was extruded at 320 °C and 300 rpm on a TSE equipped with intensive kneading zone. Nylon 6/6 (63 parts) was fed downstream in the same extruder, and the product mixture was subjected to a second intensive kneading section and pelletized. Molded test parts showed impact strength of 38 kg-cm/cm compared to 6 kg-cm/cm for a control blend containing no EP copolymer. When the above blend was prepared with no added MA or radical initiator, and using EP-g-MA prepared in a separate step (0.5-2.0% bound MA), molded test parts with impact strength of 28-58 kg-cm/cm were obtained. SMA copolymer or PS-g-(butadiene/butyl acrylate) copolymer could be used in place of PS-g-GMA. Nylon 4/6 could also be used in these blends.

Nylon blends with EPDM-g-GMA have been described by Olivier (1986b). For example, a mixture of 80 parts nylon 6 and 20 parts EPDM-g-GMA (2.8% bound GMA) was extruded under unspecified conditions and molded to give test parts with notched Izod impact strength of 15.6 ft-lb/in.

Compatibilized blends of nylon with EP rubber have been prepared by Akkapeddi et al. (1989) using EP grafted with N-methacrylyl caprolactam. For example, a mixture of 19.79 parts EP copolymer (47% ethylene), 0.2 parts N-methacrylyl caprolactam, and 0.01 parts DCP was fed to a Leistritz corotating TSE operated at 140 rpm. The extruder consisted of 10 zones held at 230-270 °C. A mixture of 70 parts nylon 6 (49 meq/kg amine end groups and 49 meq/kg carboxylic acid end groups) and 10 parts high-amine nylon 6 (72 meq/kg amine end groups and 22 meq/kg carboxylic acid end groups) was fed to the extruder at zone 4 (270 °C), and the nylon-rubber graft copolymer was formed in zones 5-10. Molded test parts showed notched Izod impact strength of 14.3 ft-lb/in. compared to 2.9 ft-lb/in. for a control blend made the same way but without N-methacrylyl caprolactam. N-Acrylyl caprolactam or N-methacrylyl laurolactam could also be used. SEBS could be grafted with N-methacrylyl caprolactam instead of EP rubber. Nylon 6/6 or 4/6 were also used.

Blondel and Jungblut (1986) claimed to form polyamide-polyolefin copolymers through grafting of olefin-terminated polyamide to polyolefin in an extruder. For example, a functionalized polyamide was prepared by copolymerization of 11-amino undecanoic acid in the presence of 12 wt % crotonic acid to give an oligomer with average MW 1100. A mixture of 400 g olefin-terminated polyamide with 9,600 g EVA (28% vinyl acetate) and 200 ppm DTB was extruded on an SSE at 150 rpm and temperature profile 200-185-220 °C at 10 kg/hr to yield flexible materials that were easily processed.

Blends of nylon with chlorinated PE were formed by Coran and Patel (1980a, 1983a) by mixing along with stabilizers in a Brabender mixer at 225 °C for 5 min following melting of the nylon. Compression molded test parts showed tensile properties higher than expected on the basis of volume-fraction-weighted averages. Selective extraction of the blends and analysis of the residues gave results consistent with the formation of a small amount of copolymer, perhaps by displacement of chloride groups by terminal nylon amine groups.

Improved mechanical properties were obtained by addition of a functionalized PP such as PP-g-MA or PP-g-carboxymethyl maleamic acid to blends of dynamically vulcanized rubbers also containing both PP and nylon (Coran et al., 1985; Coran and Patel, 1982a).

Nylon/Modified Polybutadiene or Styrene Copolymers Nylon blends with improved impact and tensile properties were prepared by Owens and Clovis (1972) by extrusion with a two-phase polymer prepared by first polymerizing an elastomeric phase followed by polymerizing a rigid thermoplastic phase containing nylon-reactive carboxylic acid groups. For example, butadiene, styrene, and methyl methacrylate were polymerized in emulsion to form a seed followed by treatment with a mixture of methyl methacrylate, ethyl acrylate, and methacrylic acid in such a manner that no new particles were formed. Molded test parts of a double-extruded blend with 75% nylon 6/6 gave notched Izod impact strength of 2.4 ft-lb/in. compared to 0.8 ft-lb/in. for unmodified nylon 6/6.

Nylon 6 blends with MA-grafted hydrogenated polybutadiene have been described by Hergenrother et al. (1984, 1985). For example, 20 parts hydrogenated polybutadiene (12% residual unsaturation level; 44% 1,2-content) grafted with 0.4 wt % MA, and 80 parts nylon 6 were coextruded on a TSE with temperature profile 232-254-288 °C at 75 rpm. Molded test parts analyzed for 35.6% graft copolymer and had notched Izod impact strength 19 ft-lb/in. compared to 0.7 ft-lb/in. for molded test parts of pure nylon 6. MA-grafted hydrogenated polyisoprene, hydrogenated styrene/butadiene/styrene copolymer, and hydrogenated high cis-polybutadiene were also used in nylon blends.

Blends of nylon with styrene/methyl methacrylate/maleic anhydride copolymer have been reported by Kasahara et al. (1983). For example, a 1:1 blend of nylon 6 and a copolymer containing 83 mol % styrene, 8 mol % methyl methacrylate, and 9 mol % MA (MW 90,000) was melt-kneaded on a TSE at 260 °C and shearing rate of 400 sec^{-1}. The product had dispersed phase particles 100-1,000 Å in size and had good impact and tensile properties. The presence of copolymer was proven by solvent fractionation and FTIR analysis. A control blend of nylon and a copolymer not containing MA had dispersed phase particle size $\geq$ 5 μm, poorer mechanical properties, and no nylon-containing copolymer. Nylon 6/6 or styrene-MA or SAN-MA copolymer could also be used in these blends.

Blends of nylon with MA-grafted hydrogenated styrene-butadiene copolymer have been described by Shiraki et al. (1986, 1987a,b). For example, an 80 : 20 mixture of nylon 6 with MA-grafted copolymer (0.5-2.3 wt % MA) was extruded on a 30 mm TSE (L/D = 28) and molded into test parts which showed markedly higher tensile and impact properties than a similar control blend without MA. Nylon 6/6 and AA-grafted hydrogenated styrene-butadiene copolymer could also be used. Improved properties in a wide variety of other thermoplastic resin blends other than those with nylon were also claimed.

Blends of nylon with either SEBS-g-MA or MA-grafted PP or a mixture of the two have been described by Lutz et al. (1989). For example, a mixture of 40 parts nylon 6/6, 20 parts SEBS-g-MA (1.6 wt % MA), and 40 parts PP-g-MA (0.16 wt % MA) was extruded on a 30 mm TSE at 270-285 °C at 300 rpm. Molded test parts showed notched Izod impact strength of 17 ft-lb/in.

Nylon copolymers with MA-grafted ABS have been claimed by Grant and Howe (1988). For example, a 1:1 blend of nylon 6 with ABS-g-MA (1.0 wt % MA) was extruded on a corotating 28 mm TSE at 260 °C. Molded test parts showed no delamination and notched Izod impact strength of 15.5 ft-lb/in. compared to blends of nylon with unfunctionalized ABS which delaminated. Nylon 6/6 and amorphous nylon were also compatibilized with ABS in this manner.

Müssig et al. (1989) formed nylon molding compositions with high impact strength and good flow through coextrusion of nylon with a polyether or fatty acid ester, and an epoxide – or anhydride – modified resin. For example, 93.2 parts nylon 6, 6 parts polyether of ethylene oxide and propylene oxide (OH number 56), and 0.8 parts of copolymer of styrene/methyl methacrylate/GMA (40/40/20; MW 15,000) were fed separately to the throat of a 53 mm TSE and extruded at 260 °C and 90 rpm. Molded test parts showed notched impact strength of 11.3 kJ/m^2 and flow length 63 cm compared to 2.3 and 55 for unmodified nylon and 8.4 and 58 for nylon blended only with polyether. Styrene/maleic anhydride copolymer or diisobutylene/maleic anhydride copolymer could be used as functionalized polymer. Nylon 6/6 and a polyamide derived from isophthalic acid and hexamethylene diamine were similarly blended.

Ide and Hasegawa (1974) described nylon 6/SMA blends containing nylon-SMA copolymer. The copolymer formed during “extrusion molding” of mixtures of the materials at 230 °C. Evidence for copolymer was obtained by selective solvent extraction and DSC. Physical properties such as tensile strength and elongation did not improve significantly compared to blends containing no copolymer.

Compatibilized blends of nylon 6 with PS formed by extrusion with SMA copolymer have been claimed by Sims (1976). For example, a mixture of 358 parts nylon 6, 90 parts PS, and 2.3 parts SMA copolymer (50% MA) was extruded at 227-232 °C. Extrudate rods showed no sign of fibrillation. Rods formed from blends without SMA showed splitting and cracking.

Glass-filled nylon 6 compositions with improved mechanical properties were obtained by coextrusion with a rubbery graft copolymer and styrene-maleic anhydride copolymer (Anonymous, 1978). For example, a blend of 9,200 g nylon 6, 200 g statistical copolymer of 70 wt % styrene/30 wt % MA, and 800 g of a rubbery copolymer obtained by grafting SAN to a 98% butyl acrylate/ 2% polyene shell was extruded on a TSE at 280 °C with addition of 4,954 g glass fibers. Molded test parts had improved impact strength compared to a comparable blend made without SMA copolymer.

Compatibilized blends of nylon 6/6 with SMA copolymer and a hydroxyethyl methacrylate copolymer have been prepared by extrusion by Woodbrey and Moncur (1982). For example, 2.8 kg nylon 6/6 and 0.7 kg of a copolymer prepared from 3.325 kg butyl acrylate and 0.175 kg 2-hydroxyethyl methacrylate were extruded on a 3.81 cm SSE at 280 °C. The material was passed through the extruder four times. The product (3 kg) was extruded under similar conditions three times with 60 g SMA copolymer (Mn 1,900; acid number = 270). Molded test parts showed notched impact strength of 742 J/m compared to 164 J/m for a control blend with SMA copolymer omitted. Poly(hexamethylene phthalamide) (40/60 tere/iso) could be used in place of nylon 6/6.

In contrast to the above studies with nylon 6 or 6/6, Dean (1985) failed to detect copolymer in blends of nylon 11 or nylon 12 with styrene-MA copolymer prepared on a Sterling SSE at 260 °C. Characterization was by DSC, DMA, selective solvent extraction, and IR spectroscopy. Improvements in physical properties of 50:50 blends were attributed to the attractive interaction between the elastomeric portion of the SMA copolymer and the amorphous regions of the nylon.

Nylon 6/6 copolymer formation with alkyl acrylate-vinyl benzylchloride copolymer has been claimed to result from displacement of reactive chloride by nylon amine end groups (Moncur, 1982). For example, 2.7 kg nylon 6/6 was mixed with 0.9 kg of a copolymer of 10-25% ethyl acrylate, 75-90% butyl acrylate, and sufficient vinyl benzyl chloride to give 0.23-0.33% reactive chlorine. The blend was extruded on a 3.81 cm SSE at 280 °C, dried and reextruded a total of 4-12 times. Molded test parts generally showed maximum impact strength after 5-8 extrusion passes. Vinyl chloroacetate was also used as source of reactive chlorine in the acrylate copolymer.

Kasahara et al. (1982) have used nylon blends with functionalized polystyrene copolymers to compatibilize nylon with PPE which is completely miscible with polystyrene. For example, a mixture of 50 parts styrene-MA copolymer (10 mol % MA) and 50 parts nylon 6 was melt-kneaded on a TSE, and the isolated extrudate pellets (40 parts) were blended with PPE (60 parts) and reextruded on a TSE at 220-280 °C. Molded test parts showed good mechanical properties with no evidence of delamination. PPE grafted with styrene monomer in an extruder was also used in these blends. Rubber-reinforced SMA copolymer, styrene/methyl methacrylate/MA copolymer, or styrene/N-phenyl maleimide copolymer could also be used. Control blends containing PPE, nylon and either PS or styrene/methyl methacrylate copolymer had poorer properties.

In another process for compatibilizing PPE with nylon (Taubitz et al., 1987d), a mixture of 40 parts PPE, 40 parts nylon 6, 10 parts styrene/butadiene/styrene copolymer, and 10 parts of a copolymer of 97% styrene and 3% chloromethylstyrene was extruded on a corotating TSE at 280 °C. Molded test parts showed markedly higher notched impact strength than molded parts from blends containing no styrene/chloromethylstyrene copolymer. Styrene/GMA and styrene/vinyl oxazoline copolymers could also be used as functionalized polystyrene copolymer.

4.4.5b Polyphenylene Ether/Nylon Blends

Ueno and Maruyama (1982a) reacted 7:3 PPE : nylon 6/6 with 1 part MA in a Brabender mixer at 250-300 °C for 5 min to give material whose molded test parts had greatly improved impact strength compared to a control blend made without MA, implying that PPE-nylon copolymer had formed as a compatibilizer.

Jalbert and Grant (1987) extruded PPE with 0.1-3.0 wt % MA on a 1.75 inch SSE at 50 rpm and 315-330 °C melt temperature to give resins with less than 0.2 wt % unbound MA. Blends of MA-functionalized PPE with nylon 6/6 (1:1) prepared on a 1.75 inch SSE gave molded test parts with maximum impact strength at only 0.25-1.00 wt % MA with wt % insolubles 9-18% (determined by successive extraction with formic acid/toluene). Grafting of MA to PPE in the presence of DCP radical initiator was shown to have detrimental effects on impact properties of the corresponding nylon blends. Itaconic acid was also used in place of MA in this disclosure.

In a related patent (Grant et al., 1988), it was shown that either preextrusion of nylon with MA followed by blending with PPE, or extrusion of PPE, nylon, and MA as a mixture resulted in blends with inferior properties. Other polyamides such as nylon 6, nylon 12, and an amorphous nylon were compatibilized with PPE-g-MA in further examples of the invention.

Similar compatibilized blends of 1:1 PPE-g-MA and nylon 6 prepared on an SSE have been reported by Akkapeddi et al. (1988). In certain examples EP-g-MA was used as impact modifier.

Blends containing PPE-g-MA, nylon, and SEBS-g-MA have been described by Nakazima and Izawa (1989). For example, a mixture of 25 parts PPE-g-MA, 50 parts nylon 6/6, and 25 parts SEBS-g-MA (0.6 wt % MA) was extruded on a 45 mm TSE (L/D = 33) at 300 °C. Molded test parts showed generally higher impact strength and higher level of chloroform/formic acid insolubles than similar blends containing either unfunctionalized PPE or unfunctionalized SEBS. Nylon 6 was also used in place of nylon 6/6.

Ueda et al. (1988) described blends of PPE and nylon compatibilized by coextrusion with MA and a styrene-butadiene block copolymer. For example, a mixture of 50 parts PPE, 40 parts nylon 6/6, 10 parts radial teleblock styrene-butadiene copolymer, and 0.5 parts MA was extruded on a 30 mm TSE at 300 °C and a shear rate of 5 cm/sec. Molded test parts showed impact strength of 30 kg-cm/cm. Blends in which the block copolymer contained no styrene block portion (e.g. polybutadiene or SBR random copolymer), or in which the styrene block was <25% or >90% of the copolymer, or in which the shear was only 2 cm/sec, had significantly lower impact strength. Nylon 6 could be used in place of nylon 6/6.

Blends of nylon with anhydride- or epoxide-functionalized EP copolymer and PPE have been described by Abe et al. (1988). For example, a mixture of 50% PPE, 0.6% MA, and 10% EP-g-MA (0.7 wt % MA) was fed to a continuous twin screw kneader while 40% nylon 6 and stabilizers were introduced at a point between the first feed hopper and a downstream air vent. The mixture was melted and kneaded at 260 °C and 380 rpm. Molded test parts showed higher impact strength than that of similar blends with unfunctionalized EP copolymer. EP-g-GMA could be used in place of EP-g-MA, and GMA monomer in place of MA. Optionally, a radical initiator such as 1,3-bis(t-butylperoxy isopropyl)benzene could be added to the PPE, MA, EP-g-MA portion of the blend and nylon 6/6 could be used in place of nylon 6 (Abe et al., 1989).

Blends of PPE and either nylon 6/6 or nylon 6 have been prepared by coextrusion with either MA or epoxidized polybutadiene (Shibuya et al., 1988a; Shibuya and Kosegaki, 1987). For example, 43% PPE, 43% nylon 6/6, 0.5% MA, and 14% styrene/butadiene copolymer were kneaded and extruded on a TSE at 280 °C. Molded test parts showed 27 kg-cm/cm impact strength. Styrene/isoprene copolymer could also be used as impact modifier.

Yates and Ullman (1988) extruded a blend of 49 parts PPE and 10 parts EPDM or EP copolymer with 0.6-3.1 parts MA at 288 °C on a counterrotating TSE at 400 rpm with vacuum venting in a segment before the die. The extrudates were blended with 41 parts nylon 6/6 and reextruded under similar conditions to give material that showed Izod impact strength of 149.5-720.9 J/m with values generally increasing with increasing MA level. Poor results were obtained when PPE/EPDM were extruded followed by reextrusion with nylon and MA. SEBS and styrene/butadiene/ styrene copolymers could be used in place of EPDM or EP copolymers.

In a related patent (Yates, 1988), a mixture of 49 parts PPE and 10 parts styrene/butadiene/styrene copolymer was extruded under the same conditions as above with 0.7-1.5 parts triethyl-, tribenzyl-, or tributylammonium fumarate. The resulting extrudate was reextruded with 41 parts nylon 6/6 under similar conditions to give blends with impact strength 123-748 J/m with values generally increasing with increasing fumarate loading.

Van der Meer and Yates (1987) grafted mixtures of PPE and SEBS with FA in the absence of radical initiators. For example, 49 parts PPE, 10 parts SEBS, and 0.7-1.5 parts FA were extruded on a counterrotating TSE at 288 °C and 400 rpm. Blends of the resulting extrudate with nylon 6/6 were prepared by extrusion on a co-rotating TSE at 300 rpm and 285 °C. Molded test parts had higher impact strength than blends of nylon 6/6, SEBS, and unfunctionalized PPE extruded with FA. Mixtures of SEBS with EPDM rubber were also used in compatibilized blends. Nylon 6 could be used in place of nylon 6/6.

Fujii et al. (1989) have demonstrated improved properties in PPE-nylon blends compatibilized with MA if the nylon component has an amine end group/carboxylic acid end group ratio of >1. Nylons with various ratios of amine to carboxylic acid end groups were prepared by mixing two grades of nylon 6, one with .084 mmol/g terminal amine and .018 mmol/g terminal carboxylic acid, and another with .046 mmol/g terminal amine and .070 mmol/g terminal carboxylic acid. A mixture of 0.5 parts MA, 50 parts PPE, and 50 parts nylon 6 was extruded at 290 °C on a TSE and molded into test parts. Blends containing nylon with ratio of terminal amine to terminal carboxylic acid >1 gave ductile molded parts with drop impact strength 464-605 kg-cm. Blends prepared either without MA or with MA and containing nylon with ratio of terminal amine to terminal carboxylic acid <1 gave brittle molded parts with drop impact strength ≤ 251 kg-cm. Nylon 6/6 with a high level of terminal amine end groups could be used in place of nylon 6.

Blends of PPE-g-MA with a mixture of crystalline and amorphous nylons have been described by Shibuya et al. (1988b). For example, a blend of 50 parts PPE, 40-4 parts amorphous nylon (e.g. poly[alkyl terephthalamide]), 10-46 parts crystalline nylon 6/6, and 0.3-0.5 parts MA was melt-kneaded on a TSE at 280 °C. Molded test parts showed higher impact strength than blends containing either of the two polyamides alone. Nylon 6 was also used in place of nylon 6/6.

Gallucci et al. (1989) prepared compatibilized PPE-nylon blends by extrusion with citric acid hydrate. For example, a blend of 40 parts PPE, 41 parts nylon 6/6, 10 parts SEBS, and 0.25 parts citric acid was extruded on a TSE at 300 °C. Molded test parts showed notched Izod impact strength of 2.6 ft-lb/in. and Gardner impact of $>$ 320 ft-lb compared to 0.8 ft-lb/in. and 18 ft-lb for the same blend made without citric acid. Malic acid could be used in place of citric acid. A variety of other polyamides including nylon 6 could be used in place of nylon 6/6.

Compatibilized blends of PPE-g-FA (or PPE-g-fumaric acid monoethyl ester) with nylon 6 or 6/6 have been described by Taubitz et al. (1989b). Low temperature impact strength was enhanced by addition of a block copolymer of styrene-butadiene-styrene partially hydrogenated to the extent of 86-89%. Optionally, high impact PS could also be present in these blends.

Yates and White (1989) prepared compatibilized PPE-nylon blends by extrusion of nylon with a PPE functionalized with carboxy groups, for example by metallation with alkyl lithium and treatment with carbon dioxide in solution. In one example, a mixture of 45 parts carboxylated PPE, 45 parts nylon 6/6, and 10 parts SEBS was extruded on a TSE at 400 rpm and 288 °C with vacuum venting. Molded test parts showed impact strength of 278 J/m compared to 37 J/m for a similar blend made using unfunctionalized PPE.

4.4.5c Polyphenylene Ether/Polyester Blends

Copolymer formation between a carbodiimide-functionalized PPE and PBT has also been accomplished in extruder reactors to give compatibilized blends. For example, Han and Gately (1987) functionalized PPE with carboxylic acid groups by extrusion with MA. The functionalized PPE was reacted in solution with an excess of a polycarbodiimide such as 4,4′-bis(4-cyclohexylcarbodiimido) diphenylmethane to give a carbodiimide-functionalized PPE through reaction of one of the carbodiimide groups with an acid group on PPE. A mixture of 45% functionalized PPE, 45% PBT, and 10% SEBS as impact modifier was extruded on a TSE and gave molded test parts with notched impact strength 774 J/m. High levels of PPE bound to PBT were demonstrated by selective extraction with suitable solvents. PET could also be compatibilized with PPE using this method.

Blends of PPE with polyesters in combination with phenoxy resins have been described by Sugio et al. (1985). For example, a mixture of 41.1 parts PBT, 2.3 parts MA, and 4.6 parts phenoxy resin (derived from condensation of a bisphenol with epichlorohydrin) was melt-kneaded in an extruder at 200-260 °C. The product (48 parts) was extruded with 32 parts PPE and 20 parts glass fibers at 280 °C. Molded test parts showed higher tensile strength and higher level of chloroform-insoluble material than control blends without either MA or phenoxy resin. Unfilled blends were also prepared. Styrene/maleic anhydride copolymer could be used in place of MA. PET could be used in place of PBT. It was believed that a transesterification reaction took place among PPE (or PPE-g-MA), polyester, and phenoxy resin leading to copolymer formation.

Impact modified PPE-PET blends with styrene-GMA copolymer have been disclosed by Sano and Ohno (1988). Presumably, copolymer formation occurs between polyester carboxylic acid end groups and epoxide groups on the styrene copolymer. The blends were claimed to possess improved properties if the SEBS or hydrogenated styrene/isoprene copolymer added as impact modifier was present in the form of a network structure rather than as discrete particles. Such network morphologies were present in increasing proportion as the molecular weight of the PPE used in the blends increased. For example, a mixture of 40:50:10:10 PPE (IV 0.48 dl/g), PET, hydrogenated styrene/isoprene copolymer, and styrene-5% GMA copolymer was melt-kneaded on a TSE and molded into test parts with excellent solvent resistance and impact strength of 17.8 kg-cm/cm. The same blend with PPE having IV 0.44 dl/g or less had significantly lower impact strength. A similar effect of PPE molecular weight in PPE-g-MA blends with nylon was also claimed.

Blends of PPE with polyesters in combination with epoxidized liquid polybutadiene have been described by Kobayashi et al. (1988). For example, a mixture of 40 parts PPE, 60 parts PET, and 0.5-5.0 parts polybutadiene with 7.5% oxirane oxygen content (MW $\approx$ 1,000) was melt-kneaded on a 30 mm TSE at 270 °C. Molded test parts showed impact strength of 4.1-7.5 kg-cm/cm compared to 1.7 kg-cm/cm for a similar blend containing unfunctionalized polybutadiene. Optionally, a styrene-butadiene block copolymer was added as impact modifier.

Hamersma et al. (1988) prepared compatibilized blends of PPE with polyesters such as PBT in the presence of a styrene-isopropenyl oxazoline copolymer (SIPO). For example, a mixture of 30 parts PPE, 40 parts PBT, 8 parts bisphenol A polycarbonate, 10 parts SIPO, and 12 parts SEBS as impact modifier was extruded at 320 °C and 300 rpm. The resultant blend had higher flow and similar impact strength to a blend prepared without SIPO. Similar blends prepared using styrene-maleic anhydride copolymer had poor impact strength.

Compatibilized blends of PPE-g-GMA with carboxylic acid-terminated polyesters have been formed by extrusion by Sybert et al. (1987). For example, a mixture of 36 parts PPE-g-GMA, 55 parts PBT, and 9 parts SEBS was extruded on a TSE and molded into test parts which showed notched Izod impact strength of 219 J/m and no evidence of delamination.

4.4.5d Polyester/Polyolefin Blends

Impact modified blends of polyesters with epoxy-functionalized polyolefins have been described by Epstein (1979b). For example, 80 parts PET was mixed with 20 parts EVA-g-GMA (5% GMA). Blends were extruded on a W&P TSE at a temperature 5-100 °C above the melting point of the polyester matrix, usually around 310 °C. The extruder was equipped with 2-5 kneading blocks and at least one reverse pitch element to generate high shear. Molded test parts showed substantially higher impact strength than blends containing no functionalized polyolefin. PBT and poly(1,4-cyclohexylenedimethylene terephthalate/isophthalate) copolymer were also used.

Blends of PBT with EPDM-g-GMA with high impact strength have been disclosed by Olivier (1986) and by Pratt et al. (1988). For example, a mixture of 80 parts PBT and 20 parts EPDM-g-GMA (1.4-2.6 wt % GMA) was extruded three times through a Killion extruder (L/D = 20) at 232 °C. Molded test parts showed notched Izod impact strength of 14.8-17.4 ft-lb/in. compared to 1.8 ft-lb/in. for an 80:20 blend of PBT with unfunctionalized EPDM and 6.6 for an 80:20 blend of PBT with EPDM-g-GMA (0.4 wt % GMA). In place of GMA there could be used glycidyl acrylate or, preferably, a mixture of glycidyl acrylate with methyl methacrylate to inhibit crosslinking during the solution grafting reaction. EPDM grafted with methyl methacrylate alone was not an effective impact modifier for PBT. A related patent describes formation and use in blends of a masterbatch in which a portion of the PBT is prereacted with epoxide-functionalized EPDM (Phadke, 1988b). A related patent (Hepp, 1985) describes blends of EPDM-g-GMA with PBT alone or in combination with bisphenol A polycarbonate or PET. The blends could be filled with glass or mica.

In a related patent (Pratt, 1988) mixtures of 78 parts PBT, 18 parts EPDM-g-GMA (6.4% GMA), and 4 parts of a third thermoplastic polymer with Tg > 100 °C could be extruded on either an SSE, or 28 mm, 30 mm, or 90 mm TSE at 232-299 °C. The third polymer was either SAN copolymer or bisphenol A polycarbonate or polysulfone, and was added to give improved heat resistance to the blend.

4.4.5e Other Examples of Graft Copolymer Formation

ABS which had been grafted with GMA in an extruder has been used to prepare compatibilized blends with PS functionalized with carboxylic acid groups (Taubitz et al., 1988a). For example, 40 parts of the product obtained from grafting ABS with 1.5% GMA was extruded with 40 parts carboxylic acid-terminated PS on a TSE with 3 min residence time at 210 °C. The product analyzed by selective solvent extraction and GPC contained only 15% unbound PS. Control blends made with unfunctionalized ABS contained 96-100% unbound PS.

Amine-terminated acrylonitrile-butadiene rubber has been effectively compatibilized with PP-g-(N-carboxymethyl maleamic acid) by melt mixing in a Brabender mixer at 185 °C (Coran and Patel, 1981a, 1982b, 1983b). Enhanced physical properties were obtained when a dimethylol phenol curing agent was added to the compositions. PP grafted with maleic acid in the presence of peroxide could also be used in these blends.

4.4.6 Interchain Copolymer Formation (Type 4): Crosslinked Graft Copolymers

4.4.6a Crosslinking through Reaction between Functionality on Each Polymer

Ethylene Copolymer + Styrene Copolymer Ethylene/vinyl acetate (8 wt %)/GMA (4 wt %) copolymer has been reacted with an equal weight of styrene-maleic anhydride copolymer (8 wt % MA) at 200 °C in a twin screw kneader extruder to give 42% grafted styrene copolymer by selective solvent extraction (Abe et al., 1984). Blends of this copolymer with PPE/PS gave improved mechanical properties compared to blends without the copolymer.

Miscible blends of PPE with high impact polystyrene (HIPS) have improved mechanical properties when both ethylene-methacrylic acid copolymer and styrene-GMA copolymer are also present (Fujii and Ting, 1988). For example, a mixture of 47 parts PPE, 30 parts HIPS, 8 parts SEBS, 15 parts ethylene-methacrylic acid copolymer, and 1 part styrene-GMA copolymer was extruded on a 28 mm W&P TSE at 290 rpm and 177-288 °C. Molded test parts had tensile elongation 110% and falling dart impact strength 340 in.-lb, compared to 29% and 243 in.-lb when neither of the two copolymers was present, and 30% and 226 in.-lb when only styrene-GMA copolymer was present at 1 part level.

Hohlfeld (1986) compatibilized blends of LLDPE with PS by coextrusion with acid-functionalized PE and oxazoline-functionalized PS. For example, a mixture of 25 parts LLDPE, 25 parts LLDPE-g-MA, 30 parts PS, and 20 parts styrene-isopropenyl oxazoline copolymer (1% oxazoline) was combined in a Brabender mixer at 280 °C and 50 rpm. Molded test parts showed 3,638 psi tensile strength and 13% elongation. A control blend of 1:1 LLDPE : PS showed 1,131 psi tensile strength and 2% elongation.

Similarly, mixing of styrene-isopropenyl oxazoline copolymer (SIPO; 1% oxazoline) with EAA copolymer (9% acid) 1:1 in a Brabender mixer at 220 °C resulted in compatible blends which had considerably improved physical properties compared to a control blend of PS with EAA (Schuetz et al., 1989). HDPE-g-MA (1%), carboxy-terminated PET, and vinylidene chloride/ methacrylic acid copolymer were also shown to form copolymer with oxazoline-functionalized PS under these conditions. Zinc chloride (0.2%) could be added to the blend containing HDPE-g-MA to catalyze copolymer formation. Styrene-acrylonitrile-isopropenyl oxazoline terpolymer (75:24:1) was shown to form copolymer with PP-g-AA (6% acid) under these conditions.

More recent studies on blends of acid-functionalized PE and oxazoline-functionalized PS have been published by Baker and Saleem (1987a,b), and Saleem and Baker (1990). Blends of SIPO copolymer (1 mol % oxazoline) with

10-90 wt % carboxylic acid-functionalized PE (9 mol % acid) were prepared using a Haake-Buchler Rheomix mixer typically at 225 °C for 6-30 min and 50-150 rpm. The blends were characterized and the presence of PS-PE copolymer was inferred from selective solvent extraction, FTIR, SEM, DSC, and mechanical testing. Blend properties were correlated with composition, shear rate, time, and temperature.

A similar study has been published by Fowler and Baker (1988) on blends of styrene-isopropenyl oxazoline copolymer with acrylonitrile-butadiene-AA terpolymer (7 mol % acid) prepared in a Haake-Buchler Rheomix mixer typically at 185 °C for 5 min. SIPO copolymer was used alone or diluted with unfunctionalized PS keeping the blend ratio at 4 parts polystyrene matrix to 1 part dispersed rubber phase. The torque rise during mixing was directly related to the concentration of oxazoline-carboxylic acid pairs. Improved rubber dispersion and molded part impact strength were observed with increasing oxazoline concentration. When the oxazoline concentration was above 5%, impact strength decreased as the effective rubber particle size was too small for efficient toughening.

Hohlfeld (1986) has prepared blends of PPE with EAA by coextrusion with styrene-isopropenyl oxazoline copolymer. For example, 35 parts PPE, 30 parts EAA, and 35 parts SIPO copolymer (1% oxazoline) were mixed in a Brabender mixer at 50 rpm and 280 °C to give torque level of 300 after 8 min. Molded test parts showed notched Izod impact strength of 1.28 ft-lb/in. and 8% tensile elongation. A similar blend made with 35 parts PS in place of SIPO showed 130 torque level, 0.6 ft-lb/in. impact strength, and 2% elongation.

Graft copolymers of EEA copolymer with peroxide-functionalized styrene-alkyl acrylate copolymers have been formed by reactive extrusion by Moriya et al. (1988). For example, a peroxide-containing polymer was synthesized by copolymerization of n-butyl acrylate, t-butylperoxy methacryloyloxyethyl carbonate, and styrene to give copolymer with Mn 87,000 and active oxygen content 0.02 wt %. Coextrusion of 100 parts peroxy copolymer with 100 parts EEA at 230 °C gave grafted copolymer with 49% efficiency. Alternatively, the two copolymers could be blended in a Brabender mixer at 180-270 °C for 5 min. Ethyl acrylate could be used in place of butyl acrylate in synthesis of the peroxy copolymer.

Dean (1986) prepared blends of sulfonated EPDM with styrene/maleimide/2-vinyl pyridine copolymer. For example, a mixture of 18 parts sulfonated EPDM, 82 parts copolymer containing 74.9% styrene, 13.2% maleimide (or other monomer with active hydrogen with pKa 7.5-10.5), and 11.9% 2-vinyl pyridine was extruded at 226 °C. Molded test parts showed notched Izod impact strength of 8.6 ft-lb/in. compared to 0.6 ft-lb/in. for parts made from styrene copolymer alone. The functionality on the EPDM was described as a sultone group (85 meq/100 g EPDM). Presumably, the maleimide group ring-opens the sultone to a sulfonic acid during extrusion with formation of a covalent bond.

The patent describes compositions which have simultaneously a covalent and a zwitterionic bond between the EPDM and styrene copolymer.

Ethylene or Styrene Copolymer + Polypropylene Blends of PP-g-MA with polyester and EVA-g-GMA have been described by Mashita et al. (1986). For example, a mixture of 66 parts PP-g-MA (0.11 wt % MA), 29 parts PBT, and 5 parts EVA-g-GMA (10 parts GMA) was melt-kneaded on a 65 mm vented extruder at 240 °C. Molded test parts had markedly higher impact strength and tensile properties than blends containing either unfunctionalized PP or no EVA-g-GMA. Selected blends were also made with added PP homopolymer, with EP-g-MA in place of PP-g-MA, and with PET in place of PBT. Ethylene-GMA copolymer (6 parts GMA) could also be used in these blends.

Graft copolymers of PP with peroxide-functionalized styrene-alkyl acrylate copolymers have been formed by reactive extrusion by Moriya et al. (1989). For example, a peroxide-containing polymer was synthesized in the presence of 700 g dispersed PP by copolymerization of 300 g styrene, 6 g t-butylperoxy methacryloyloxyethyl carbonate and 1.5 g BOP in aqueous suspension to give PP impregnated with copolymer having active oxygen content 0.13 wt %. Selective extraction showed that no grafting to PP took place under these conditions. Extrusion of the mixture on an SSE gave grafted copolymer with 62% efficiency. Omission of the peroxide-containing monomer led to a blend with 1.9% grafting efficiency. Methyl methacrylate, 21:9 styrene/acrylonitrile, or 21:9 styrene/butyl acrylate could be used in place of styrene in synthesis of the peroxy copolymer.

Natural Rubber + Polyethylene A study by Choudhury and Bhowmick (1989) described compatibilized blends of epoxidized natural rubber (ENR) with acid- or anhydride-functionalized polyolefins. Blends were prepared using a Brabender mixer at 150 °C, 60 rpm, and 5-6 min residence time. In certain cases DCP was added as a curative and mixing was continued until a significant torque increase was observed. Molded test parts containing NR, ENR, PE, and PE-g-MA showed 45% improvement in tensile strength over a control blend containing only NR and PE.

Polyvinyl Alcohol + Polyethylene Blends of polyvinyl alcohol (PVA) with PE useful as gas barrier materials have been prepared by Schmukler et al. (1986a,b). For example, a mixture of 50 wt % PVA (DP = 330, 35 mol % hydrolyzed) and 50 wt % HDPE grafted with 1.5 wt % methylbicyclo[2.2.1.]-hept-5-ene-2,3-dicarboxylic anhydride was reacted in a Brabender mixer at 325 °C for 5 min at 120 rpm. The product showed no gross phase separation, and had an IR spectrum which showed a decrease in anhydride bands and new bands at 1750 cm^{-1} and 1229 cm^{-1} not attributable to acetate.

Polyphenylene Ether + Polypropylene or Ethylene Copolymer or Styrene Copolymer Togo et al. (1988) have prepared compatibilized blends of PPE with PP by extrusion of acid-functionalized PPE with epoxide-functionalized PP. For example, PPE-g-MA was prepared by extrusion of 3 kg PPE with 90 g MA and 15 g DCP on a TSE at 300-320 °C to give material with 1.1 wt % grafted MA. PP-g-GMA was prepared by extrusion of 3 kg PP with 90 g GMA, 300 g styrene, and 15 g DCP on a TSE at 180-220 °C to give material with 6.2 wt % grafted styrene and GMA. A 1:1 mixture of PPE-g-MA and PP-g-GMA was melt-kneaded in a laboratory mixer at 60 rpm and 270 °C for 10 min and pressed into a sheet which had tensile strength of 360 kg/cm^2 compared to 180-250 kg/cm^2 for the same blend made using either unfunctionalized resins or PP grafted only with styrene monomer. Use of PP grafted only with GMA gave a PPE blend with 100 kg/cm^2 tensile strength. PP grafted in an extruder with hydroxyethyl acrylate/styrene could be used in place of PP-g-GMA.

Taubitz et al. (1988e) have described blends of acid-functionalized PPE with functionalized shell-core polymers. For example, a mixture of 4.5 parts PPE-g-FA, 2.3 parts SAN (25 wt % AN), 1.2 part styrene-butadiene-styrene block copolymer, and 2 parts of a material with crosslinked butyl acrylate core and a styrene-acrylonitrile-GMA shell (72/25/3) was extruded on a TSE at 280 °C with 3 min residence time. Molded test parts showed impact strength of 17 kJ/m^2 compared to 2 kJ/m^2 for a control blend made using unfunctionalized PPE and ABS in place of functionalized shell-core polymer. Hydroxypropyl acrylate could be used in place of GMA to synthesize the functionalized shell-core polymer in which case a similar blend had impact strength of 18 kJ/m^2.

Weiss (1989) prepared blends of PPE with good impact strength by extrusion of an acid-functionalized PPE with an epoxide-functionalized EPDM. For example, PPE was coextruded with 0.7 wt % fumaric acid (FA) on a vacuum vented, corotating TSE at 300 °C to produce PPE-g-FA. In a separate step 100 parts EPDM was extruded with 7.4 parts GMA and 0.74 parts peroxide initiator on a 30 mm corotating TSE at 160 rpm and 200 °C to give material with 5.4 wt % GMA incorporation and 29% gel content. Extrusion of 90 parts PPE-g-FA with 10 parts EPDM-g-GMA on a corotating TSE at 300 °C followed by molding gave test parts with no delamination and notched Izod impact strength of 7.5 ft-lb/in. Control blends containing either unfunctionalized PPE or unfunctionalized EPDM either delaminated or had lower impact strength.

4.4.6b Crosslinking of Two Polymers through Addition of a Third Reagent

It is usually difficult to make copolymer by addition of a poly-functional reagent (e.g. a polyelectrophile) to a polymer blend containing two functionally terminated polymers (e.g. nucleophile terminated) since the polyelectrophile often dissolves preferentially in one of the two polymers and homogeneous chain extension occurs instead of coupling across a phase boundary. The success of this compatibilization strategy often relies on high concentration of pendant rather than terminal functionality on the two polymers to be crosslinked. Alternatively, a polyfunctional reagent reactive with the main chain of both polymers may be used, but in all cases a certain amount of homopolymer crosslinking is to be expected.

The examples are categorized by type of crosslinking reagent.

Quinone Methide Precursor Hartman (1975) prepared graft copolymers of ethylene-butene-1 copolymer (2% butene; 0.3% unsaturation) with butyl rubber (98% isobutylene/2% isoprene; 1.4% unsaturation) by mixing with p-t-butylphenol-formaldehyde resin in a Brabender mixer 325 °C for 15 min. For example, 36 parts ethylene copolymer, 4 parts butyl rubber, 1.5 parts phenol-formaldehyde resin, and 0.5 parts stannous chloride dihydrate were combined to give product with 9.7% rubber grafted to ethylene copolymer.

Nitroso Compounds Ueno and Maruyama (1982b) prepared graft copolymers of PPE with PS or styrene copolymers by coextrusion with oxime or nitroso compounds. For example, a mixture of 400 g PPE, 600 g SAN copolymer, and 2 g p-quinone dioxime was extruded on a 30 mm extruder at 250 °C with about 5 min residence time. The resulting product had improved mechanical and optical properties compared to a control extruded without dioxime. In a second example, PPE (1 part) mixed with 1.5 parts PS and either .0075 parts p-quinone dioxime or .002 parts polydinitrosobenzene in a Brabender mixer at 250 °C for 10 min showed evidence of complete grafting according to the product's solubility properties. Comparative examples with no oxime or nitroso compound, or with CHP as additive showed no evidence for graft copolymer formation.

Peroxides Yonekura et al. (1988) prepared thermoplastic elastomers by coextrusion of peroxide-crosslinkable olefin copolymers and peroxide-decomposable polyolefins in the presence of peroxides. For example, a mixture of 55 parts EPDM (78% ethylene), 20 parts ethylene/4-methyl-1-pentene copolymer (96.5% ethylene), 25 parts PP, and a solution of 0.2 parts L130 peroxide and 0.3 parts divinylbenzene in 0.5 parts mineral oil was extruded on a TSE at 230 °C under nitrogen and formed into a sheet through a T-die. The sheet had good tensile and film properties compared to control blends made without ethylene/methylpentene copolymer. In a related patent (Otawa et al., 1987), elastomeric compositions were prepared by coextrusion of EPDM, MA, divinyl benzene, L130, and either PP or PE.

Van Ballegooie and Rudin (1988) prepared blends of o-vinyl benzaldehyde-styrene copolymer with LLDPE by coextrusion with DCP and triallyl isocyanurate (TAIC). o-Vinyl benzaldehyde comonomer (4.7-5.4 mol %) was incorporated to make the PS more susceptible to radical grafting. Mixtures of 1:1

LLDPE and the styrene copolymer were extruded either alone or with DCP or with DCP and TAIC on a counterrotating TSE with a flat temperature profile of 200 °C. Mean residence time of material was 5 min. The amount of graft copolymer formed through radical initiated coupling of PS and PE by TAIC was determined by selective extraction of unbound PS with toluene and analysis of the residue by IR spectroscopy. The copolymer levels ranged from 1% to 8%. A finer microstructure was observed in the morphology of blends made using TAIC compared to control blends made without it. Small improvements in mechanical properties were also observed in the blends containing copolymer.

Nitrene Precursors Krabbenhoft (1986) prepared graft copolymers of polycarbonate with SEBS by extrusion with a bis(sulfonyl azide). For example, a mixture of 1 part bisphenol A polycarbonate, 2 parts SEBS, and 1 wt % 4,4′-biphenyl disulfonyl azide was extruded at 177-182 °C on an SSE. Selective extraction with dichloromethane showed 47% grafted PC. The product was blended with EEA and additional PC to give a final blend composition of 82.5% PC, 7.5% SEBS, and 10% EEA. Molded test parts showed less delamination and improved double-gate notched Izod impact strength compared to control blends of the same resins without disulfonyl azide.

Bis-Nucleophiles Togo et al. (1988) prepared compatibilized blends of PPE with PP by extrusion of acid-functionalized PPE with acid-functionalized PP in the presence of a diamine. For example, PPE-g-MA was prepared by extrusion of 3 kg PPE with 90 g MA and 15 g DCP on a TSE at 300-320 °C to give material with 1.1 wt % grafted MA. PP-g-MA was prepared by extrusion of 3 kg PP with 90 g MA, 300 g styrene, and 15 g DCP on a TSE at 180-220 °C to give material with 6.5 wt % grafted styrene and MA. A 1:1 mixture of PPE-g-MA and PP-g-MA was melt-kneaded with 1 wt % p-phenylenediamine in a laboratory mixer at 60 rpm and 270 °C for 10 min and pressed into a sheet which had tensile strength of 430 kg/cm^2. BPA or a polyepoxide resin were also used in place of p-phenylenediamine.

Taubitz et al. (1988d) have extruded immiscible pairs of acid- or anhydride-functionalized polymers with a difunctional reagent capable of reacting with acid or anhydride on each polymer. For example, a mixture of 5 parts 95 : 5 PPE-g-1% FA/PS was extruded with 4 parts ABS-g-FA (1% FA) and 0.03 parts 1,6-hexanediol and 0.95 parts of an impact modifier such as SEBS. Molded test parts had higher impact strength at -23 °C than control blends either without hexanediol or with either unfunctionalized PPE or unfunctionalized ABS in place of their FA grafts. Other difunctional reagents employed included hexamethylenediamine and glycerindiglycidyl ether.

4.4.7 Interchain Copolymer Formation (Type 5): Ionic Bond Formation

Graft copolymers have also been made in which the graft linkage between polymers is an ionic bond instead of a covalent bond. Usually two polymers, each with a low concentration of bound ionic groups, are coextruded in the presence of a divalent metal cation such as zinc or aluminum. The two polymers become associated through ionic bonds mediated through the divalent cations. Monovalent cations such as Na or K may also be used in which case association is mediated through ion-dipole attraction. With either type of cation a morphology is formed in which there are concentrated domains of associated ionic species in a matrix of the (usually) incompatible homopolymers.

In most cases sulfonated or carboxylated polymers are used. The acidic functionalities are most often pendant to the main polymer chain and are prepared either through copolymerization or through subsequent grafting. They are usually preneutralized in a separate step with a divalent metal salt before melt blending since polymers bearing unneutralized sulfonic acid groups may not be stable to normal melt processing temperatures. When two neutralized ionic polymers are coextruded with intensive mixing, redistribution of intramolecular ionic crosslinks to intermolecular ionic crosslinks is promoted since ionic crosslinks are generally reversible at elevated temperature. Depending on the nature of the two polymers, the type and concentration of ionic groups, and the nature of the divalent metal cation, a stable, compatibilized blend of the two polymers may result.

Impact modified blends of PPE could be prepared with ionomeric polyolefins or polysiloxanes if an ionomeric PPE or PS was present as compatibilizing agent and both zinc stearate and triphenyl phosphate were added as plasticizers (Campbell et al. 1986, 1989; Golba and Seeger, 1987). For example, a mixture of 52 parts PPE, 13 parts sulfonated PPE zinc salt, 23.3 parts PS, 11.7 parts

$SO_3^-Zn^+_{1/2}$ + EPDM-$SO_3^-Zn^+_{1/2}$ ⟶

sulfonated PS
representative structure

SO_3^--Zn^{++}-^-O_3S-EPDM

copolymer with ionic crosslinks

sulfonated EPDM zinc salt, 18.2 parts triphenyl phosphate, and 11.7 parts zinc stearate was extruded on a TSE at 160-270 °C.

Molded test parts showed impact strength of 534 J/m. A control blend without sulfonated PPE gave test parts with 10.7 J/m impact strength. Sulfonated PS zinc salt could be used in place of sulfonated PPE as compatibilizing agent and was in certain blends more effective than the corresponding sodium salt. A sulfonated polydimethylsiloxane and the sodium salt of a butadiene/acrylonitrile/AA terpolymer were also effective as impact modifiers. The mechanism of compatibilization was believed to involve ionic crosslinking between sulfonated PPE or PS and the sulfonated impact modifier to form a graft copolymer.

Brown and McFay (1986, 1987) formed a graft copolymer by coextrusion of the zinc salt of sulfonated polystyrene with an EP rubber functionalized with phosphonate ester groups. Addition of a small amount of zinc stearate to the polymer mixture was necessary to promote hydrolysis of phosphonate ester to phosphonic acid zinc salt during extrusion so that ionic crosslinks could form with sulfonated PS salt. The polyolefins functionalized with phosphonate esters served as masked ionic polymers which were more easily processed than the salts but could be converted to salt form under controlled conditions in the melt.

$$EPR\text{-}g\text{-}\overset{O}{\overset{\|}{P}}(OEt)_2 \xrightarrow[\text{zinc stearate}]{\text{melt hydrolysis}} EPR\text{-}g\text{-}\underset{OEt}{\underset{|}{\overset{O}{\overset{\|}{P}}}}\text{-}O^{-}\ \overset{++}{Zn}\ {}^{-}O\overset{O}{\overset{\|}{C}}C_{17}H_{35}$$

$$\xrightarrow{PS\text{-}g\text{-}SO_3^{-}\ \overset{+}{Zn}_{1/2}} EPR\text{-}g\text{-}\underset{OEt}{\underset{|}{\overset{O}{\overset{\|}{P}}}}\text{-}O^{-}\ \overset{++}{Zn}\ {}^{-}O\underset{O}{\underset{\|}{\overset{O}{\overset{\|}{S}}}}\text{-}g\text{-}PS$$

copolymer with ionic crosslinks

Since PPE is miscible with PS, the compatibilization of PS with EP or EPDM rubber through ionic crosslinks could be used to compatibilize PPE with EP or EPDM. PPE blends were obtained with outstanding impact strength, improved stress crack resistance, and no tendency to delaminate compared to control blends.

Coran and Patel (1981a, 1983b) prepared compatibilized blends of PP with carboxylated acrylonitrile-butadiene rubber by mixing at 180-190 °C in a Brabender mixer 90 parts PP, 10 parts PP-g-MA, and 0.15 parts triethylenetetramine, followed by mixing with an equal amount of carboxylated rubber. Molded test parts had 250% tensile elongation compared to 13% for a similar blend containing unmodified PP. The compatibilization mechanism may involve imide formation between PP-g-MA and polyamine followed by protonation of a PP-attached amine group by carboxylated rubber with concomitant ionic association.

Compatibilized blends of neutralized sulfonated EPDM elastomers with styrene-4-vinylpyridine copolymer have been described (Peiffer et al., 1986; Agarwal et al., 1987; Lundberg et al., 1987). For example, a mixture of 80 parts zinc salt of a sulfonated EPDM (sulfonation level = 20 meq per 100 g polymer) and 4 phr zinc stearate was combined with 20 parts styrene-4-vinylpyridine copolymer (8.5 mol % vinylpyridine) in a Brabender mixer at 100 rpm for 10 min at 185-200 °C. The blends were characterized by IR spectroscopy, melt viscosity, DMA, DSC, and SEM. Enhanced physical properties were observed compared to blends containing PS in place of styrene-4-vinylpyridine copolymer. Compatibilization was believed to involve coordination of pyridine groups to Zn cation associated with sulfonate on EPDM. Sulfonated EPDM neutralized with Na or Mg exhibited a lower degree of compatibility with styrene-4-vinylpyridine copolymer.

4.5 Coupling/Crosslinking Reactions

4.5.1 Introduction

Coupling reactions involve reaction of a *single polymer* with a condensing agent, a polyfunctional coupling agent, or a crosslinking agent to build molecular weight by chain extension or branching, or to build melt viscosity by crosslinking. Suitable polymers have end groups or side chains capable of reacting with the condensing, coupling, or crosslinking agent. The difference between a condensing agent and a coupling agent may be explained as follows. Condensing agents are used to chain extend polymers which have *two different functionalities* as end groups, e.g., nylon or PET. A typical condensing agent may be mono- or polyfunctional, and reacts with only one of the two types of end group, typically giving a low molecular weight, volatile by-product such as water in the process. This modified end group is now activated for subsequent reaction with the other type of end group. When the two end groups react, the condensing agent acts as a leaving group and *is not incorporated into the extended polymer chain.*

Coupling agents are always polyfunctional and may be used to chain-extend or branch polymers with *either one or two different types of functionality* as end group. A coupling agent reacts with two (or more) polymer chains and *is incorporated into the final chain-extended or branched polymer.* The functionalities suitable for polyfunctional coupling agents are the same types that were useful for Interchain Copolymer Formation in Section 4.4 : for example, cyclic

phite (TPP) resulted in increased intrinsic viscosity of the polymer as a function of amount of TPP, temperature, residence time, and end group concentration. Evidence was presented that chain extension proceeded by reaction of TPP with hydroxyl end groups to give an alkyldiphenylphosphite ester and phenol by-product followed by reaction of the new phosphite ester with carboxylic acid end group to give chain extended polyester and diphenylphosphite as shown in the scheme on p. 165.

4.5.3 Coupling Reactions through Polyfunctional Coupling Agents

Coupling agents are polyfunctional reactants which react with polymers to give either chain extension or branching. The coupling agent itself is incorporated into the coupled polymer chain. Intensive mixing is again required to insure adequate reaction within realistic extruder residence times. Stoichiometry matching between substrate polymer and coupling agent is important since under ideal reaction conditions an excess of coupling agent will result in capping of the majority of functional groups and little chain extension or branching will occur. In the ideal case a polymer with n moles of functional groups would react in an extruder with n/2 moles of a bifunctional coupling agent for complete reaction to chain extended product. In practice the relative reaction rates of end capping and coupling must be considered, and a statistical mixture of products often results. A slight excess of coupling agent is frequently used because of limited residence time in the extruder which may not allow all end groups to encounter a limited amount of coupling agent. Also some families of polymers, particularly polyesters, may generate more end groups during extrusion. Examples in which a relatively large excess of coupling agent is used so that end capping, rather than coupling, is the predominant reaction are covered under Polymer Functionalization and Functional Group Modification, Section 4.7.

In all coupling reactions a viscosity increase was observed in the final polymer product. It can be seen that in all the examples polymers with nucleophilic end groups react with a coupling agent which has two or more electrophilic sites. As was pointed out in the section on Interchain Copolymer Formation, the most common electrophiles with suitable reactivity for coupling within the time limits of a typical extrusion process are epoxide, oxazoline, isocyanate, and carbodiimide. Polyfunctional reagents bearing these functionalities are most commonly used in the examples which follow.

Polyamides + Polyepoxides Polyamides such as nylon 6 and 6/6 have been chain extended by reaction with polyepoxides (Anonymous, 1971). For example, extrusion of 1 kg nylon 6/6 with 7 g epoxydicyclopentylglycidylether at

275 °C on a 30 mm extruder gave material with flow index 0.3 compared to an initial value of 60. Other polyepoxides used were vinylcyclohexenediepoxide, 1,4-hexadienediepoxide, and 1,3,7-nonatrienetriepoxide.

Blends of nylon 6/6 with BPA polycarbonate showed improved properties in the presence of a polyepoxide such as 3,4-epoxycyclohexylmethyl-3,4-epoxycyclohexane carboxylate (Maresca, 1989). For example, a blend of 25 parts nylon, 75 parts PC, and 0.1 parts diepoxide was extruded on a W&P ZSK-28 TSE at 296-307 °C. Molded test parts showed improvement in mechanical properties and solvent resistance compared to control blends without epoxide.

Polyamides + Polyisocyanates Polyamides have been chain extended by reactive extrusion with diisocyanates by Nelb et al. (1987). For example, a polyamide prepared by reaction of azelaic acid, adipic acid, and 4,4′-methylenebis (phenyl isocyanate) (MDI) could be extruded on a corotating 28 mm TSE at 272-282 °C at 100-120 rpm in the presence of 1 wt % MDI. The final product had inherent viscosity 0.627 dl/g. A control sample extruded under the same conditions without MDI had inherent viscosity 0.519 dl/g. The reaction was also performed using a counterrotating 34 mm TSE at 259-299 °C.

Polyether ester amides have been coupled by reactive extrusion with diisocyanates by Droescher et al. (1985a). For example, 9,900 g of a polyether ester amide prepared from 25 parts laurolactam, 75 parts dodecanoic acid, and polyoxytetramethylene glycol (Mn 1,000; equivalent amount based on dodecanoic acid) was extruded on a TSE at 196 °C with 2 min residence time along with 100 g isophorone diisocyanate oligomer (isocyanate functionality = 3.2). The final product had relative viscosity 2.07 compared to initial relative viscosity 1.81. Similar extrusion of a polymer with 61:39 ratio laurolactam : dodecanoic acid at 220 °C and 4 min residence time gave material with relative viscosity 2.17 compared to 1.86 initially.

Polyamides + Polycarbodiimides Thomas et al. (1978a) have coupled polyamides such as nylon 6/6 using polycarbodiimides on a W&P ZSK TSE with an intensive mixing screw configuration with kneading blocks at 295 °C. For example, extrusion of 1 wt % polycarbodiimide derived from 4,4′-diphenylmethane diisocyanate with nylon 6/6 gave material with relative viscosity 309, carboxylic acid end groups 18, and amine end groups 31, compared to initial values of 48, 74, and 47, respectively.

Polyamides + Polylactams A brief report by Akkapeddi and Gervasi (1988) described chain extension of polyamides by reactive extrusion with terephthaloylbislaurolactam (TBL). For example, nylon 6 was mixed with 0.5% dioctyl adipate (coating medium) and 2% TBL, and extruded on a Killion 1 inch SSE (L/D = 30) at 260 °C, 40 rpm, and 8 lb/hr. The product had relative viscosity = 3.0 and 0.021 meq/g residual amine groups. A control sample extruded without chain extender had relative viscosity = 1.85 and 0.055 meq/g residual amine groups. Terephthaloylbiscaprolactam was also effective as a chain extender. Nylon 6/6 and an amorphous nylon could also be chain extended using this process.

Polyesters + Polyepoxides Borman and Rock (1977) coupled PBT by reaction with polyepoxides. For example, extrusion of PBT with 0.5 wt % bis(3,4-epoxycyclohexylmethyl) adipate using a 1 inch Wayne extruder at 232-282 °C gave material with melt viscosity 9,400 poise compared to 6,950 poise initially.

Other polyepoxides which have been disclosed for branching/chain extension of polyesters by reactive extrusion include N,N-diglycidyl benzamide and related diepoxides (Arai and Tanaka, 1985a); N,N-diglycidyl aniline derivatives (Arai and Tanaka, 1985b); N,N-diglycidyl hydantoin, uracil, barbituric acid, or isocyanuric acid derivatives (Arai and Tanaka, 1984a; Kodama et al., 1979); N,N-diglycidyl diimides (Arai, 1984); N,N-diglycidyl imidazolones (Arai and Tanaka, 1984b) and epoxy novolaks in combination with phosphonium salt catalysts (Thomas, 1978; McNally et al., 1986).

PBT in blends with PBT/caprolactone copolymer has been chain extended with mixtures of phenyl glycidyl ether and diethyleneglycol diglycidyl ether in a 40 mm TSE (Kobayashi et al., 1986).

Aoyama et al. (1977) prepared flame retardant blends containing PBT or PBT copolymer with tetrabromobisphenol A by coextrusion with a polyepoxide such as Epikote 815 (diglycidyl ether of bisphenol A-epichlorohydrin oligomer) on a 65 mm extruder at 240 °C. Molded parts of blends containing 0.3-3.0 parts polyepoxide had markedly higher tensile elongation than control blends without epoxide.

Coleman (1980) prepared blends of PBT with a butyl acrylate-based core-shell resin and 1.5-3.0 parts diglycidyl ether of bisphenol A-epichlorohydrin oligomer using a 1.5 inch extruder operated at 240-270 °C and 75-150 rpm. Molded test parts showed improved retention of impact strength after exposure to boiling water.

Polyesters + Polyoxazolines Inata and Matsumura (1986) have described chain extenders for polyesters in a series of papers. In one example in which reactive extrusion was employed, PET was chain extended with 2,2′-bis(2-oxazoline) by coextrusion on a 30 mm extruder at 290 °C and 3 min residence time. PET with initial viscosity 0.78 and 41 meq/kg carboxylic acid groups had viscosity 1.07 and 10 meq/kg carboxylic acid groups after extrusion with 0.5 wt % bis(oxazoline).

Polyesters + Polyisocyanates The melt strength of polyesters such as PBT and PET can be increased by processing with polyisocyanates. For example, Dijkstra et al. (1971) showed that PET viscosity increased from 0.68 dl/g to

0.88 dl/g upon extrusion with 0.8 wt % MDI on a 1.75 inch extruder at 260 °C and 20 rpm. When 1.5 wt % MDI was used, product viscosity was > 1.1 dl/g.

Gebauer et al. (1979) extruded PBT with initial reduced viscosity of 0.9 dl/g at 70.5 kg/hr with 1.78 wt % MDI using a 60 mm SSE (L = 30D) at 80 rpm and 245-280 °C temperature profile. The product had reduced viscosity of 1.51 dl/g. It was shown that chain extension involved reaction of diisocyanate with hydroxyl end groups since the carboxyl end group level was unchanged after extrusion.

Kolouch and Michel (1983) also used MDI to build viscosity in PET in a TSE. For example, PET resin with initial inherent viscosity 0.70-0.74 dl/g was extruded with 0.85 wt % MDI at 4.5 kg/hr using a 28 mm TSE with corotating, intermeshing screws operated at 40 rpm. The TSE screw configuration consisted of six zones. The blend was fed to feed zone 1 at ambient temperature and conveyed to melting and plasticization zones 2 and 3 at 160-250 °C followed by intensive mixing zone 4 and devolatilization zone 5 at 280-320 °C. Zone 5 was maintained at ≤ 1.33 kPa pressure for efficient removal of volatiles and carbon dioxide. The material passed through a final metering zone at 260-280 °C and exited the die at 255-265 °C. The final product was free of gas bubbles and had inherent viscosity 1.02-1.08 dl/g. The viscosity of a ternary blend of PET with ethylene/propylene/1,3,5-hexatriene (EPH) copolymer and EPH-g-maleic anhydride was also improved using this process.

Kriek et al. (1985) prepared molding compositions of PET with glass fibers by coextrusion with 0.5-8.0 wt % of blocked isocyanate-terminated polyurethane. For example, a mixture of 64.1 parts PET (IV = 0.6), 30 parts glass fibers, and 4.8 parts of a prepolymer prepared from hydroxy-terminated polypropylene oxide, propylene oxide-ethylene oxide copolymer, trimethylol propane, toluene diisocyanate, and caprolactam was extruded on a 2 inch vented extruder at 90 rpm and 250-280 °C. Molded test parts had impact strength of 136 J/m compared to 86 J/m for a blend without blocked isocyanate-terminated polyurethane.

A brief report by Akkapeddi and Gervasi (1988) described chain extension of PET by reactive extrusion with a blocked diisocyanate derived from reaction of MDI with caprolactam. For example, PET (IV = 0.7 dl/g) was mixed with 0.5% dioctyl adipate (coating medium) and 1.9% blocked diisocyanate, and extruded on a Killion 1 inch SSE (L/D = 30) or a Wayne 1 inch SSE (L/D = 24) at 280 °C and 60 rpm. The product had IV = 0.87 dl/g and 0.01 meq/g residual carboxylic acid groups. A control sample of PET extruded without chain extender had IV = 0.66 dl/g and 0.03 meq/g residual carboxylic acid groups. Terephthaloylbislaurolactam was also effective as a chain extender for PET probably through reaction with hydroxy end groups. Other coupling agents such as m-phenylene bis(2-oxazoline), BPA diglycidyl ether, or an oligomeric carbodiimide were not as effective for building IV in PET under these conditions.

Reaction of 48 g PET with 2 g polyphenylene polyisocyanate (average of 2 isocyanate groups per molecule) in a Brabender apparatus at 270 °C for 5 min gave material with viscosity 98,000 poise compared to an initial value of 9,000 (Costanza and Berardinelli, 1980). Use of a TSE with 90-120 sec residence time at 220-280 °C was also mentioned. Increased reaction rate was obtained by addition of triphenylphosphine catalyst.

Droescher et al. (1985b) used isophorone diisocyanate oligomer (isocyanate functionality = 3.2) to branch/chain extend PBT or PBT/PET blends in a TSE at 250 °C and 2 min residence time.

Polyesters + Polycarbodiimides PBT has been chain extended by reaction with polycarbodiimides such as that derived from self-reaction of toluene diisocyanate end capped with p-chlorophenylisocyanate. Thomas et al. (1978b) extruded PBT with 2% polycarbodiimide at 250 °C and 90 sec residence time on a W&P ZSK TSE with an intensive mixing screw configuration with kneading blocks. Material was produced with intrinsic viscosity 1.4 and < 2 meq/kg carboxylic acid compared to initial values of 0.75 and 50-55 meq/kg. Filled and flame retarded blends containing polycarbodiimide and PBT copolymer with tetrabromo- or tetrachlorobisphenol A were covered in a related patent (Berardinelli and Edelman, 1977).

Polycarbodiimides derived from oligomerization of aromatic diisocyanates such as toluene diisocyanate have been coextruded with PET to give chain extension. For example, a 70:30 blend of PET with glass fiber extruded on a 28 mm W&P ZSK TSE at 254 °C and 100 rpm with 1-2 wt % polycarbodiimide (based on PET) gave material with unnotched Izod impact strength 19.8 ft-lb/in. compared to 14.1 ft-lb/in. for material extruded without polycarbodiimide (Thomas et al., 1978c).

Polyesters + Polyphthalimides Okuzumi (1977) has coupled PET in an extruder using N,N-terephthaloyl bis(phthalimide). For example, extrusion of PET (initial IV = 0.592) with 2 wt % bis(phthalimide) on a 1 inch extruder with 1 min residence time gave material with IV = 0.790.

4.5.4 Coupling Reactions through Crosslinking Agents

Dynamic Vulcanization Dynamic vulcanization refers to crosslinking ("vulcanization"; "curing") of elastomers during melt processing. Usually the term is used in connection with crosslinking of elastomers in the melt during mixing with thermoplastic resins to give blends which have the properties of thermoplastic elastomers. When mixed in the proper proportions, these blends have a rubbery dispersed phase of crosslinked elastomer and a continuous phase of thermoplastic resin. Crosslinking the rubbery phase stabilizes the rubber particulate morphology which prevents reagglomeration of rubber particles before the thermoplastic matrix can crystallize or cool to a glassy state, and during any subsequent processing steps. Numerous examples of different crosslinked elastomer- thermoplastic resin blends prepared by dynamic vulcanization have been reported in both the open literature and in patents. In certain cases there has been evidence of copolymer formation between thermoplastic matrix and crosslinked elastomer phase and these examples are included under Interchain Copolymer Formation in Section 4.4.

Coran and Patel (1980b), and Coran et al. (1978) prepared blends of vulcanized EPDM with PP with dynamic vulcanization of the EPDM. For example, a blend of 100 parts EPDM, 66.7 parts PP, 5 parts zinc oxide, and 1 part stearic acid was melted in a Brabender mixer at 100 rpm and 180-190 °C. After 2-3 min, x/2 parts tetramethylthiuram disulfide and x/4 parts 2-benzothiazyl disulfide were added followed after 30 sec by x parts sulfur (usually 2 parts). An increase in mixing torque was observed and mixing was continued for 2-3 min before the product was removed and passed through a cold roll mill to give sheet. The sheet was cut into small pieces which were remelted and mixed in the Brabender for an additional 2 min before the product was removed and again pressed into sheet before compression molding into test parts. It was found that the crosslink density was directly dependent upon the amount of sulfur added and that tensile strength increased as the crosslink density increased. Tensile set improved sharply with a small amount of crosslinking and then leveled off. The average rubber particle size was 1-2 μm. Control blends containing rubber particles of larger size were prepared by vulcanizing EPDM with various amounts of crosslinking agent in the absence of PP and then blending with PP in the Brabender mixer. It was found that the mechanical properties were inversely proportional to rubber particle diameter. Carbon black and extender oil could be added to these blends. Polyethylene could be used in place of PP. Blends were also reported in which the EPDM was cured with a mixture of DCP and phenylene bismaleimide, or with DBPH alone.

EPDM blends with PP with superior physical properties were also prepared using a dimethylolphenol as curing agent in a similar melt process (Abdou-Sabet and Fath, 1982). The blends were stated to contain a negligible amount of EPDM-PP copolymer.

In related patents dynamically vulcanized blends prepared in a similar manner in Brabender equipment were disclosed for combinations of natural rubber, styrene-butadiene rubber, acrylonitrile-butadiene rubber, or polybutadiene rubber with PP or PE (Coran and Patel, 1978a); EVA rubber with PP or PE (Coran and Patel, 1978b); isobutylene-isoprene rubber with PP or PE (Coran and Patel, 1978c), styrene-butadiene, acrylonitrile-butadiene, EPDM, or polybutadiene rubber with PBT, polycarbonate, polyetherester, or polycaprolactone; or nylon 6 in combination with PBT or polycarbonate (Coran and Patel, 1979a); chlorosulfonated PE with PP or PE (Coran and R. Patel, 1979b); polybutadiene, styrene-butadiene, polyisoprene, or natural rubber with PP cured using either a mixture of tetramethylthiuram disulfide, N-cyclohexyl-2-benzothiazole sulfenamide, and 4,4-dithiodimorpholine; or zinc dimethyldithiocarbamate in combination with a urethane curative derived form p-nitrosophenol, and dicyclohexylmethane-4,4-diisocyanate; or dimethylol-p-octylphenol with zinc oxide (Coran and Patel, 1981b); epichlorohydrin rubber with either PBT or a polyester terpolymer of 1,4-butanediol, 1,3-propanediol, and terephthalic acid (Coran and Patel, 1981c); and epichlorohydrin rubber with PMMA (Coran and Patel, 1981d).

Abdou-Sabet and Shen (1986) prepared dynamically vulcanized EPDM-PP blends using a TSE with shear rate at least 2,000 sec^{-1}. For example, mixtures of varying proportions of PP, EPDM, additives, and methylol phenolic curing resin were fed to either a corotating ZSK 53 or 83 mm TSE at 83-116 kg/hr and 350 rpm with average residence time 32-42 sec. The screws contained roughly equal numbers of conveying and kneading elements. The resins were melted and mixed in the first third of the extruder and a curing accelerator such as stannous chloride was added at this point via an inlet port. The remaining portion of the extruder at 180-230 °C acted as dynamic vulcanization zone. A vent was present immediately before the exit die for removal of any volatiles. The product showed improved physical properties compared to similar material made at low shear rate on a Banbury mixer. The process is especially useful for preparing blends with > 50 wt % rubber loading.

Coran and Patel (1979c, 1980c) have prepared dynamically vulcanized blends of acrylonitrile-butadiene rubber (22-45% AN) with nylon in a Brabender mixer using a process similar to that described above for EPDM-PP blends. Nylons used included 6, 6/9, 6/6, 6-6/6 copolymer, and poly(hexamethylene isophthalamide). Self-curing and non-self-curing rubbers were used, and in the former case the effect of added crosslinking agents was minimized since the properties of the blends improved simply from melt mixing. Curing agents used included the sulfur + additives system described above, DBPH, or m-phenylene bismaleimide + 2-benzothiazyl disulfide. The addition of a dimethylol phenolic compound gave improved blend properties either due to improved rubber curing (see above) or perhaps through reaction with nylon to give chain extension so that a better viscosity match between nylon and crosslinked rubber was obtained. Also, some copolymer may have been formed between nylon and rubber which would have led to improved compatibilization. Plasticizers and fillers could also be added to these blends.

Dynamically vulcanized blends of EPDM, EVA, or nitrile-butadiene rubbers with PP, PS, SAN, or nylon have been described by Coran and Patel (1981e). For each combination a 40:60 ratio of thermoplastic to rubber was used and the blends were prepared using the Brabender procedure described for EPDM-PP blends. Curing additives contained DBPH with and without m-phenylene bismaleimide. Blend mechanical properties were related to the crystallinity and stress at yield or break of the thermoplastic matrix, and the tensile modulus and critical surface tension for wetting each material. In general the ultimate mechanical properties of the blends increased with increasing crystallinity of the thermoplastic phase and with increasing similarity in critical surface tension of the two phases. This work was extended with similar conclusions by Coran et al. (1982) to blends based on combinations of PP, PE, PS, ABS, SAN, PMMA, PBT, nylon 6/9, or polycarbonate thermoplastic resins with rubbers from the group butadiene rubber, isobutene-isoprene rubber, styrene-butadiene rubber, acrylonitrile-butadiene rubber, chloroprene, chlorinated PE, poly-trans-pentenamer, EVA, EPDM, and natural rubber.

A study by Romanini et al. (1987) appeared on dynamic vulcanization of EPDM rubber in the presence of PP using an unspecified paraformaldehyde resin. For example, a blend of 60% EPDM, 40% PP, and paraformaldehyde resin was extruded on a TSE at 230 °C to give material with maximum crosslink density when 9-12% paraformaldehyde resin was used. Morphology as well as mechanical, rheological, and chemical resistance properties of blends with various ratios of EPDM:PP were reported.

Coran and Patel (1980a, 1983a) have described blends of nylon 6, 6/9, or a terpolymer of nylon 6, 6/6, and 6/10 with chlorinated PE dynamically vulcanized using DBPH with or without either trimethylolpropane triacrylate or m-phenylene bismaleimide.

Blends of acrylonitrile-butadiene rubber with EEA copolymer have been dynamically vulcanized with peroxide by Aldred and Fogg (1986). For example, 50 parts rubber (36 wt % AN), 50 parts EEA copolymer, 2.2 parts DCP, and 1 part stearic acid were mixed in a Brabender mixer at 120 rpm. Over 15 min, the instrument torque at 190 °C had risen in two distinct stages indicating first crosslinking of the rubber, then crosslinking of the EEA. No evidence for copolymer formation between the two components was presented. After further processing, molded sheet showed tensile strength of 13 MPa compared to 4.4 MPa for a comparable blend made without DCP.

Other Crosslinking Reactions in the Presence of Peroxides Summaries have been published on PE crosslinking by coextrusion with peroxides by Dorn (1985) and Henman (1983).

Early work by Gregorian (1965) showed that polyethylene could be cross-linked through melt reaction with peroxide in a Brabender Plastograph. For example, 90.9 parts PE (melt index 0.7; density 0.958) was reacted for 15 min at 170 °C with a mixture of 9 parts petroleum wax (mp 145 °C; spec. grav. 0.925) and 0.1 parts DCP. The final product had melt index 0.24 and improved environmental stress crack resistance. t-Butyl perbenzoate and 2,7-dimethyl octane-2,7-di-t-butyl peroxide were also used.

A study of PE crosslinking using benzoyl peroxide has been published by Kampouris and Andreopoulos (1987). Various LDPE grades were mixed with

BOP at 120 °C in a Brabender mixing head. The torque measured after 10 min increased with increasing concentration of BOP up to 5 phr. Gel content measured by solvent extraction also increased to a maximum of about 55% with increasing BOP concentration.

Schuddemage et al. (1977) prepared EPDM-modified SAN in an extruder through coextrusion of EPDM with SAN containing a radical initiator with a half-life of at least 5 hrs at 120 °C. The radical initiator is conveniently added to a polymerization reaction mixture of the components prior to extrusion. For example, styrene and acrylonitrile were prepolymerized in the presence of EPDM (50% propylene; 10 ethylidene norbornene units per 1,000 carbon atoms; MW $\approx$ 200,000) in an autoclave with di-t-butyl peroxide, dimeric alpha-methyl styrene and white oil additives. The reaction mixture was then subjected to suspension polymerization in a second autoclave in the presence of water and partially saponified polyvinyl acetate. A radical initiator such as dibenzoyl peroxide was added at this stage. The isolated, dried product was extruded at 40 rpm on a Leistritz TSE containing a shearing zone at 180 °C for reducing rubber particle size to < 1 μm, and a crosslinking zone at 230-250 °C. Molded test parts showed notched Izod impact values 2-3 times higher than those obtained when the final product was crosslinked by extrusion on an SSE with short residence time either with or without preliminary crosslinking during suspension polymerization.

Blends of EPDM with PP have been crosslinked by reactive extrusion with divinyl benzene (DVB) and peroxide by Otawa et al. (1988). For example, 70 parts EPDM, 30 parts PP, 1 part MA, and 0.5 parts DVB were extruded with 0.3 parts L130 using a corotating TSE at 220 °C. Mineral oil (30 parts) and butyl rubber (10 parts) were also added. The product had 98% gel content compared to 2% gel content for material extruded without DVB and peroxide.

Ethylene/octene-1 copolymer has been crosslinked in the presence of 2% loading of a variety of other polymers by extrusion with 2,500 ppm L130 using a SSE in a blown film process (Boivin and Zelonka, 1987). Clear films with higher melt strength than controls were obtained depending upon the identity of the additive polymer. Additive polymers included EVA and polyethylene- or polypropylene-g-MA.

Related work by White (1988) disclosed crosslinking of ethylene/butene-1 or ethylene/octene-1 copolymers by reactive extrusion with peroxide concentrates with carriers such as ethylene/butene-1 copolymer. Triallyl isocyanurate could be present as co-curing agent in the concentrate.

Narkis and Wallerstein (1986) studied polycaprolactone crosslinking in the melt with peroxides. Polycaprolactone (Mw = 63,000; Mn = 26,000) was mixed with known concentrations of DCP or t-butyl cumyl peroxide at 100 °C in a Brabender mixing head at 30 rpm for 15 min after which the temperature was raised to 180 °C to decompose the peroxide. Measured gel content increased with increasing concentration of either peroxide above a critical level of 0.75

wt %. At 3.0 wt % of peroxide, gel content was 65% using DCP and 90% using t-butyl cumyl peroxide.

Crosslinking Using Oxidizing Agents Christensen and Voss (1985) achieved crosslinking of polyphenylene sulfide by reactive extrusion in the presence of air. For example, PPS was fed at 26.5-41 lbs/hr to a 53 mm W&P ZSK TSE with corotating, intermeshing screws. The melt temperature was 285-405 °C and the screw speed was 239-301 rpm. Air was injected into the molten polymer at three points in the extruder at a rate of 690-978 SCF. A devolatilization zone was present immediately before the die and material residence time was about 30-36 sec. PPS with initial melt flow of about 2,500 g/10 min (at 316 °C) had a final melt flow of 205-332 g/10 min.

An earlier patent claimed a process for increasing melt viscosity of PPS by treatment with oxygen in an extruder but no examples were given (Moberly, 1985).

Chan and Venkatraman (1986), and Chan and Tanous (1986) crosslinked polyether ether ketone (PEEK) and polyether ketone (PEK) through extrusion on either corotating or counterrotating twin screw extruders (no details) at 400 °C in the presence of 0.25-2.0 wt % elemental sulfur with vacuum venting to remove hydrogen sulfide. Evidence for crosslinking came from rheological, DMA, and gel content measurements. The process was believed to involve initial PEEK chain scission followed by crosslinking through covalent bond formation with sulfur.

Crosslinking Initiated by Mechanical Stress Porter and Casale (1985) have published a summary of polymer reactions caused by mechanical stress which includes examples of polyolefin degradation and crosslinking during melt extrusion. The presence of oxygen is often critical for high levels of crosslinking.

Bergström et al. (1979) studied LDPE degradation and gel formation in a Brabender mixer under a nitrogen atmosphere at 180° and 220 °C. Melt temperature and mixing time had the most effect on gel formation while mixing efficiency had little effect. The presence of an antioxidant led to decreased gel formation.

Crosslinking Using Bifunctional Coupling Agents Park (1985) crosslinked EAA copolymer (6.5 % AA) by reactive extrusion with a polyepoxide or with an epoxysilane in the presence of 50 wt % LDPE. A blowing agent was injected into the extruder and the final product was a cellular foam. For example, an EAA/LDPE blend was extruded on a 1.25 inch extruder at 120-190 °C with 0.125 parts epoxy novolac resin to give a stable foam. In the absence of crosslinking agent an unstable foam was produced.

In a related patent (Corbett and Bearden, 1984), SAA copolymer was crosslinked by reactive extrusion with either a polyepoxide, a divinyl ether, or a diisocyanate. For example, 100 parts SAA (8% AA) was coextruded on a 1.25 inch extruder with 12 parts blowing agent, 1 part talc, 0.2 parts epoxy novolac resin (epoxy equivalent weight 175), and 0.05 parts MgO catalyst to give a crosslinked foam with melt tension (220 °C) 12 g compared to 3 g for the same material made without epoxy resin. In addition to MgO, a large number of other catalysts were examined with SAA and epoxy resin in a Brabender mixing head.

Coupling reactions have also been mediated through silane crosslinking. For example (Schmid and Hoppe, 1984), the end groups of nylon 12 or nylon 6 have been capped by reactive extrusion with either 1-isocyanato-3- triethoxysilylpropane or 1-(2,3-epoxypropyl)-3-trimethoxysilylpropane in an extruder at 60 rpm and 230 °C. The extruded product was stored at > 40% humidity and then heated at 225 °C for 45 min to affect crosslinking.

Charles and Gasman (1981) crosslinked PBT by reaction with 0.02 wt % bis(azidoformylethylenedioxy) phthalate in a 1.5 inch extruder at 200-250 °C and 120-180 sec residence time. Material with initial IV of 0.8 had IV > 2 after extrusion.

Nogues (1988a) reacted PP-g-MA in either a Brabender mixer or a 30 mm SSE at 200 °C with diols such as 1,6-hexanediol, polyoxytetramethylene glycol (MW 650), or polyoxyethylene glycol (MW 600 or 2,000). Reaction of diol with grafted anhydride could be followed by monitoring the appearance of the IR band at 1730 cm^{-1} due to the new ester group. The products reacted with diol had significantly improved adhesion to aluminum compared to PP-g-MA. Increasing adhesion was obtained using increasing equivalents of hydroxyl per MA group up to 1-2 equivalents. Statistically, at least a portion of the PP-g-MA is crosslinked at these stoichiometries. Use of a mono-alcohol such as 1-octanol had no beneficial effect on adhesion. EP-g-MA could be used in place of PP-g-MA. The anhydride groups on PP-g-MA could also be converted to amine groups by reaction with alpha, omega diamines such as a polyamide 11 oligomer (MW 1,050) obtained by polycondensation of 11-aminoundecanoic acid in the presence of hexamethylenediamine. In this case extent of reaction could be monitored using the IR band at 1650 cm^{-1} due to the amide group.

In a related patent (Nogues, 1988b) PP-g-MA or EP-g-MA reacted in a Brabender mixer at 180 °C with 1,6-hexanediol, polyoxytetramethylene glycol (MW 650), polyoxyethylene glycol (MW 600 or 2,000), or polyamide 11 oligomer (MW 1,050) showed significantly increased tensile strength and elongation in molded test parts compared to control parts with no polyol.

4.5.5 Ionic Crosslinking

The categories are by type of metal salt or base used in the crosslinking process. In some of the following patents more than one type of metal salt was given in examples.

Zinc Salts Polyisoprene rubber grafted with MA has been crosslinked by mixing in the melt with zinc stearate (Luijk et al., 1972). For example, 100 parts polyisoprene-g-MA (0.7 phr MA content) was combined in a Brabender mixing apparatus with 5 parts zinc oxide and 3 parts stearic acid at 180 °C for 20 min. Molded test parts showed markedly higher tensile properties and lower tension set at break than a similar blend containing no stearic acid.

Carboxylic acid groups on EAA copolymer have been neutralized/crosslinked by coextrusion with aqueous zinc or sodium acetate solution (Ziegler et al., 1986). For example, 50 kg/hr EAA (8 wt % AA) was introduced into a 5.7 cm TSE (L/D 33) at 140 °C and conveyed to a reaction zone where it was treated with 7 l/hr 20 wt % aqueous zinc acetate solution at 180 °C followed by degassing at 220 °C to remove acetic acid and water. The final product had 31% neutralized carboxylic acid groups and a melt index of 1.9 g/10 min compared to an initial melt index of 15 g/10 min. Ethylene-butyl acrylate-AA terpolymer has also been neutralized using this procedure.

The preparation of blends of 0.1-25 parts EAA with ultrahigh molecular weight polyethylene by extrusion with zinc stearate has been disclosed by Herten and Louies (1986). Blends were prepared on a Brabender SSE with temperature profile 45°, 80°, 260°, and 275 °C (die). Typically, 97.5-98.5 parts polyethylene were coextruded with 1-2 parts EAA and 0.5 parts ZnSt. The purpose of the invention was to allow processing of ultrahigh molecular weight polyethylene on conventional extrusion equipment. It was believed that during melt processing there is some neutralization of carboxylic acid-containing polymer which results in a certain amount of ionic crosslinking, and that crosslinked material acts as a lubricant for the ultrahigh molecular weight polyethylene.

Ethylene-methacrylic acid copolymer has been neutralized/crosslinked in either a 2 or a 3.5 inch extruder containing a melting section, a mixing section, a devolatilization section, and a final metering section (Rees, 1968). For example, 15-22 lbs/hr copolymer (15% methacrylic acid) and 0.4-1.5 lbs/hr zinc oxide were fed separately to the feed throat of the extruder. The insoluble zinc oxide is distributed in the molten polymer in the melting zone. At the beginning of the mixing zone an aqueous solution of a low molecular weight carboxylic acid such as acetic acid is injected at 4-9 cc/min into the extruder at 156 °C and the melt temperature is raised to 210 °C before devolatilization and exit. Copolymer with initial melt index of 71 g/10 min had melt index <2 g/10 min following this treatment. Ethylene/vinyl acetate/methacrylic acid terpolymer was also crosslinked using this process. Decrease in copolymer melt index was also achieved by extrusion in the presence of aqueous sodium hydroxide or dry mixtures of metallic salts such as magnesium hydroxide.

In a related patent (Bush and Milligan, 1972), a similar extruder reactor was used for neutralization/crosslinking of ethylene-methacrylic acid copolymer using a concentrate of zinc oxide in an ethylene-methacrylic acid copolymer with melt index substantially less than that of the copolymer to be crosslinked.

For example, pellets of copolymer containing 15% methacrylic acid (melt index 60 g/10 min) were fed to the throat of an extruder along with 7.5 wt % pellets of a concentrate containing 60% zinc oxide and 40% ethylene-methacrylic acid copolymer (melt index 500 g/10 min). Water (5 wt % based on copolymer) was fed to the molten polymer stream at the beginning of the mixing zone at 225 °C at which point the pressure was 2,000 psi. The extrudate was passed through a devolatilization zone before exiting the extruder to give product with melt index 1.1 g/10 min. A lubricant such as zinc stearate could also be present in the concentrate.

Ethylene-methacrylic acid (EMA) copolymer has also been neutralized/crosslinked by extrusion with a concentrate of the copolymer containing zinc oxide, zinc stearate, and either zinc acetate or zinc formate (Powell and Prejean, 1976). For example, a blend of 92.5 parts EMA (10 wt % acid) and 7.5 parts concentrate containing 54.4% EMA, 42% zinc oxide, 3% zinc acetate, and 0.6% zinc stearate was fed to a 3.5 inch plasticating extruder with initial plasticating section 7 diameters long at 120 °C and mixing section 13 diameters long at 240-280 °C. The product was clear and had melt index 1.1 compared to 35 initially. Ethylene/methacrylic acid/isobutyl acrylate copolymer has also been crosslinked by this process.

Magnesium Salts EAA copolymer has been neutralized/crosslinked by coextrusion with a 1:1 concentrate of MgO in ethylene/1-octene (7%) copolymer (Neill et al., 1987, 1988). For example, 50 parts/hr copolymer and 6 parts/hr MgO concentrate are fed through two separate feeders into a 53 mm W&P TSE operated at 200 rpm and 185-285 °C temperature profile with a devolatilizing zone to give clear material with melt flow index 2.27 and tensile strength 4,240 psi. For comparison, heterogeneous material which had MFI 2.86 (foamed due to further reaction of MgO with EAA) and tensile strength 3,890 psi was obtained under comparable conditions with addition of pure MgO. MgO concentrates were also made with LLDPE, LDPE, and HDPE.

Similarly, high density polyethylene filled with MgO was reported to have better physical properties when combined with EAA (Herman et al., 1983). For example, coextrusion on a TSE at 148-221 °C of 100 parts HDPE with 20 parts MgO and 5 parts EAA (3% AA) gave molded parts with marginally higher Izod impact strength than the corresponding blend without EAA.

Aluminum Salts Crosslinking of carboxylic acid groups on ethylene/n-butyl acrylate/methacrylic acid copolymer has been achieved by coextrusion with aluminum salts. For example, Statz (1988) extruded 2,400 g copolymer (60 parts ethylene, 28 parts n-butyl acrylate, 12 parts methacrylic acid) with 1,028 g nylon 6, 178 g aluminum acetylacetonate, and 255 g N-ethyl-p-toluene sulfonamide plasticizer on a W&P TSE at 150 rpm at 277 °C melt temperature with devolatilization. A blend resulted with compression set at 70 °C (22 hrs) of

55%. The teachings of this patent indicate that aluminum ionomers are not processable using standard thermoplastic processing techniques. Therefore, melt neutralization in an extruder is one of the few practical ways to make such materials.

A related patent (Strauss, 1987) disclosed the use of $Al(OH)_2(CH_3COO)$ or related salts for neutralization/crosslinking of carboxylic acid groups on ethylene/n-butyl acrylate/methacrylic acid copolymer in blends with HDPE using an 8.9 cm extruder at 270-292 °C. Output from this first extruder was fed directly to a 5.1 cm extruder with two devolatilization zones for removal of volatiles. Material with initial melt flow index about 800 had MFI about 180 following neutralization and crosslinking. When glacial acetic acid was injected into the molten copolymer-aluminum salt mixture, MFI decreased to 0.43. The process was most effective when the aluminum salt was added in the form of a concentrate with the copolymer. Hydrated alumina could also be used as aluminum source when glacial acetic acid was added. Ethylene-methacrylic acid copolymer was also crosslinked.

Garagnani et al. (1985) grafted polypropylene with 0.5 wt % N-carboxymethyl maleamic acid or related maleamic acids in a TSE at 190 °C in the presence of 0.05 wt % initiator such as 1,3-bis(t-butylperoxyisopropyl)benzene. The grafted polymer (100 parts) was reextruded at 200 °C on a TSE with 4 parts of a concentrate containing 50 parts LDPE and 50 parts aluminum isopropylate to give material which was crosslinked as determined by resistance of molded test parts to creep. EP, HDPE, and EVA were also treated in this manner. Aluminum isopropylate could also be introduced in pure form.

Potassium Hydroxide Saito et al. (1981b) grafted styrene-butadiene-styrene block copolymer (15% butadiene) with MA followed by crosslinking with metal hydroxide. For example, a mixture of 100 parts SBS copolymer, 0.6 MA, 0.5 butyl hydroxytoluene, and 0.1 phenothiazine was extruded on a 40 mm SSE (L/D = 24) at 195-205 °C. Copolymer with initial melt index 5.6 had MI = 3.3 g/10 min following grafting. The product was reacted with 0.69 parts KOH in a Brabender Plastograph for 5 min at 200 °C to give crosslinked material with MI = 0.02 g/10 min.

Ammonium Salts through Addition of Amines Polymers containing carboxylic acid groups have also been ionically crosslinked through ammonium salt formation by extrusion with polyamines (Rees, 1969). For example, ethylene-methacrylic acid copolymer (10 wt % acid) was fed to a 2 inch extruder at 30 rpm and 170-180 °C. Hexamethylene diamine (3-10 wt %) was injected into the mixing section of the extruder and the product was extruded and formed into sheets which had markedly higher tensile strength and stiffness than copolymer extruded without diamine. When the reaction temperature exceeded about 300 °C, an intractable, crosslinked product was obtained arising from conversion of ammonium salt linkages to amide linkages with loss of water. The crosslinking reaction could also be carried out in solution and using polyammonium salts instead of polyamines.

4.6 Controlled Degradation

Controlled degradation of polymers in extruder reactors generally involves lowering of molecular weight to meet some specific product performance criterion, or, in the case of biological polymers, degrading to release valuable low molecular weight species. Brief review articles which include some examples of chemical changes in polyolefins and other polymers during extrusion have been published (see for example Carlsson and Wiles, 1986).

4.6.1 Polypropylene and Other Polyolefins

Porter and Casale (1985) have summarized polymer reactions caused by mechanical stress including examples of polyolefin degradation and crosslinking during melt extrusion.

Dorn (1985) has briefly summarized PP degradation occurring during coextrusion with peroxides.

Henman (1979) has reviewed melt stabilization of PP during extrusion. The effects on PP molecular weight of extrusion with or without peroxides were described. The mechanisms of PP degradation were considered and the efficiencies of various stabilizers were determined. Butylated hydroxytoluene, quinone methide precursors, and benzoquinones were most effective, probably due to their ability to trap alkyl radicals.

In an early patent Greene and Pieski (1964) passed PP with a melt index of 0.5 through a 2 inch Egan extruder (L = 48 inches) where it was reacted with a 3.8% solution of t-butyl hydroperoxide in benzene (concentration 800 ppm relative to polymer) at 230 °C increasing to 266 °C at the die head. The product had a melt index of 10. Polybutene and polypentene were treated in a similar manner.

PP has also been degraded in the presence of peroxides in a 32 mm TSE operating at 10 rpm and 230 °C with residence time about 1 min (Anonymous, 1966). A radical forming compound such as di-t-butyl peroxide, lauryl peroxide, benzoyl peroxide, or potassium persulfate was added to the molten polymer as a hexane solution. Melt seals were formed in the extruder which prevented volatiles from entering the feed section and exit ports. The reaction mixture was devolatilized before exiting the die. PP with initial reduced viscosity 5.5 gave final reduced viscosity 4.9 in the absence of peroxide and 2.1-4.2 in the presence of 0.005-0.02% of the various peroxides listed above.

PP has also been premixed with radical initiator before extrusion for degradation (Anonymous, 1976). For example, PP with melt flow index 0.2 mixed with 0.1 wt % 3,3,6,6,9,9-hexamethyl-1,2,4,5-tetraoxanonane and stabilizers was extruded on a TSE at melt temperature 240 °C with 2 min residence time to give product with melt flow 16.0-20.7, depending upon the stabilizer package.

Aliphatic peroxides which release t-butyl alcohol as byproduct such as L101 have been employed for PP degradation (Baba et al., 1975). For example, PP was extruded at 230 °C with the above peroxide and 0.20 parts 2,6-di-t-butyl-4-methylphenol using an extruder with 65 mm screw diameter. Polymer with initial intrinsic viscosity 7.13 dl/g had final viscosity 4.66 and melt flow index 0.048 dg/min in the absence of peroxide and 2.57-0.93 dl/g and 0.70-94.8 dg/min after extrusion in the presence of 0.01-0.20 parts peroxide, respectively.

Castagna et al. (1976) have degraded PP in the presence of peroxide and air in an extruder. For example, PP with melt flow of 20 was extruded with 0.04 wt % t-butyl peroxy isopropyl carbonate at 260 °C on an 8 inch extruder with air charged to the feed hopper at 3 ft^3/min to give material with melt flow of 40.

Morman and Wisneski (1984) prepared polypropylene-peroxide concentrates by extrusion of PP at 190 °C and 2 min residence time in the presence of 0.275 wt % L130. Material with initial melt index less than 1 gave final product with 78% unreacted peroxide remaining and melt index 40-45 (corrected for degradation in the melt indexer). Reextrusion at 238 °C and 3 min residence time gave material with melt index about 550.

Ehrig and Weil (1987) employed peroxides which do not release t-butyl alcohol for PP degradation using either a 1 inch Killion or a 2.5 inch Davis Standard vented extruder. Examples of such peroxides include 2,2-di(t-amyl) peroxy propane and 3,6,6,9,9-pentamethyl-3-n-propyl-1,2,4,5-tetraoxacyclononane.

Fritz and Stöhrer (1986) studied peroxide-initiated PP degradation in a TSE. A 30 mm TSE (L/D = 20) with devolatilizing section was linked to an on-line rheometer and a computer system for product monitoring and rapid adjustment of processing conditions to produce constant quality of material.

A number of papers have appeared on mathematical modeling of peroxide-promoted PP degradation in an SSE. Model predictions have been compared with experimental data generated using a 38 mm SSE (L/D = 24).

The effects of DBPH peroxide concentration, temperature, and screw speed on molecular weight distribution, molecular weight averages, and MFI have been reported (Tzoganakis et al., 1988a,b,c; Suwanda et al., 1988a,b).

Earlier work of a similar nature was performed using a Brabender 19 mm SSE also with DBPH as peroxide (Hudec and Obdrzalek, 1980).

A recent study by Pabedinskas et al. (1989) described a process control scheme for PP degradation on an SSE using die pressure drop as measured variable and peroxide concentration as manipulated variable.

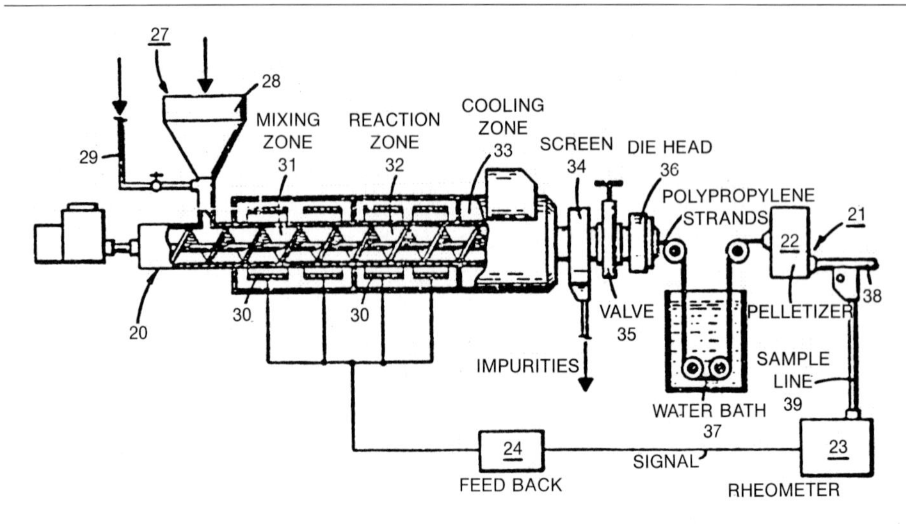

Figure 4.17 Extruder reactor for degradation of polypropylene (Kowalski et al., 1971)

Tzoganakis et al. (1989a) published a study on polymer residence time distribution (RTD) in a 38 mm SSE (L/D = 24) during peroxide initiated PP decomposition. A radioactive tracer method was used and the effect of screw speed, temperature, and peroxide concentration on RTD was examined. Screw speed had little effect on distribution while increased peroxide concentration gave a broader distribution and increased temperature gave a narrower distribution.

Trieschmann et al. (1966) showed that PP degradation could be achieved simply by shear heating in a 3.48 cm SSE under a nitrogen atmosphere. After a total residence time of only 3-4 sec at 256-263 °C, a PP with IV 1.7-1.8 dl/g and molecular weight distribution 2.5 was produced from material with initial IV 5.5 dl/g and MWD 7.3.

Work at Exxon resulted in controlled rheology PP products from extrusion of PP in the presence of air. Kowalski (1986) has summarized the history of this process. Fig. 17 shows a representative single screw extruder reactor used in this process (Kowalski et al., 1971; see also Staton et al., 1971; Kowalski, 1971). PP powder was thoroughly mixed in the feed hopper with air introduced into the feed hopper through pipe 29 and fed to a mixing zone 31. The material was melted, subjected to shear mixing at a melt temperature in excess of 290 °C, cooled, and passed through an adjustable back-pressure valve 35 and a die. The extrudate was quenched in a water bath, pelletized, and a portion passed through a melt rheometer 23 where the melt viscosity at a specified temperature was measured. The results from the melt rheometer were employed through a feed back loop to raise or lower the temperature in the extruder to produce either lower or higher viscosity material, respectively. Using the above reactor, the molecular weight distribution of a PP as represented by a die swell of 14.5 could be reduced to give a die swell of 4.8 at 290 °C.

An improved procedure was reported by Watson et al. (1975) which involved injection of air under a pressure of 1,800 psig into the reaction section of the extruder at 200-230 °C at a rate of 1,800 cc/lb of PP. This latter process allowed preparation of material with MFR 38 and die swell 2.0 from starting material with MFR 0.3 in a single extrusion pass, in contrast to 1-4 passes required for the earlier process where air was mixed with PP in the feed hopper.

Related patents by Keller et al. (1973) and Buntin et al. (1974, 1976) reported that PP which had been partially degraded in a prior extrusion step could be degraded further in a second extrusion to yield material in the form of filaments suitable for production of nonwoven mats. For example, partially degraded PP with melt flow rate 33.6 was fed to an extruder held at 383 °C where it was conveyed to a die held at 316 °C. Air was introduced into the die at 327 °C and 72.5 lbs/min/in^2 die slot where it was mixed with molten polymer to yield filaments with apparent viscosity 63.6 poise. When the extruder temperature was lowered to 372 °C, unacceptable material with grains of high viscosity PP and apparent viscosity 115 poise was produced.

Mueller-Tamm et al. (1981) have degraded polyisobutylene by extrusion with a specific antioxidant to prevent carbon black formation. For example, a polyisobutylene with initial MW 4,000,000 was mixed with 100 ppm alpha-tocopherol at 180 °C in the presence of air in a Brabender Plastograph at 50 rpm to give material with molecular weight 48,000-52,000 and no carbon black. A control blend containing 2,6-di-t-butyl-4-methylphenol as antioxidant failed to prevent carbon black formation.

Edwards and Padliya (1986) degraded butyl rubber containing 98.4 mol % isobutylene and 1.6 mol % isoprene in the presence of air. For example, a butyl rubber with initial MW 514,000 was extruded as a mixture with talc or calcium carbonate (30-40 wt %) on a 30 mm corotating, intermeshing TSE at 200 rpm. Dry air was injected at 1.4 kg/cm^2 downstream of the polymer feed and upstream of a vacuum vent maintained at 71 cm Hg. Dynamic melt seals prevented backmixing of air or volatiles with feed. Polymer residence time varied from 0.9 min at a throughput of 6.8 kg/hr (reaction zone temperature 220 °C) to 1.2 min at a throughput of 4.55 kg/hr (reaction zone temperature 160 °C). The extruded product had MW 34,000-104,000 depending on the reaction temperature and air injection rate. The molecular weight obtained by extrusion in the absence of air was 2-3 times higher.

Polyethylene with initial melt index 1.6 and density of 0.951 has been degraded by Smith et al. (1968) by melting in a 1.25 inch SSE at 225°-275 °C with single flight metering screw which fed directly into 1 inch metal tube, the center of which was fitted with a metal rod 15/16 inch in diameter. Material residence time in the metal tube assembly was about 28 sec at 368 °C. The final polyethylene had melt index 18 and density 0.953.

4.6.2 Polyesters

Reactive extrusion has been used to reduce the IV of PET used in fiber production.

$$\left(\overset{O}{\overset{\|}{C}}-C_6H_4-\overset{O}{\overset{\|}{C}}-OCH_2CH_2-O\right)_n + HO\text{-}CH_2CH_2\text{-}OH$$

$$\downarrow$$

$$\left(\overset{O}{\overset{\|}{C}}-C_6H_4-\overset{O}{\overset{\|}{C}}-OCH_2CH_2-OH\right. + \left.HO\text{-}CH_2CH_2\text{-}O\overset{O}{\overset{\|}{C}}\right)$$

For example, Watkins and Dean (1982) coextruded PET (initial IV .58 dl/g) with 0.191% ethylene glycol injected into the extruder throat at the feed inlet. The product showed a decrease in IV to .46 dl/g when the extruder was run at 265-273 °C. PET extruded under the same conditions without ethylene glycol showed a decrease in IV only to .56 dl/g.

Dyer and Meyer (1974) extruded PET in the presence of aluminum trihydrate to reduce IV. Water released from the trihydrate during extrusion at 300 °C degraded PET from initial IV 0.6 dl/g to final IV 0.38 dl/g.

4.6.3 Polyamides

Nylon 6/6 (500 kg) was treated by Follows et al. (1986) in a polymerization vessel at 50 °C with 0.375 kg water and discharged to give material with relative viscosity 88.5. Remelting the polyamide in a nonvented extruder with residence time 5 min gave material with relative viscosity 87.2. Repeating the process without water addition gave material with relative viscosity 97.2. Water could also be injected directly into the molten polyamide in the extruder in which case the product viscosity before exiting the extruder was measured by on-line rheometry connected to a feed-back loop to control the amount of water injected.

4.6.4 Biological Polymers

Biological polymers have been subjected to controlled degradation in extruder reactors. Fulger et al. (1985) treated coffee extraction residue consisting of 70% moisture with 0.5-2.0 wt % sulfuric acid in two single screw extruders in series at 107-200 °C to produce a solution of mannan oligomer (DP = 1-10). Phosphoric acid, acetic acid, or carbon dioxide could be used in place of sulfuric acid.

Waste sawdust or paper pulp has been partially degraded to glucose monomer in an extruder (Rugg and Brenner, 1982, 1983; Rugg and Stanton, 1982a,b, 1983, 1986; Huber et al., 1988). For example, sawdust (150 lbs/hr) and water (30 lbs/hr) were fed via hopper 10 of Fig. 18 to the 53 mm TSE operated at 300 rpm. Excess water could be taken off at drain 23 in the feed zone where the material was preheated to 93 °C. Following the feed zone, a dynamic seal at 24 consisting of reverse flighted screw elements surrounding an unthreaded, radially recessed screw element prevented escape of pressurized volatiles through the feed throat. The material was treated at 232 °C with 1 wt % sulfuric acid introduced at 34 at 120 lbs/hr, and steam introduced at 42 to produce about 50 lbs/hr glucose in a mixture of other materials.

Den Otter (1984) converted coal to a mixture of fluid hydrocarbons by reaction with hydrogen under pressure in a SSE. For example, coal with 7.5 wt % moisture and 2 wt % ash was treated with hydrogen at a maximum pressure of 230 bar at 450 °C to yield 25% gaseous product, 29% liquid product, and 46% solid residue containing 19% ash and 27% carbon. A portion of the fluid hydrocarbon product was recycled to one or more spots in the extruder to act as a lubricant for coal which had not yet become plastic.

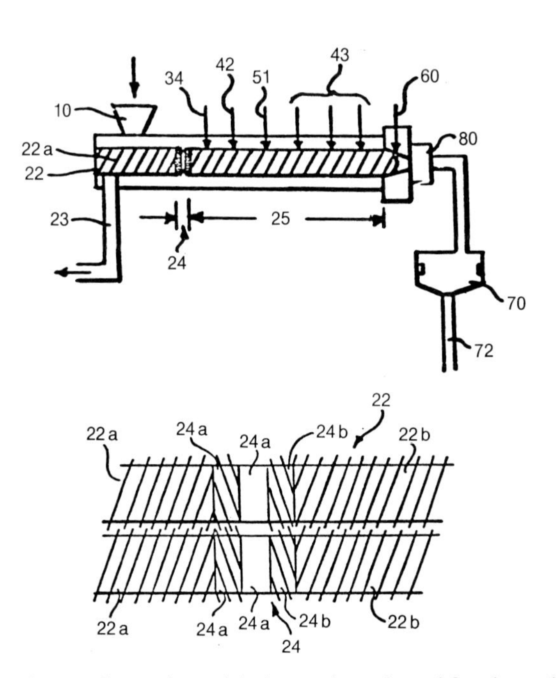

Figure 4.18 Extruder configuration with dynamic melt seal for degradation of biomass (Rugg and Stanton, 1982a)

Coal desulfurization has been claimed by Ryason (1980) in a process in which coal is extruded at 390-490 °C with injection of an aqueous solution of a sodium or potassium salt such as acetate, formate, carbonate, bicarbonate, or halides. Reaction times may vary from a few seconds to 15 min. No examples of desulfurization were given.

4.7 Polymer Functionalization and Functional Group Modification

Reactive extrusion has been used to introduce a variety of functional groups into polymers and to modify existing functional groups. Functionalization of polymers by grafting of *potentially polymerizable monomers* was covered primarily in the section on Graft Reactions although certain examples are included here.

4.7.1 Halogenation

The most sophisticated example of polymer functionalization performed in an extruder reactor is chlorination or bromination of polyolefins developed by workers at Exxon (Newman and Kowalski, 1983, 1984, 1985; Kowalski et al., 1985a,b,c, 1986). Conceptually, this process presents a number of difficult problems such as adequate mixing of molten polymer with chlorine gas, efficient removal of hydrogen chloride by-product, prevention of corrosion by reactant or by-product, and prevention of side reactions such as polyolefin degradation or dehydrohalogenation. Kowalski discusses aspects of this work in an earlier chapter of this book.

4.7.2 Sulfonation

Siadat et al. (1980) sulfonated oil-extended EPDM in a Brabender 1.91 cm SSE at residence times of 6.4-19.0 min through reaction with sulfuric acid and acetic anhydride fed as two separate streams. The final product was steam-stripped and neutralized to yield material with sulfur content 6-9 meq/100 g.

Boocock (1989) and White (1988) used sulfur trioxide/trimethylamine complex to sulfonate polyolefins. For example, ethylene/butene-1 copolymer has been reacted with the sulfur trioxide complex using a 1.9 cm Brabender at 273 °C with residence time 2.5-5 min and extruded into a film. The sulfonated product was dyeable in a dye bath. Polypropylene could also be sulfonated using this procedure. Sulfonation could also be carried out using a masterbatch prepared from 1,500 g ethylene/butene-1 copolymer and 45 g sulfur trioxide/trimethylamine complex.

Reactive Extrusion by Marino Xanthos

Errata

The following are corrections for errors on page 57, 206, 207, and 236.

p. 57 paragraph 3 line 1: "Currently" should read "Concurrently"

p. 206, caption for fig. 5.2: The caption should read "Enthalpy of common polymers"

p. 207, caption for fig. 5.3: The caption should read "Effect of screw speed on net energy removal, G = power input, q = heat transfer out, $\sum$ = net energy out"

p. 236: Table 6.1 is corrected as follows:

Table 6.1 Some Geometric Relationships

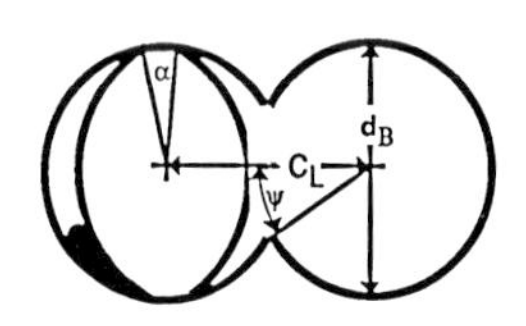

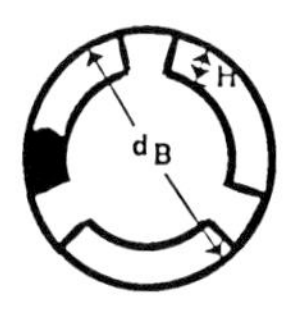

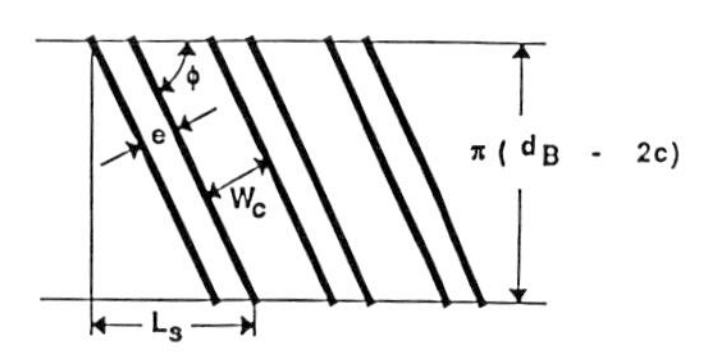

Twin Screw (Booy, 1978)		Single Screw
$S = (A_0/N_C)\sin\phi$	$A_0 = A_B - A_S$	$L_s = \pi(d_B - 2C)\tan\phi$
$A_B = 2\{R_B^2(\pi - \psi) + (R_B C_L/2)\sin\psi\}$		$N_C(W_C + e) = L_s\cos\phi$
$A_S = 2\{2[\psi C_L^2 - C_L R_B \sin\psi] + \alpha(R_B^2 + (C_L - R_B)^2]\}$		$L_B = L\sin\phi$
$\cos\psi = C_L/d_B$	$\psi = \pi/4 - \alpha/2$	

4.7.3 Introduction of Hydroperoxide Groups

Korber (1979) introduced hydroperoxide groups into polyethylene by coextruding the polymer with air at 190 °C on a 20 mm (L = 40 D) counterrotating TSE with 3 min residence time at 200 rpm. Air is introduced into an induction zone of the extruder at 25 l/hr at 75 bar and into an oxidation zone at 100 l/hr at 55 bar. A PE with 3,200 ppm hydroperoxide groups could be produced. The process has also been applied to EVA.

Graft reactions with a monomer such as butyl acrylate may be performed on the hydroperoxide-containing polymer in the same extruder by introducing the monomer after the oxidation zone and removing unreacted monomer by vacuum venting (Binsack et al., 1981).

4.7.4 Introduction of Carboxylic Acid or Trialkoxysilyl Groups through Grafting with Sulfonyl Azides

Udding (1988) introduced carboxylic acid groups into SEBS by melt reaction with 3-azidosulfonylbenzoic acid. For example, 50 g SEBS (Kraton G 1652) and 2 g azido compound were mixed in a Brabender mixer at 190°-210 °C, 30 rpm and 5 min residence time. Analysis by NMR spectroscopy showed 80% grafting of azido compound of which 80% was grafted to the aromatic groups. A molded test sheet showed increased tensile strength and elongation compared to unfunctionalized SEBS.

Trialkoxysilane arylsulfonyl azides have been used to introduce crosslinkable silane functionality into polymers (Barnabeo, 1985). For example, 75 g ethylene-butene-1 copolymer (8 wt % butene) was treated in a Brabender mixer at 60 rpm and 140 °C with 0.6 g antioxidant and 25 g of ethylene-butene-1 copolymer which had been premixed with 4 g of a 50 wt % solution of 2-(trimethoxysilyl)-ethylbenzenesulfonyl azide in methylene chloride and dried. The reaction temperature was raised to 180 °C for 20 min after which 0.03 g dibutyltin dilaurate was added. Molded test plaques after moisture curing showed a Monsanto rheometer value of 31 lb-in.

Trialkoxysilane alkylsulfonyl azides have also been used to introduce crosslinkable silane functionality into polymers (Gillette, 1989). For example, HDPE was treated with 1 wt % triethoxysilylhexane sulfonyl azide (TSHSA) and 1 wt % triethoxyvinylsilane in a Brabender mixer at 200 °C for 5 min and the product was molded into test plaques which were treated with steam. Extraction with decalin showed that the plaques contained 95% gel. A stated advantage of this procedure was that it could be used to functionalize polymers which degraded in the presence of the usual peroxide radical initiators which must be employed when functionalizing with vinyl silanes. Reaction of PP with 2 wt % TSHSA and 1 wt % triethoxyvinylsilane under similar conditions followed by molding and steam treatment gave plaques with 63% gel. Ultrahigh molecular weight PE, polysulfone, and polyoxymethylene were also functionalized using this procedure.

4.7.5 Capping of Carboxylic Acid Groups

Reactive extrusion with a number of different capping agents has been used for capping carboxylic acid end groups on polyesters to improve their thermal stability. Capping agents are usually monofunctional, and contain the same electrophilic groups which were used to couple or crosslink polyester homopolymers in Section 4.5, usually, epoxide, carbodiimide, isocyanate, and oxazoline.

Epoxide In one example, Schwarz (1971) capped PET by coextrusion in a TSE with epoxides. PET with relative viscosity 65 and 19.1 meq/kg residual carboxylic acid groups was fed to a TSE at 295 °C. Ethylene oxide was injected into the extruder at 150 psig at a rate to give 0.8% by wt in the polymer. The PET throughput rate was adjusted to give a residence time of 14 min between epoxide injection and spinneret exit. A melt filter before the spinneret showed a pressure of 1,500 psig. The final product was obtained as fiber free from voids with relative viscosity 37 and 1.0 meq/kg residual carboxylic acid content. In the absence of ethylene oxide a pressure of 2,000 psig was observed at the melt filter before the fiber spinneret, and the extruded PET fiber had relative viscosity 38 and 33.3 meq/kg residual carboxylic acid groups. Propylene oxide could also be used in this process.

Titzmann et al. (1972) capped PET by coextrusion on a TSE with 1 wt % phenyl glycidyl ether at 285 °C and 26 min residence time. The initial carboxylic acid end group concentration of 12.4 meq/kg was reduced to 3.0 meq/kg. It was stated that an extruder was less economical to use in this process than a tumbling drier.

Rothwell et al. (1983) capped PET by extrusion with N-2,3-epoxypropyl phthalimide using a 3 inch extruder (L/D = 24) between 200° and 300 °C. PET treated in this manner had 6 meq/kg remaining carboxylic acid end groups compared to material with 26 meq/kg obtained by extruding PET without epoxide. Other capping agents used were mixtures of either ethylene carbonate or dodecyl glycidyl ether with potassium iodide/triphenylphosphite mixture. With these latter capping agents the final product had to be devolatilized before final use.

Related patents disclose the use of N-2,3-epoxypropyl benzamide for end capping of PET during extrusion on a 1 inch extruder at 300 °C (Lazarus and Chakravarti, 1978), and the use of ethylene carbonate/alkali or alkaline earth salt mixtures for the same purpose using a Brabender extruder with 5 min residence time (Lazarus and Lofquist, 1982).

N-glycidyl isocyanuric acid derivatives have also been used to end cap PET during reactive extrusion (Arai et al., 1985).

Korver (1978) achieved capping of PET with phenyl glycidyl ether in the presence of lithium acetate using either a 1 or 2 inch extruder. PET was premixed with lithium salt and dried before extrusion into fiber. The initial carboxylic acid end group concentration of 10 meq/kg was reduced to 7 meq/kg after extrusion with the epoxide compared to 17 meq/kg obtained for extrusion without epoxide.

Inata et al. (1980, 1981) extruded PET with diallyl glycidyl isocyanurate or related olefin epoxides at 280 °C to give capping of carboxylic acid end groups with retention of olefinic functionality. Curing is effected by UV or electron beam to produce crosslinking through olefin groups. Filament prepared from this capped, functionalized PET showed an IV increase, improved elongation, and less shrinkage compared to untreated PET.

Carbodiimide Barnewall and Scheibelhoffer (1976) capped PET containing 11 meq/kg carboxylic acid end groups by extrusion on a 1 inch extruder with N,N′-di-o-tolylcarbodiimide to give material with 4 meq/kg acid end group. The PET was fed to the extruder at 8 lb/hr while the carbodiimide was injected into the feed throat at 18 cc/hr.

Tintel (1988) studied capping of PET with an unspecified polycarbodiimide. For example, PET with IV 1.01 dl/g and 18.5 mmol/kg carboxylic acid end group concentration was extruded with 1.25-2.50% polycarbodiimide on an SSE equipped with a Barr screw. With increasing level of polycarbodiimide molded test parts showed increasing improvement in retention of tensile strength upon exposure to saturated stream. PET was also capped by extrusion with 25 mmol/kg 1,4-butanediol diglycidyl ether in the presence of 3.2 mmol/kg of a catalyst such as triphenyl phosphine or sodium benzoate. Carboxylic acid level decreased to 10-12 mmol/kg using triphenyl phosphine and 7-8 mmol/kg using sodium benzoate although in the latter case a sharp increase in viscosity was also observed presumably through crosslinking reactions involving a sodium alkoxide intermediate from ring-opened epoxide. No capping was observed in the absence of catalyst.

Cordes and Sterzel (1977) capped carboxylic acid end groups on PBT using carbodiimides that are soluble in PBT. A 10 % concentrate of carbodiimide with PBT was made on a ZDSK 28 mm TSE at 240 °C before backblending with additional PBT using a static mixer. Suitable carbodiimides included 2,2′,6,6′-tetrachlorodiphenyl-, 2,2′-dimethyl-4,4′-dinitrodiphenyl-, and 2,2′-dinitro-4,4′-dichloro-diphenylcarbodiimide.

Isocyanate Gilliam and Paschke (1979) capped PET carboxylic end groups using MDI by extrusion on 1 inch SSE (L/D = 24) at 220-315 °C. The modified polyester had greatly improved peel adhesion compared to unmodified material.

Oxazoline and Related Compounds Matsumura et al. (1982) used oxazoline and oxazine capping agents (0.8-1.4 parts relative to polymer) to cap PET or PBT at 240 °C in an extruder with average residence time 3 min. Initial PBT meq/kg carboxylic acid of 43 was lowered to 3-7 by extrusion with 2-alkoxy-, 2-alkylamino-, or 2-alkylcarbamoyl-2-oxazoline or related compounds. For comparison, PBT extruded without capping agent showed carboxylic acid content of 48 meq/kg. Masterbatches of 2-phenylcarbamoyl-2-oxazoline at 10-15 part loadings in PET, PBT, or in BPA polycarbonate could be prepared by extrusion for backblending with PET.

Ortho Ester Hallden-Abberton et al. (1988) treated polyglutarimides with ortho esters in an extruder for removal of residual carboxylic acid and anhydride groups. For example, polyglutarimide was fed at 60 g/min to a 20 mm TSE with non-intermeshing screws consisting of a plastication zone six diameters long, a closed barrel section six diameters long for pumping and pressure generation, a reaction zone 31.5 diameters long held at 286-313 °C containing an injection port and a vent, and finally a low pressure devolatilization zone. Trimethylorthoformate was injected to the beginning of the reaction zone at 8.2 g/min. The final product had 0.033 meq/g acid and anhydride units compared to 0.585 meq/g in the starting material. Residual water in the polymer reacted with some of the orthoformate necessitating drying before extrusion or use of excess orthoformate. Triethylorthoformate, dimethyl carbonate (optionally with triethylamine catalysis), and 2,2-dimethoxypropane were also used to react with acid and anhydride groups. Polymethyl methacrylate could be converted to polyglutarimide by reactive extrusion with methylamine and treated with orthoformate in the same extrusion step (see below).

4.7.6 Cyclization of Pendant Carboxylic Acid or Ester Groups

Anhydride Formation Pendant carboxylic acid and/or ester groups on acrylic copolymers have been partially converted to cyclic anhydrides during extrusion simply by thermal dehydration (Hunt and Maslen, 1977). For example, extrusion of methyl methacrylate/methacrylic acid copolymer at 280 °C with devolatilization gave product with 10-25% anhydride content (as wt % of acid content before extrusion) as determined by IR.

Cyclization of pendant carboxylic acid and/or ester groups on acrylic copolymers to anhydride may be facilitated by adding a basic ring-closing promoter during extrusion (Kato et al., 1988). Examples of copolymers so treated were 90:10 methyl methacrylate-methacrylic acid, 90:10 styrene-methacrylic acid, and 87:10:3 methyl methacrylate-methacrylic acid-AA. Ring-closing promoters were solid sodium hydroxide, sodium methoxide, sodium stearate, sodium or potassium carbonate, potassium acetate, 2-phenyl-4-methylimidazole, triethylamine, and 1,8-diazabicyclo[5.4.0]undec-7-ene (DBU). Premixed blends of copolymer and promoter were extruded on a 40 mm devolatilizing extruder at 50 rpm and melt temperature 280 °C. The amount of anhydride formation was calculated by measuring IR absorbance at 1,805 cm^{-1}. The amount of ring-closure was critically dependent upon the amount of promoter present since too much promoter resulted in ionic crosslinking of the polymer when an ionic promoter was used.

Imide Formation Pendant carboxylic acid or ester groups on polymers have been cyclized to imides by reactive extrusion with amines (Kopchik, 1981). For example, polymethylmethacrylate (PMMA) was fed at 23 g/min to a counterrotating TSE consisting of five zones each about 84 cm in length with an average barrel temperature of 310 °C operated at 150 rpm. Ammonia gas was introduced at 8 g/min under 49 atm pressure to the beginning of a reaction section separated from the feed zone by a vapor seal consisting of nonflighted screw elements. The reaction product passes through at least one other melt seal into a devolatilization zone containing a vacuum vent for removal of unreacted ammonia and volatile byproducts. The final product was not crosslinked as shown by its solubility in DMF, and contained 3.26% nitrogen in the form of 2,6-piperidinedione (glutarimide) units formed by ring closure of adjacent carboxylic acid units in the chain. In other examples additional high pressure and low pressure vents were present in different sections of the extruder. For example, a 70/30 copolymer of MMA with EA was fed at 2,724 kg/hr to a 51.4 cm counterrotating TSE where it was melted and passed through a melt seal into a reaction zone. Ammonia was added to the latter section of the reaction zone at 272.4 kg/hr and 33.42 atm. Unreacted ammonia and volatile reaction products exited through a vent upstream of ammonia addition. The vapor stream was thus countercurrent to the polymer stream. The reaction product passed through a second melt seal to an extruder segment with an atmospheric vent and then to a vacuum vented devolatilization zone for removal of remaining volatiles. The final product was produced at 2,084 kg/hr, and had 9.0% nitrogen and a molecular weight within 10% of that of the starting polymer. Copolymers of methylmethacrylate with ethylene; ethyl methacrylate; methyl, ethyl, or butyl acrylate; AA; methacrylonitrile; or vinyl acetate have also been imidized using this process. Other amines used besides ammonia were methyl, ethyl, isopropyl, n-butyl, cyclohexyl, dodecyl, aniline, and trichloroaniline at extruder barrel temperatures of 260-310 °C depending upon the amine.

In a related patent (Hallden-Abberton et al., 1988) polymethyl methacrylate (45 g/min ; MW 119,000-138,000) was treated with methylamine (14.6-17.9 g/min) at 302 °C and 890-1,050 psig in the first 61 cm section of a 20 mm TSE with non-intermeshing screws. The product was treated in a subsequent zone with 3.10-8.73 g/min trimethyl orthoformate (see above under Capping of

Carboxylic Acid Groups) followed by devolatilization. The final product had 6.8-7.3% nitrogen and 0% residual carboxylic acid or anhydride groups. Product prepared without treatment with orthoformate had 0.575-0.678 meq/g residual carboxylic acid or anhydride groups, and had poorer chemical resistance and weatherability.

Lambla (1988) has described conversion of polymer-bound cyclic anhydride groups to imide groups by reaction at 180 °C in an SSE with ammonia which had been presaturated into the polymer at high pressure. The exact identity of the polymer backbone was not specified. A second process was also described in which the polymer was treated in the extruder with an unspecified compound which releases ammonia upon heating. Higher conversion of anhydride to imide was obtained using the first process (69-79%) instead of the second one (25-54%).

Proximate carboxylic acid groups on PP-g-AA have been reported by Flood and Plank (1976) to cyclize in an extruder with o-aminophenol to produce polypropylene with pendant benzoxazole groups, although no details of the process were given.

4.7.7 Carboxylic Acid Neutralization

Hydrophobic thermoplastics in the form of stabilized aqueous dispersions have been prepared by coextrusion of the thermoplastic with a surfactant, aqueous potassium hydroxide, and a second polymer bearing carboxylic acid or anhydride groups (Tanaka and Honma, 1987). For example, a mixture of 100 parts LDPE, 10 parts polyethylene-g-maleic anhydride, and 2 parts oleic acid was fed to a corotating TSE ($L/D = 20$) at 112 parts/hr. Aqueous potassium hydroxide (23%) was injected at 3.5 parts/hr at an intermediate point in the extruder, and the mixture was continuously extruded at 160 °C feeding into a water-jacketed static mixer at 90 °C before exiting. The final product could be dispersed in water to give spherical particles with average size 0.45 μm. EVA, EPDM, or SEBS copolymer could be used in place of LDPE.

In a related patent (Homma et al., 1988) a 10:1 mixture of ethylene-butene-1 copolymer was similarly extruded with EAA copolymer (15 wt % acid; 2.4 meq/g carboxylic acid) on a TSE at 110 parts/hr with injection of 6 parts/hr 20% aqueous potassium hydroxide at a barrel segment downstream of the feed throat. The product had 2.1 meq/g carboxylate salt and formed a good dispersion in water with average particle size 1.5 μm.

4.7.8 Ester Saponification or Transesterification

Pendant ester groups on ethylene-vinyl acetate copolymer have been hydrolyzed under controlled conditions in extruder reactors.

$$\left[\!-CH_2-CH(OCCH_3)-\right]_n \;(\text{C}=\text{O}) + NaOCH_3 \longrightarrow \left[\!-CH_2-CH(O^-Na^+)-\right]_n + CH_3\underset{\underset{O}{\|}}{C}OCH_3$$

Ratzsch et al. (1983) injected sodium methoxide in methanol into a molten stream of EVA in a TSE to achieve different degrees of saponification depending upon temperature, concentration of sodium methoxide, and whether methyl acetate byproduct was removed by vacuum venting. In one example, 2 kg/hr EVA (31.5% vinyl acetate) was reacted in a 28 mm TSE ($L = 929$ mm.) at 403-473°K with 7.5% sodium methoxide in methanol injected at 160 ml/hr into the polymer melt. The material was devolatilized at a zone 130 mm past the point of methanol injection to remove 60% free methanol after 20 sec reaction time. EVA with initial melt index 175 (at 463°K) had melt index 27-35 with 34% saponification following reaction.

Ester groups on EVA have been hydrolyzed by transesterification with an alcohol either in a discontinuous process in a Haake-Rheocord mixer, or in a continuous process in a 30 mm TSE with corotating, intermeshing screws (Lambla et al., 1987a,b; Bouilloux et al., 1986). In a representative example an EVA copolymer (5-28% VA) was fed to the throat of the TSE which was operated at 170 °C and 150 rpm. Sodium methoxide (0.27-1.16 phr) in 1-octanol was fed to the extruder at a section 1/3 of the distance down the barrel. The product was devolatilized before exiting the extruder. The extrudate was analyzed by IR and NMR spectroscopy and by DSC, and showed 14-62% conversion of acetate groups to alcohol. Among the other alcohols examined were methanol, butanol, dodecanol, 1,4-butanediol, 1,12-dodecanediol, and sorbitol, all of which gave 52-75% conversion at 3 phr sodium methoxide, except sorbitol which gave only 12% presumably because of its poor solubility in molten EVA.

Saxton (1982) achieved alcoholysis of polyvinyl acetate by premixing polymer with methanol/sodium methoxide in a static mixer followed by feeding to a ZSK 53 or 90 mm TSE operated at 130-150 rpm and 28-58 °C with 0.5-4.0 min residence time. The extrudate was fed directly to a neutralization tank for isolation. The final product had 82-97 mol % conversion to alcohol.

Akaboshi et al. (1963) mixed 260 parts per hour 22% polyvinyl acetate in methanol in < 10 sec in a mixing tower with additional 0.1 part methanol and 13.7 parts per hour 23% aqueous sodium hydroxide in the form of thin films. The mixture was introduced into a corotating, self-wiping TSE at 40 °C and, after a residence time of 3 min, material was obtained which was 99.5% saponified.

4.7.9 Destruction of Unstable End Groups

Polyacetal End Group Thermolysis Polyacetal-polyether copolymers can be stabilized against thermal degradation if the copolymer is first extruded to destroy unstable hydroxymethylene end groups which form formaldehyde on heating.

$$—(OCH_2)_x—(OCH_2CH_2)_y—OCH_2OH \quad \text{unstable end group}$$

$$\downarrow \text{base}$$

$$—(OCH_2)_x—OCH_2CH_2OH \; + \; CH_2O$$

stable end group

Extrusion is usually carried out in the presence of base to promote condensation of the released formaldehyde to water soluble paraformaldehyde either under homogeneous conditions in an SSE (Clarke, 1967; Orgen, 1968) or heterogeneous conditions in a TSE (Golder, 1974a,b). For example (Golder, 1974b), a copolymer of trioxane with ethylene oxide was fed to a self-wiping TSE (L/D 43) which consisted of a feeding and melting zone (L/D 12.5), a hydrolysis zone (L/D 20.5), and a devolatilization zone (L/D 10). An aqueous solution of < 1% triethylamine was injected into the hydrolysis zone at 100 psi where reaction takes place at 190-240 °C. Under these conditions the reactant vaporized and reacted with molten polymer under heterogeneous conditions. The hydrolysis zone was separated from the other two zones by melt seals formed by reverse thread screw elements. Remaining reactant and other volatile material were removed through a vent following the hydrolysis zone. Copolymer which had been subjected to hydrolysis in the extruder showed < .017% wt loss per min at 230 °C compared to .035-.070% wt loss per min for untreated copolymer.

Sextro et al. (1988) melted a copolymer of 96.6 : 3.4 1,3-dioxolane/s-trioxane in a TSE at 190 °C and treated with 2 wt % water solution containing 1,000 ppm triethylamine and 10 ppm sodium carbonate as deactivators. After devolatilization with removal of formaldehyde derivatives and isolation, the product showed improved thermal and hydrolytic stability compared to product mixed with deactivators before melting in the extruder.

Acyl Fluoride Hydrolysis Morgan and Sloan (1985) hydrolyzed acyl fluoride end groups on hexafluoropropylene-tetrafluoroethylene copolymer by coextrusion with 1 % water on a corrosion resistant 28 mm corotating, intermeshing TSE. The output was fed directly to a 38 mm SSE designed to generate pressure at low shear rate for filtration of the final product through a screen pack. Copolymer extruded without water had 37 meq/kg COF while material extruded with water had 0 meq/kg COF after extrusion.

4.7.10 Conversion of Anhydride to Alcohol or Amine

PE-g-MA has been reacted with alpha, omega-aminoalcohol in an extruder to yield imide alcohol, although no characterization of the materials was given (Gallucci, 1986, 1987). The net result represents conversion of an electrophilic anhydride group to a nucleophilic alcohol group. Either p-aminophenol or 2-aminoethanol at 1.5-1.7 % loading were extruded with PE-g-MA on a .75 inch SSE at 50 rpm and 180-260 °C.

4.7.11 Binding of Stabilizers to Polymers

Munteanu (1987) prepared a review on "Polyolefin Stabilization by Grafting" which contains many examples of melt functionalization of polyolefins. Among the methods covered were binding of stabilizer molecules to polyolefins either by direct grafting of olefin-containing stabilizers; reaction of epoxide-containing stabilizers with carboxylic acid functionalized polyolefins; and reaction of acid-containing stabilizers with epoxide functionalized polyolefins

Scott and coworkers have shown that certain thiol-containing stabilizers can become covalently bound to polymers during melt processing. A review on "Mechanochemical Modification of Polymers by Antioxidants and Stabilizers" has appeared (Scott, 1987). In most of the examples which follow, the most important parameter for determining extent of binding of stabilizer to polyolefin was the polymer viscosity which varies with temperature and which determines the extent of radical formation by shear scission of the polymer chain.

In one example (Ghaemy and Scott, 1981), a masterbatch of 5 parts stabilizer in 100 parts ABS was prepared by mixing in a variable torque rheometer at 170 °C and 72 rpm for 3 min. Stabilizers used included 4-ethoxymercaptoaceta-to-2-hydroxybenzophenone and 2,6-di-t-butyl-4-hydroxybenzyl mercaptan. Air was excluded from the mixing chamber during processing, and the polymer was discharged into cold water. The product was diluted to 1 part additive per 100 parts ABS and compression molded at 190 °C for 3 min into 0.2 mm thick film. Unbound additive was removed by continuous extraction of the film with hexane until no more additive was detected in the extract (3-5 days). The level of covalently bound additive was estimated by comparison to calibration curves by UV or IR spectroscopy as appropriate. At 5 parts stabilizer per 100 parts ABS, 42-47% of stabilizer added became covalently bound to the polymer. The effectiveness of stabilizers for preventing UV and thermal degradation was directly proportional to the amount of stabilizer bound to the polymer. A mechanism was proposed involving radical chain addition of thiol to ABS through addition at unsaturated sites in the butadiene portion of the copolymer. Thiol radicals may be generated through hydrogen abstraction by mechanochemically formed radicals on the polymer.

A similar study has been published on covalent binding of 3,5-di-t-butyl-4-hydroxybenzyl mercaptan and 4-mercapto-acetamido diphenylamine to nitrile-butadiene rubber (34.5% acrylonitrile) by mixing in a torque rheometer at 55 °C (Ajiboye and Scott, 1982).

In a related paper (Scott and Tavakoli, 1982), antioxidants containing a thiol group were shown to bind to natural rubber and to styrene-butadiene rubber during mechanical processing in a torque rheometer at 70 °C.

Al-Malaika et al. (1986) achieved binding of dibutyl dithiophosphoric acid and its ammonium salt, and dibutyl dithiophosphoryl disulfide to natural rubber in the melt at 70°-140 °C using a torque rheometer. As much as 22% of added stabilizer was bound and the addition of hydroperoxides had no beneficial effect.

Using similar procedures Scott and Setoudeh (1983) and Scott (1980) showed that 4-mercaptoacetamido diphenylamine or 3,5-di-t-butylbenzyl mercaptan can become bound to saturated polymers such as LDPE and PP by mixing in a torque rheometer at 150° and 180 °C, respectively. The primary mechanism for binding was proposed to involve formation of hydroperoxides from polyolefin radicals generated by mechanical shearing in the presence of air followed by decomposition of hydroperoxide and combination with thiol radicals.

A brief report has appeared describing binding to PP of hindered piperidine stabilizers functionalized with maleate or acryloyl groups for grafting (Al-Malaika, 1988).

4.7.12 Displacement Reactions on PVC

Mijangos et al. (1984, 1989) studied the reaction of PVC in the melt with either sodium benzenethiolate or sodium benzothiazole-2-thiolate. Reactions were performed in a Brabender mixer at 160 °C and 40 rpm. Aliquots were removed periodically and analyzed by GPC with simultaneous RI and UV detectors. Samples were further analyzed by ^{13}C NMR spectroscopy. In the presence of a plasticizer such as dioctyl phthalate conversion increased to a maximum of 50% with increasing amount of nucleophile. A maximum of 0.7% conversion was obtained without plasticizer.

A related work (Mijangos et al., 1986) described reaction of functionalized plasticizers with PVC in a Haake Rheomix mixer. For example, PVC was mixed with 0.018-0.154 mol fraction of sodium 2-ethyl hexyl thiosalicylate at 160 °C for 10 min to give material with 5.3-8.6% chlorine atoms substituted as characterized by NMR and IR spectroscopy, and DSC.

4.8 Summary of Principal Trends

This survey has attempted to collect representative examples of reactive extrusion processes in a logical organization based on the primary purpose of the extruder process. The emphasis has been on the chemical aspects rather than the engineering aspects of the processes. From a chemical viewpoint a principal consideration in any reactive extrusion process is kinetics of the desired reaction. This is the common but unspoken theme in all six general types of reactive extrusion processes surveyed. If a major advantage of reactive extrusion processes is rapid production of polymer from monomer, or production of chemically modified polymers, or compatibilized polymer blends, then a key to success is designing a chemical reaction that takes place efficiently within the residence time of available extrusion equipment. As was emphasized in the text, certain general types of chemical reaction have been traditionally utilized for performing reactive extrusion processes. Specifically, the largest number of reactive extrusion patents and published papers involve grafting reactions of saturated or unsaturated polyolefins or polyphenylene ethers with unsaturated monomers, particularly maleic anhydride or vinyl silanes. Also, particularly useful for polymer modification or copolymer formation in polymer blends have been reactions of amine or carboxylic acid moieties with electrophilic reactants such as anhydrides, epoxides, isocyanates, carbodiimides, oxazolines or their analogs, and orthoesters. Despite the vast number of examples from the patent literature, very little information has been published on the kinetics of these types of reactions under extrusion conditions. Consequently, one of the only guides for designing new chemistry for reactive extrusion processes is literature precedent, the presentation of which is the primary purpose of this survey.

In designing new chemistry for reactive extrusion processes or in applying known chemical reactions to new polymer systems or new processing conditions, there are some *general requirements for success.* First, the chemical reaction must have activation energy low enough to occur under desired extrusion conditions. This is usually not a problem at the melting point of most thermoplastics. For slow chemical reactions or for reactions on low-melting polymers, the extrusion temperature may only be raised as high as that point at which polymer or reactive additives start to degrade (except when polymer degradation is the desired reaction). For a low-melting polymer, the use of a catalyst is often successful in bringing about a desired slow chemical reaction. In the opposite case of polymers which melt at a higher temperature than that required to initiate the desired chemical reaction, plasticization or lubrication is often used to lower the processing temperature of the polymer to a range suitable for the reaction to occur at a controlled rate. *In every case,* mixing in the extruder must be sufficient for reactive groups to encounter one another.

Intensive mixing in twin screw extruders may be necessary for reactions across phase boundaries (i.e. with immiscible reactants/ heterogeneous conditions). For homogeneous reactions (i.e. miscible reactants), single screw extruders often give sufficient mixing. *In every case,* efficient reaction must take place within the time constraints of extruder. Furthermore, the chemical reaction must take place completely. It is extremely difficult to sell a commercial product based on a reactive extrusion process that continues to react when it is in the hands of a customer. *In every case,* any new covalent or ionic bond which is formed in the product of a reactive extrusion process must be stable to subsequent processing conditions (e.g. further extrusion, molding, paint oven, etc.) Finally, in any reactive extrusion process, there must be successful removal of byproducts and any noxious fumes by efficient devolatilization and ventilation.

List of Abbreviations

AA = Acrylic acid
ABS = Acrylonitrile-butadiene-styrene copolymer
AIBN = 2,2′-azobis(isobutyronitrile)
BBPD = Bis(t-butylperoxy-p-diisopropyl)benzene
BOP = Benzoyl peroxide
BPA = Bisphenol A
CHP = Cumene hydroperoxide
DBPH = 2,5-dimethyl-2,5-di(t-butylperoxy)hexane
DCP = Dicumyl peroxide
DTB = di-t-butylperoxide
EAA = Ethylene-acrylic acid copolymer
EBA = Ethylene-butyl acrylate copolymer
EEA = Ethylene-ethyl acrylate copolymer
EMMA = Ethylene-methyl methacrylate copolymer
EP = Ethylene-propylene copolymer
EPDM = Ethylene-propylene-diene modified copolymer
EVA = Ethylene-vinyl acetate copolymer
FA = Fumaric acid
GMA = Glycidyl methacrylate
HDPE = High density polyethylene
IA = Itaconic anhydride
IV = Intrinsic viscosity
L101 = Lupersol 101 (trade name for 2,5-dimethyl-2,5-di(t-butylperoxy)hexane)
L130 = Lupersol 130 (trade name for 2,5-dimethyl-2,5-di(t-butylperoxy)hexyne-3)
L/D = Extruder length / diameter ratio
LDPE = Low density polyethylene
LLDPE = Linear low density polyethylene
MA = Maleic anhydride
MDI = Methylenebis(p-phenyl isocyanate)
N = Nylon
PBT = Polybutylene terephthalate
PE = Polyethylene
PET = Polyethylene terephthalate
PIB = Polyisobutylene
PMMA = Polymethyl methacrylate
PP = Polypropylene
PPE = Polyphenylene ether
PS = Polystyrene
SAA = Styrene-acrylic acid copolymer
SAN = Styrene-acrylonitrile copolymer
SEBS = Styrene/ethylene-butylene/styrene copolymer
SIPO = Styrene-isopropenyl oxazoline copolymer
SSE = Single screw extruder
TBP = t-butylperoxy benzoate
TSE = Twin screw extruder
VTMOS = Vinyl trimethoxysilane

Part III

Engineering Fundamentals of Reactive Extrusion

Chapter 5

Features of Extruder Reactors

By David B. Todd, Polymer Processing Institute at Stevens Institute of Technology, Castle Point on the Hudson, Hoboken, NJ 07030

5.1 Introduction

Carrying out reactions with polymeric materials whose viscosities are typically in the 10 to 10,000 Pa s (100 to 100,000 poise) range is generally not possible in conventional chemical reactors. Extruders offer some attractive features which can be used to advantage in reactive extrusion. Many features are common to all types of extruders, but the variations which do exist may have specific advantages for a particular process. Extruders also have limitations which must be considered when selecting equipment.

5.2 Process Considerations

Extruders have been found suitable for reactive processing because of several inherent capabilities:
- Ease of melt feed preparation

- Excellent dispersive and distributive mixing
- Temperature control
- Control over residence time distribution
- Reaction under pressure
- Continuous processing
- Staging
- Unreacted monomer and by-product removal
- Post-reaction modification
- Viscous melt discharge

The limitations in broader application of extruders for reactive processing are attributed to:

- Difficulty in handling large heats of reaction
- High cost for long reaction times

Before selecting extrusion equipment for a projected reactive extrusion process, there must be a reasonable understanding of what the process needs are likely to be, and a realistic understanding of what the equipment capabilities are. The final selection of the most appropriate equipment may involve some compromises. Early in the process study, there should be an awareness of the ultimate needs of the projected commercial plant so that critical variables can be explored in the pilot stage. A knowledge of how single and twin screw extruders work will also be beneficial in estimating overall size and selecting special features which may be required to produce the desired product.

Polymer processes which may be candidates for continuous reaction in an extruder can be classified primarily by the:

- Residence time requirements
- Energy requirements
- Nature of the starting feeds and end product

5.2.1 Residence Time Requirements

Reasonable reaction times for extruder considerations are in the range of a few seconds to about 20 min. Processes may well be technically feasible beyond these ranges, but will probably not be economically justifiable. The estimate of reaction time will have to be verified in continuous equipment, and may differ considerably from batch data.

Thus, a polymerization process which exceeds an hour in a batch reactor may still be attractive in a continuous extruder because of the more intensive mixing obtainable in the latter. In one instance, a process previously requiring over 4 hrs in a batch kettle is successfully being conducted in a twin screw reactor with a residence time of about 10 min.

5.2.2 Energy Requirements

Addition polymerizations may involve such a large heat of reaction so as to be ruled out for reactive extrusions. Breaking a RHC = CH_2 double bond to form a -RCH-CH_2- chain involves a net heat release of about 16 to 20 kcal/mole. The smaller the molecular weight of the pendant R group, the larger the heat of polymerization on a weight basis. Thus, the heat of polymerization for polystyrene is about 160 cal/g, but 350 cal/g for acrylonitrile or 480 cal/g for propylene.

With an average specific heat of 0.6 cal/g K, an adiabatic polymerization could involve a temperature rise of several hundred degrees. The desired polymer product structure cannot be maintained and the desired molecular weight cannot be achieved unless temperature can be kept under control. Thus heat removal, always a difficult process for very viscous materials, will remain a very critical element in any polymerization process.

If the polymer product is formed from ring monomers, the heat of polymerization may be only about that of the heat of solidification as the polymer precipitates out of its monomer sources. Examples of the latter are the formation of polyamide-6 from caprolactam, and polyoxymethylene from trioxane (Todd, 1989).

Ring opening polymerizations involve forming a bond in the polymer similar to the one opened in the ring, thus providing offsetting heat effects.

Condensation polymerizations may be candidates for reactive extrusion if the heats of reaction are low enough. In some instances, vaporizing and remov-

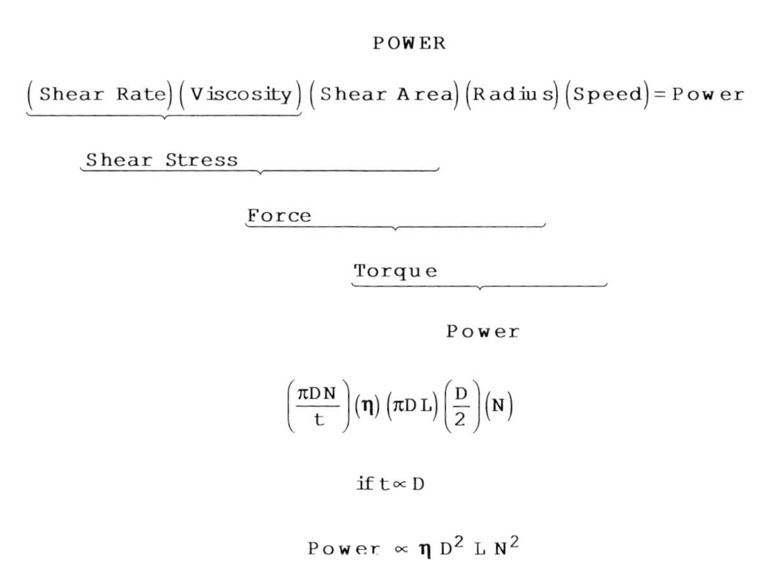

Figure 5.1 Elements of power consumption in extruders

time, length-to-diameter (L/D) ratio, and mixing characteristics, reaction volume will increase with the cube of extruder diameter, but surface increases only with the square. Thus, the surface-to-volume ratio decreases inversely as diameter is increased. Consequently, any process which, if feasible only with massive heat exchange in pilot equipment, will be very difficult to scale up.

Supplying heat is generally not too great a problem in reactive extrusion as the viscosity of the mixture is usually high enough that heat can be generated by conversion of work energy from the drive motor. Cooling may be a serious problem because most reactions are exothermic, and creating the desired intensity of mixing will add more heat.

Very little data have been published on heat transfer in extruders. In batch mixers with close fitting agitators (anchor, helical ribbons, or screw) the data have generally been correlated in terms of Nusselt number (Nu) as a function of Reynolds number (Re), Prandtl number (Pr) and viscosity ratio:

$$Nu = K_2 Re^a Pr^b (\eta_b/\eta_w)^c \tag{3}$$

The correlations have not extended into the low Reynolds numbers encountered in most applications where screw extruders are used ($Re < 10$).

Recent heat transfer data reported by Todd (1988b) for twin screw extruders indicate a decreasing dependency on rotor speed (Reynolds number) and little effect of screw clearance or degree of fill. The overall correlation:

$$HD/k = 0.94 (D^2 N \varrho/\eta)^{.28} (c_p \eta/k)^{.33} (\eta_b/\eta_w)^{.14} \tag{4}$$

represented the data over the Reynolds number range from 10^{-3} to 10, in close agreement with Chavan (1983) for tanks with screw type mixers.

Since power input may be increasing with the $(1+n)$ power of screw speed, but heat removal only with the 0.28 power, trying to lower temperature by increasing screw speed is often counter-productive, as illustrated in Fig. 3.

5.2.4 Form of Feed and Reactants

Starting a reactive extrusion process with crude monomer is rarely possible because of heat liberation problems, as noted above. However, modification of a basic polymer chain by subsequent reaction or grafting is an ideal candidate for reactive extrusion. If this reaction stage is occurring with already melted polymer, there is little energy consideration to worry about except possibly overheating arising from the mixing effort required to achieve the necessary intimacy of contact.

Basic polymer modification is more likely to start with the raw polymer in solid form as powder or pellets. Since the desired reaction will usually require that the polymer be in fluid form, the equipment selection will require an allowance for the energy and length requirements to get the polymer in the desired melt state. The overall enthalpy versus temperature required for various conventional polymers is shown in Fig. 2. Overall average heat capacities are primarily about 0.4 to 0.8 cal/g °C. Typical reaction temperatures may be in the 150 °C to 300 °C range, requiring specific energies of about 0.1 to 0.2 kWh/kg to bring the polymer to reaction temperature.

At these processing temperatures, there may be a substantial pressure requirement to introduce reactants at intermediate points down the barrel, particularly if the reactant must be maintained in liquid form. Even though the equilibrium partial pressure might be relatively low because of only a small amount of reactant to be added, the introduction of reactant will require that the pressure be in excess of the reactant's full vapor pressure. Thus, for an additive with a boiling point of 120 °C to be added to a polymer at 250 °C, the additive pump will have to develop over 2×10^6 Pa (300 psi) to be introduced into the extruder. An additional requirement is that the extruder be configured to develop a melt seal that will contain this pressure both upstream and downstream of the reactant injection point.

Solid additive reactants, if not preblended with the initial polymer, may be added through a top or side port into a starved section of the extruder barrel. Such downstream addition may have to be considered with low melting point reactants when introduction with the initial feed would so lubricate the feed mixture as to make initial incorporation very difficult.

After effecting the desired reaction, by-products or unreacted components may have to be removed in a devolatilization zone. More than one zone may be required if a very high vacuum is dictated by the process requirement (Biesenberger, 1983).

Additional non-reacting additives (stabilizers, colorants, etc.) may need to be incorporated prior to discharge of the final product. Upstream addition of these components may not be possible because of incompatibility with the reaction or devolatilization steps.

Final discharge will usually require pressure development in order to extrude the product through a strand die or other pelletizing system.

Thus, listing the sequence of process events to be accomplished will help identify the zones that the extruder must have in order to complete these sequential tasks. A knowledge of the reaction kinetics will be necessary to establish the length of the zone required for the key reaction step; the other steps can likely be predicted from conventional compounding experience.

5.3 Reactive Extrusion Equipment

The extruders which are available for continuous reactive extrusion processes differ in the features employed to bring about the desired process ends. Single screw extruders are the simplest (Fig. 4). Twin screw extruders are generally classified as being tangential or intermeshing, and further subdivided as being counter or co-rotating.

5.3.1 Single Screw Extruders

Basic performance of single screw extruders has been thoroughly covered in literature, such as the books by Bernhardt (1959), Tadmor and Gogos (1979), and, more recently, Rauwendaal (1986). Most of the theoretical treatment has been concerned with performance as a compounding extruder and for pressure development, rather than operation in a starved mode, as may frequently be required in reactive extrusion.

For reactive extrusion, the basic feed may well be in solid form. Reaction is unlikely to occur until the polymer has been melted. Thus, the initial portion of the extruder would be devoted to solids feed transport and melting. The requirements for this initial section are essentially the same as for conventional compounding, and the energy consequences similar to those depicted in Fig. 2.

After melting, contact of the polymer melt with reactants may take place in either completely filled channels or in partially starved ones. After reaction, and possibly devolatilization, the product will likely require a pressure generation step for extrusion.

In all extruders, the melt is dragged along the barrel by the rotation of the screw(s). The drag mechanism is usually best visualized by “unwrapping” the screw flight as a continuous straight stationary channel, and sliding the barrel in relative motion diagonally over the top of the channel. As noted, the extruder is also being used to generate pressure to force the viscous melt through some shaping element, such as a sheet or strand die. The pressure required for extrusion also causes a back flow (pressure flow) back down the channel.

The drag flow (Q_d) depends upon screw speed (N), diameter (D), flight width (w), flight depth (h), and helix angle (ϕ). For relatively narrow depth/diameter ratios (Bernhardt, 1959):

$$Q_d = F_d \pi D w h N \cos\phi / 2 \tag{5}$$

Another interpretation is that the drag flow is equivalent to the volumetric displacement of the melt as a consequence of the relative movement of the screw and barrel. This displacement per turn of the screw is equal to one-half

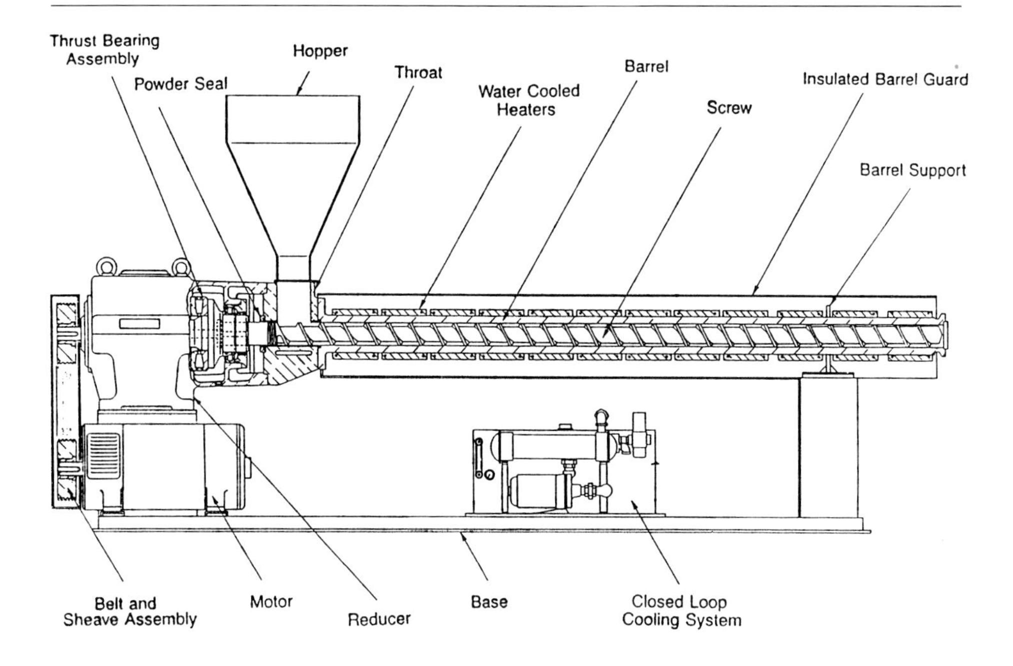

Figure 5.4 Single screw extruder

of the volume contained in one turn. Since the volume is the open cross-section (A) of the flight times the lead (Z):

$$Q_d = A\,Z\,N \cos\phi/2 \tag{6}$$

The pressure flow, (Q_p), can be ideally calculated on the basis of the driving force extrusion pressure (P), causing the flow and the resistance to flow as measured by the viscosity (η) of the melt and the geometry of the unwrapped flow channel.

$$Q_p = F_p W h^3 \Delta P \sin\phi/12\,L\eta \tag{7}$$

The net flow Q is merely the difference between drag and pressure flows:

$$Q = Q_d - Q_p \tag{8}$$

At constant screw speed and melt viscosity, the relationships between flow and pressure are linear, as indicated in Fig. 5.

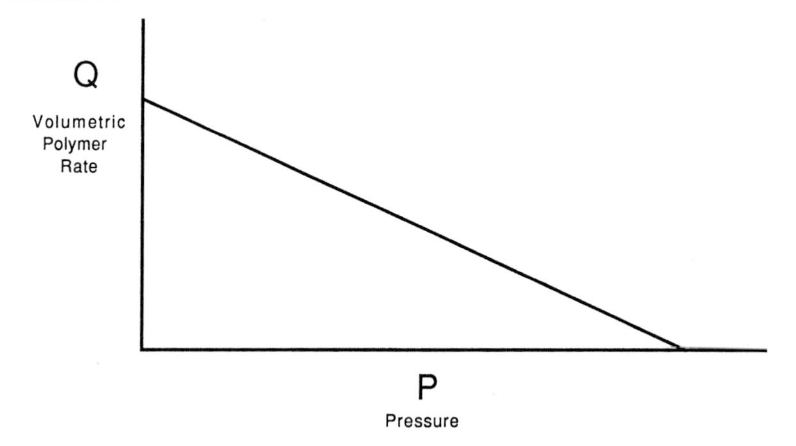

Figure 5.5 Flow vs. discharge pressure for an extruder at constant speed

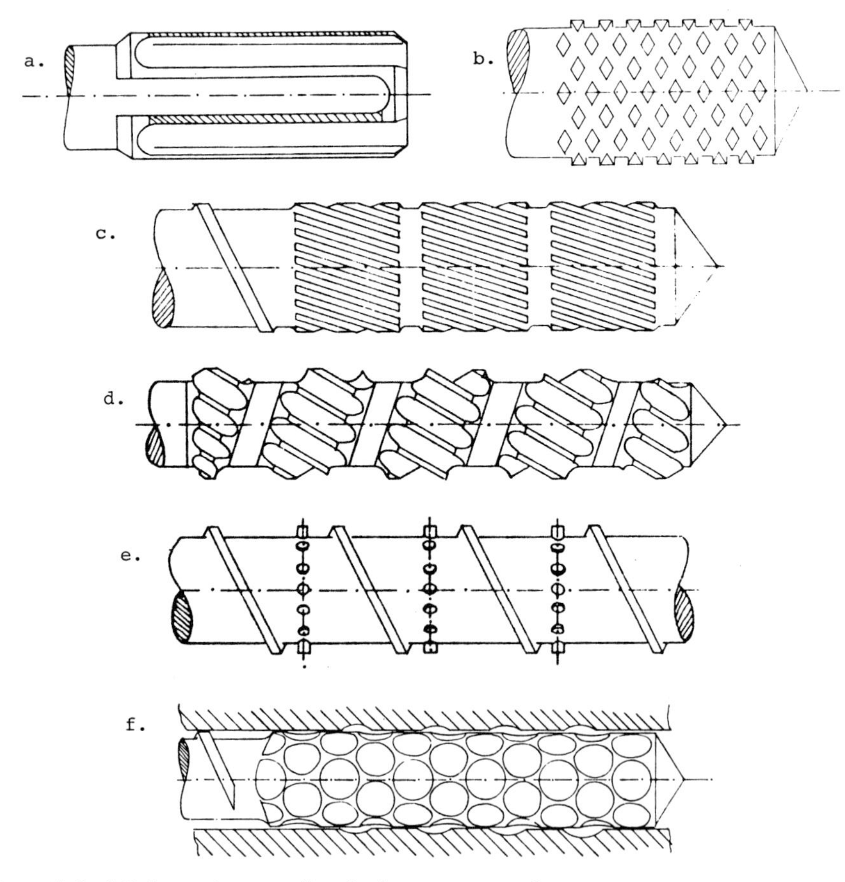

Figure 5.6 Mixing enhancers for single screw extruders:
a = Maddock (Union Carbide) b = Pineapple c = Dulmage d = Saxton e = Pin
f = Cavity transfer

The extent of mixing that occurs in a simple extruder is a consequence of the drag and pressure flows. The cross channel mixing generally results in an extended laminar layering (frequently still evident in the extrudate). Various devices have been used in single screw extruders to reorient the laminar layers for greater homogeneity (Fig. 6).

When operated starved, the flight is filled in proportion to the ratio of the actual Q to the potential Q_d. The melt is somewhat in the form of a rolling bead. Bubbles may be present even in the absence of devolatilization as the rolling bead induces both entrapment and rupture. Staging between starved zones can be affected by reducing the channel depth of the screw such that complete filling is assured.

5.3.2 Twin Screw Extruders

The development of technology for twin screw extruders has recently been reviewed by White et al. (1987). Twin screw extruders have been gradually increasing in importance in plastics compounding because of superior mixing capabilities and better control over residence time distribution. These same attributes are even more valuable in reactive extrusion.

Intermeshing twin screw extruders can vary in the degree of intermeshing to the extent of being fully self-wiping. The self-wiping feature is of particular value if a long residence time tail is deleterious to the reactive extrusion process, as material in the root of the screw may otherwise become stagnant.

Single screw extruders are generally flood fed; thus, feed rate is determined by screw speed. Twin screw extruders are generally starve-fed, thereby permitting an extra degree of control of rotor speed (and mixing) independent of feed rate. Individual feeders for each major constituent can often be used, thereby

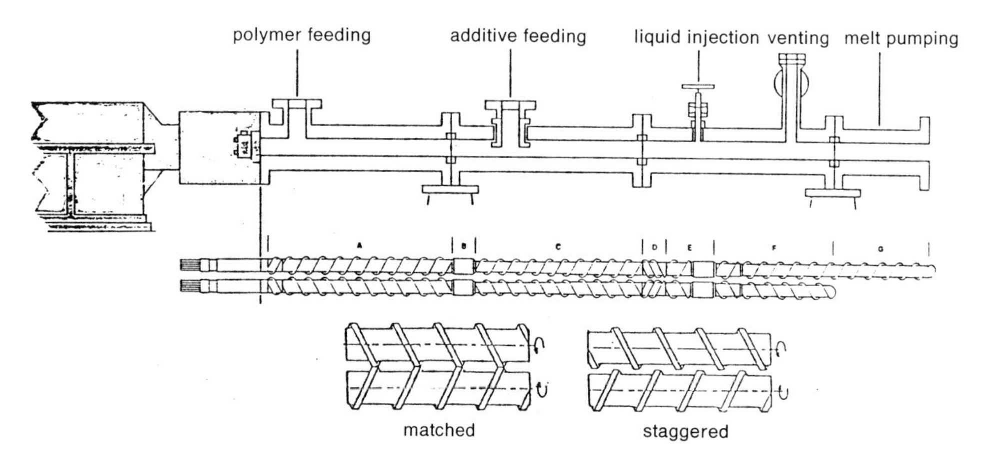

Figure 5.7 Tangential counter-rotating twin screw extruder

avoiding the necessity of preblending, but at the expense of providing more accurate multiple feeders.

Twin screw extruders also have the advantage of the additional mixing which accrues from the interaction of one screw with the other one. A great variety of interchangeable slip-on mixing elements are offered by the manufacturers to provide a great range of mixing effects.

Tangential counter-rotating twin screw extruders (Fig. 7) can be arranged with the flights either matched or staggered. The staggered array causes more interchange between the two screws, with beneficial reorientation of the flowing polymer.

The intermesh cusps of the barrel are truncated for structural reasons to avoid the weakness of feather edges. With the truncated sections, there is incomplete sealing between screws and barrel, as well as easier backmixing from one screw to the other. Thus, the flow versus pressure performance characteristics of the tangential twin screw extruders, as diagrammed in Fig. 8, reflect additional axial mixing. Two non-connected twin screws in parallel can develop twice the maximum flow at the same discharge pressure as can a single screw extruder. The greater sensitivity to pressure flow, however, allows the generally desirable increased axial flow for distributive mixing.

This type of twin screw extruder has the capability of large vent ports for degassing. There is practically no limit to the barrel length that can be provided with tangential twin screw extruders, and units with 100 L/D are in operation.

Pressure for discharge is developed in a single screw extension of one of the screws, which thus transmits the major thrust to only one of the two screws in the gear box.

Intermeshing counter-rotating twin screws (Fig. 9) offer some interesting performance characteristics. Each channel is essentially sealed from its neighbors, although there is a progressively larger leak path as helix angle is increased. The residence time distribution approaches plug flow, as each C-section is carried down the barrel with little interaction with adjacent or opposed C-sections. If additional axial mixing is desired, screw components can be selected which are less than fully intermeshing. Special gear mixing elements can also be used for multiple reorientation of the polymer flow.

With intermeshing screws, the channel depth is fixed by the amount of overlap of the screws. Degree of fill can be controlled by varying the helix angle of the screw segments, as shown in Fig. 10. High pressure can be generated at low screw speed with intermeshing counter-rotating twin screw extruders - a feature which is beneficial particularly for heat sensitive products, or where the product must be forced through a tight screen pack.

Intermeshing co-rotating twin screw extruders are available in single, double, and triple start configurations (Fig. 11). The double start variant is most commonly used in reactive extrusion, as it offers the largest reactive volume combined with the minimum shear work input.

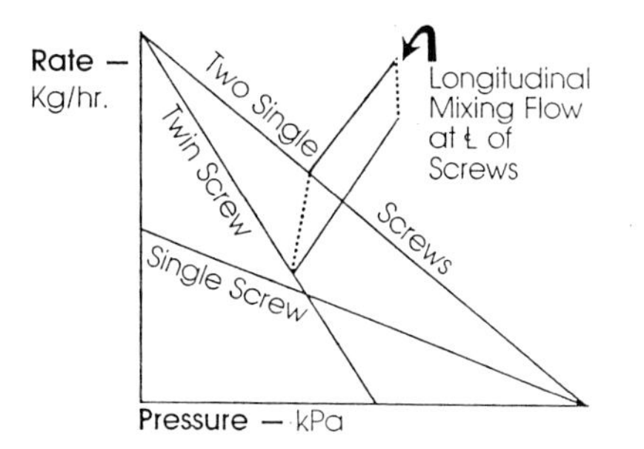

Figure 5.8 Flow characteristics for a tangential counter-rotating twin screw extruder

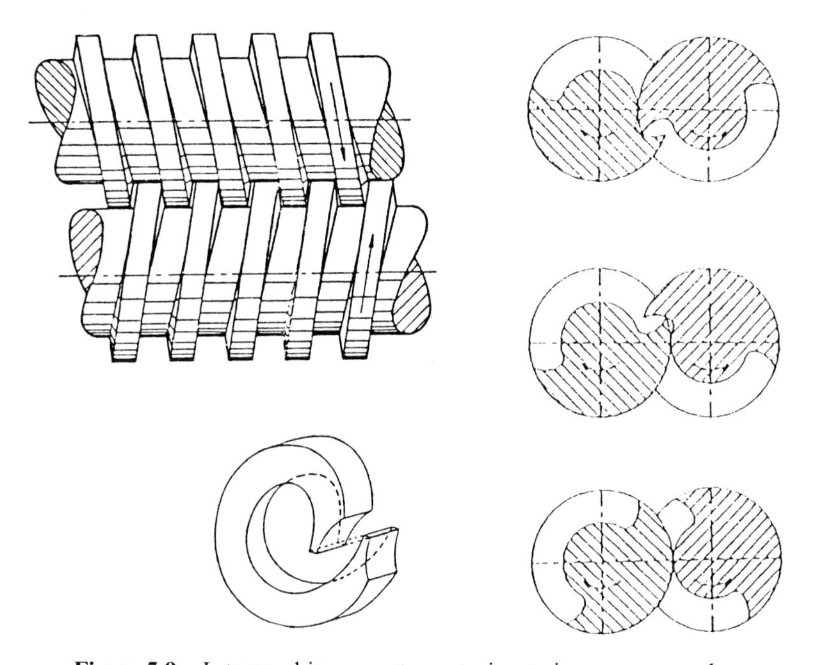

Figure 5.9 Intermeshing counter-rotating twin screw extruder

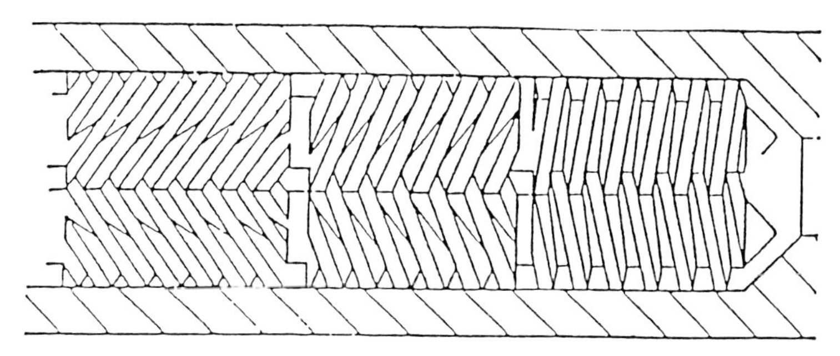

Figure 5.10 Intermeshing counter-rotating twin screws with decreasing helix angle

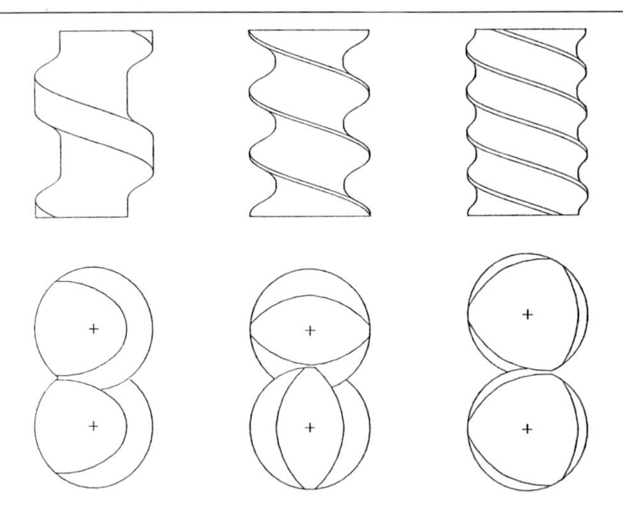

Figure 5.11 Single, double, and triple lobe intermeshing co-rotating twin screws

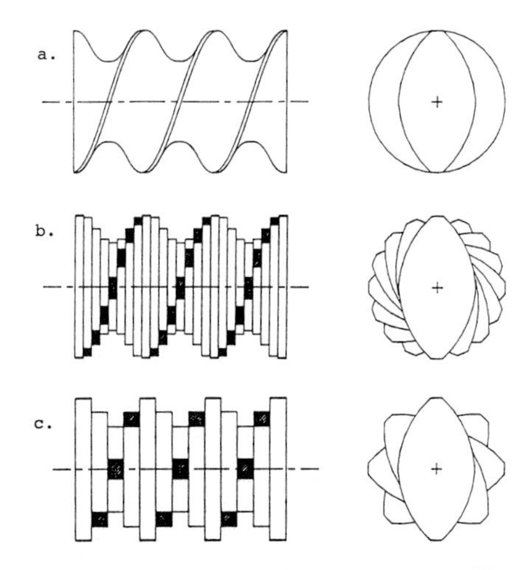

Figure 5.12 Arrangement of kneading paddles with screw-like characteristics:
a = square pitch bilobe screw b = same pitch with D/16 thick paddles at 22.5° offset
c = same pitch with D/8 thick paddles at 45° offset

Most of the mixing, both dispersive and distributive, is achieved with kneading paddles. The kneading paddles have the same meridional profile as the screws, but 90° helix angle. Depending upon axial thickness and angular offset, arrays of kneading paddles have some of the characteristics of screws, as shown in Fig. 12. The drag flow can be quite similar, but the pressure flow is much more sensitive because of the leakage paths in the gaps between offset paddles.

Dispersive mixing is effected by the smearing effect of the paddle tip against the barrel wall (Fig. 13) and also the opposite paddle. Co-rotating paddle

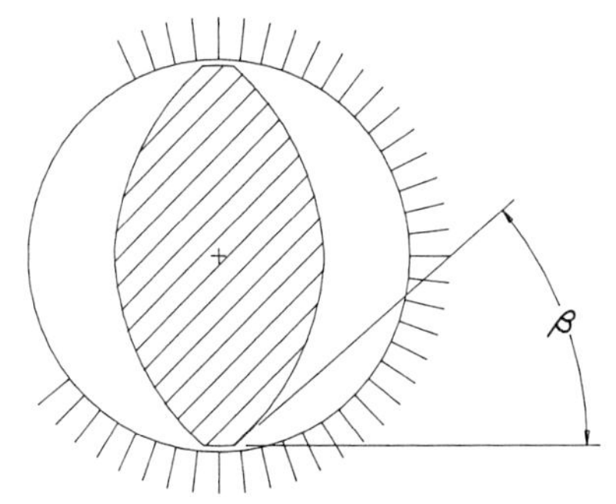

Figure 5.13 Smearing action of a kneading paddle; $\cos \beta = C/D$

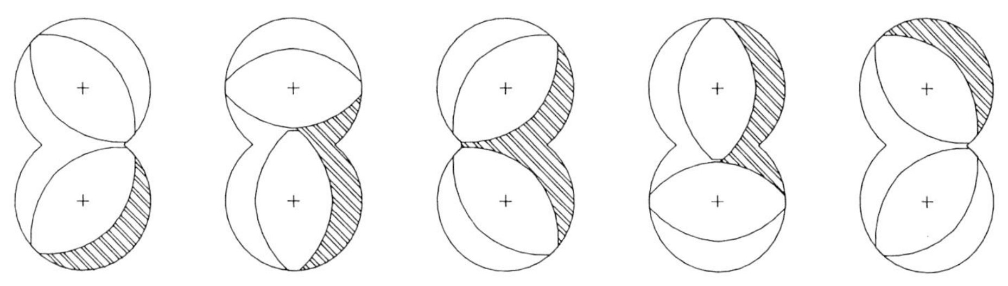

Figure 5.14 Compression/expansion mixing effect with intermeshing co-rotating twin screws

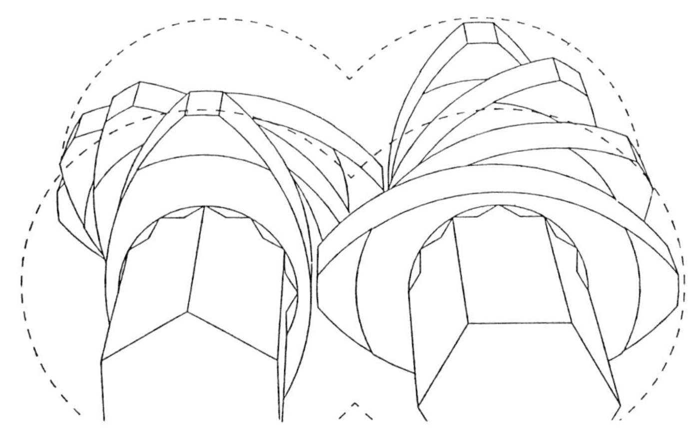

Figure 5.15 Cutting mixing action of kneading paddle arrays

elements provide a unique compression/expansion mixing effect, as shown in Fig. 14. As the shafts rotate, the crescent-shaped portions expand and then are recompressed. The contents are squeezed and generally forced axially, either upstream or downstream, depending upon the orientation of the next adjacent paddles.

There is an additional mixing effect as the paddle on one shaft slices between the next upstream and downstream paddles on the opposite shaft, as illustrated in Fig. 15.

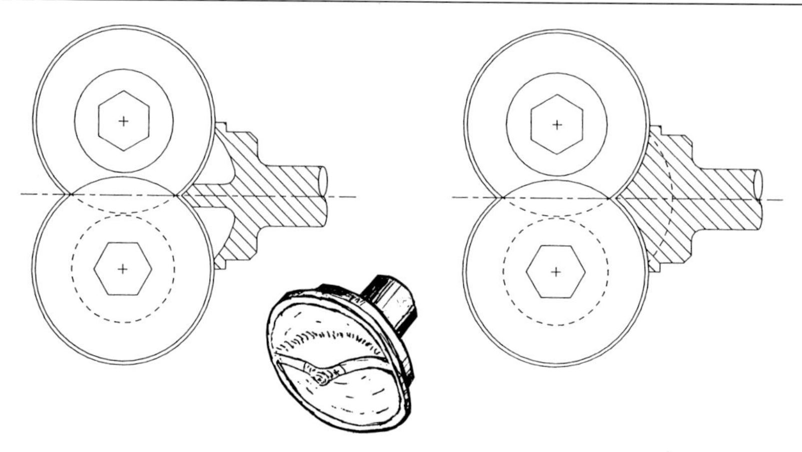

Figure 5.16 Barrel valve and blister rings combined to provide a variable flow resistance

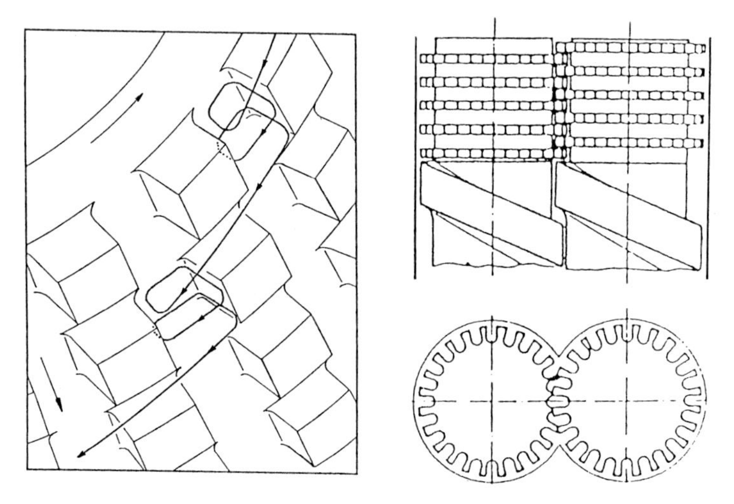

Figure 5.17 Gear type mixing elements

The mixing action described in Figs. 14 and 15 is most effective when that cross section of the barrel is completely full. Thus, fill control becomes another element of overall process control. Fill control can be achieved by opposite hand screw or paddle sections, or by use of blister rings. In conjunction with blister rings, variable restriction can be imposed by valve mechanisms in the barrel. A rotary type is shown in Fig. 16. Barrel valves, by varying the restriction, increase the pressure drop past the blister rings. Because of the pressure sensitivity of the paddle arrays, the degree of fill is readily controlled.

Gear type mixing elements (Fig. 17) can be used for multiple reorientation and two phase contacting at the expense of giving up some of the self-wiping features.

5.3.3 Equipment Response

Regardless of configuration, the flow in twin screw extruders still possess elements of drag and pressure flow. Comparison of the various designs available should be based on how the different geometric features affect the performance characteristics of importance in reactive extrusion.

5.3.4 Extruder Sizing

Overall, length will be determined by the number of tasks to be accomplished, namely:

- feed preparation (including melting if not melt fed)
- reactant incorporation (one or more stages may be required)
- reaction
- devolatilization
- pressure development

The required reaction zone length can be estimated from knowledge of the feed rate and probable reaction time. The available volume per unit length is a function of barrel diameter and flight depth, and is essentially independent of helix angle.

Single screw extruders would typically have a channel depth (h) of 0.07 to 0.23 screw diameter (D) in the reaction zone. The screw flight thickness would likely be about 0.1 Z (screw lead - the axial distance for one helical turn). Under these conditions, the total volume V, expressed as liters/meter would be shown in Fig. 18 as a function of extruder diameter and channel depth. With any extruder, it may be necessary to operate only partially full to allow room for offgassing of by-products or unreacted components. Thus, only part of the calculated extruder volume shown in Fig. 18 may be available.

Tangential counter-rotating twin screws (Fig. 7) will have twice the available volume that is shown in Fig. 18. Fully intermeshing counter-rotating twin screws (Fig. 9) have essentially the same available volume as a single screw of the same diameter. With full intermeshing, the channel depth is constant, whereas there is some flexibility in channel depth for both single screws and tangential twin screws.

Fully intermeshing co-rotating twin screw extruders (Fig. 11) can utilize deep flights without fear of root stagnation. As such, large reaction volumes are available, as shown in Fig. 19. The channel of one screw does communicate freely with that of the other screw, which enhances homogenization but at the expense of a somewhat broader residence time distribution and greater pressure sensitivity.

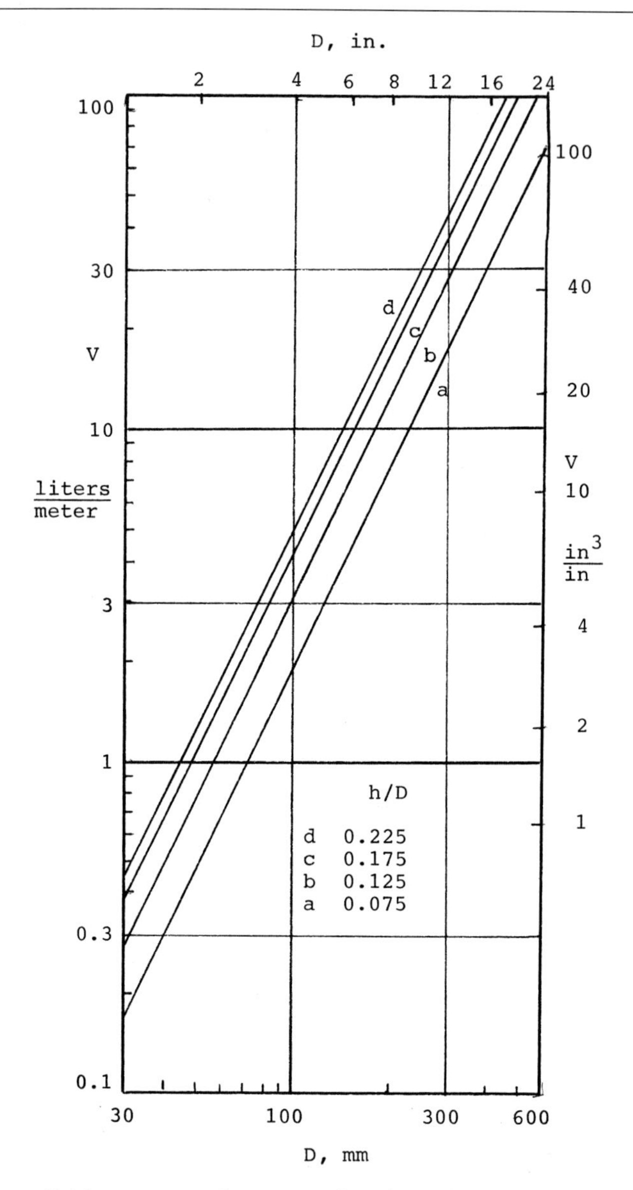

Figure 5.18 Available process volume as a function of (h) and (D) for single screws and for intermeshing counter-rotating twin screws

5.3.5 Required Volume

The process volume (V) required in an extruder to carry out a desired reactive extrusion process is simply:

$$V = W\theta/\varrho f \tag{9}$$

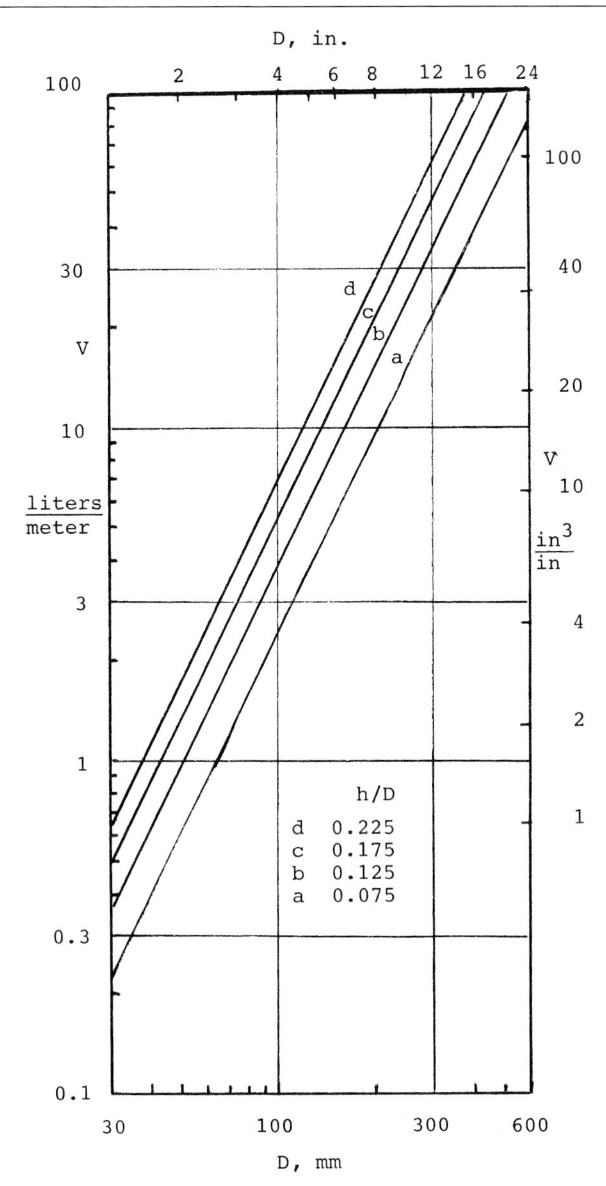

Figure 5.19 Available process volume as a function of (h) and (D) for intermeshing co-rotating twin screws

where (W) is the weight rate of flow, (θ) the required reaction time, (ϱ) the melt density, and (f) the volumetric fraction full. This volume (V) would be in addition to that required for the other necessary stages, such as feeding, melting, devolatilization, and exit pressure development.

In some instances, it may be known that a volatile by-product must be removed during reaction, so that (f) will have to be less than 1.0. It may also be desirable to keep (f) less than unity to limit the greater power input that accompanies full channels, and the consequent adiabatic product temperature rise.

When operating in the starved mode, the fraction full (f) is merely the ratio of net flow (Q) to drag flow (Q_d):

$$f = Q/Q_d \tag{10}$$

5.3.6 Residence Time

The maximum residence time is, of course, equal to the available holdup divided by the volumetric throughput rate. In practicality, however, the available cross-sections are only partially full in the solids feed zone, in the devolatilization zone(s), and perhaps in the reaction zone(s) and part of the discharge pressure zone.

If the extruder is running starved, the residence time θ is essentially independent of degree of fill, and for a barrel length L is approximately:

$$\theta = 2L/ZN \tag{11}$$

The lead length Z may range from 0.25 to 1.5 D. Too deep a channel depth compared to lead length (h/Z) is undesirable because of possibly stagnant flow in the root of the channel. With constraints on h and Z, a similar constraint is then imposed on screw speed N - will a slow enough N required for the desired reaction time θ produce enough intimate mixing for the desired reactant contact?

If the reaction must be conducted under pressure in a full zone, there will be increased backmixing and a possible consequence of a pressure flow term. The latter can be estimated from conventional extrusion equations if sufficient knowledge of intermediate stage viscosities exist.

Example 1. What size single screw extruder would be required for a reaction extrusion process for 250 kg/hr, 1.1 kg/liter melt density, gas by-product removal, and an estimated reaction time of 4 min? (Assume $f = 0.65$ for adequate venting, $h/D = 0.175$ channel depth, and maximum 25 L/D for reaction.)

$$V = W\theta/\varrho f \tag{9}$$

$$V = \frac{(250\ \text{kg/hr})(4\ \text{min})}{60(1.1\ \text{kg/liter})(0.65)} = 23.31\ \text{liters}$$

At h/D = 0.175, the process cross-section (A) is:

$$A = 0.9\pi(1 - .65^2)D^2/4 = 0.408\,D^2$$

At L/D = 25, the volume (V) is:

$$V = LA = (25\,D)(.408\,D^2)$$
$$V = 10.2\,D^3$$
$$D^3 = 23{,}310\ cm^3/10.2$$
$$D = 13.14\ cm\,(5.17\ in)$$
$$L = 3.3\ meters$$

With a screw lead (Z) of D/3, what would be the maximum screw speed permitted to retain the 4 min residence time?

$$N = 2\,L/qZ$$

$$N = \frac{2(25\,D)}{4(0.333\,D)} = 37.5\ rpm$$

Example 2. What would be the relative sizes of the reaction zones in different type extruders for a reactive extrusion rate of 1000 kg/hr requiring 3 min residence time with vapor contact? What rotor speed would be required? Assume Z equals 0.3D, h equals 0.175D, a melt density of 0.9 kg/liter, and f = 0.5.

From Equation (9):

$$V = \left(\frac{1000\ kg}{h}\right)\left(\frac{h}{60\ min}\right)\left(\frac{3\ min}{0.5}\right)\left(\frac{liter}{0.9\ kg}\right) = 111\ liters$$

This volume can be realized by a small diameter very long barrel at high screw speed, or a large diameter shorter barrel at lower speed. Using the volumes from Fig. 18, the required length and rotor speeds can be calculated for various assumed diameters for a single screw or a tangential counter-rotating twin screw extruder, as summarized in Table 1.

For a co-rotating intermeshing twin screw, a deeper flight (h = 0.225D) could be utilized because of the self-wiping feature. The required length and rotor speeds can be similarly calculated using Fig. 19, and the results are also summarized in Table 1.

With a reasonable L/D of 10 to 30 for the reaction zone length, the probable size range can be bracketed. Thus, a single screw extruder would need to be in the 200 to 300 mm size range. The twin screw extruders would be in the 170 to

250 mm diameter range. The implied maximum rotor speed would be 22 rpm for the larger diameter and 67 rpm for the smaller diameter sizes for both single and twin screw extruders if the 3 min reaction time criterion is to be held.

Exercises similar to the preceding can be used to help define the probable range of interest prior to contacting the extrusion equipment vendors, and then to determine what standard sizes the vendors have would meet the possible process needs. The final decision as to which type extruder may have to include other considerations as well, such as adequacy of mixing action, ease of venting, desirability of self-wiping, etc.

Table 5.1 Reactive Extrusion Length and Screw Speed for Example Problem

Diam., mm	Vol. (liters/meter)	Zone Length, meters	L/D	N, rpm
Single Screw at h = 0.175 D				
100	4.3	25.8	258	578
200	17	6.53	32.6	72
300	37.5	2.96	9.9	22
400	66	1.68	4.2	9
Counter-Rotating Tangential Twin Screw at h = 0.175 D				
100	8.6	12.9	129	284
200	34	3.27	16.3	36
300	75	1.48	4.5	10
400	132	.84	2.1	5
Co-Rotating Intermeshing Twin Screw at h = 0.225 D				
100	6.9	16.1	161	354
200	27	4.1	20.6	45
300	61	1.8	6.1	13
400	106	1.05	2.6	6

Nomenclature

A Open cross-section of barrel (less screw)
c_p Heat capacity
C Centerline distance of twin shafts
D Barrel, screw diameter
f Volume fraction full
F Correction factor for channel shape
G Power consumption
h Channel depth
H Heat transfer coefficient
k Thermal conductivity
K Proportionality constant
L Barrel length
n Slope of shear stress vs shear rate
N Screw speed
P Pressure
q Heat transfer rate
Q Volumetric flow rate
S Specific energy
t Flight clearance
V Volume
w Channel width
W Weight flow rate
Z Lead length of screw
β Dispersion face angle of kneading paddle
$\dot{\gamma}$ Shear rate
η Effective polymer viscosity
ϱ Melt density
ϕ Helix angle of screw
θ Residence time
$\sum$ Net energy removal

Subscripts
b Bulk
d Drag
p Pressure
w Wall

Chapter 6

Principles of Reaction Engineering

By Joseph A. Biesenberger, Polymer Processing Institute at Stevens Institute of Technology, Castle Point on the Hudson, Hoboken, NJ 07030

6.1 Reactor Types

6.1.1 A Review

It is instructive to view all continuous-flow chemical reactors as belonging to one of two extreme categories based upon the mixing histories they impart to a reacting fluid. We designate the first, for lack of a better term, a linear reactor; and the second, a backmixed reactor. Their respective characteristics are summarized in Fig. 1 and their flow dynamics are illustrated in Fig. 2.

The linear reactor is so named because of its large aspect ratio, L/H. Its large longitudinal dimension, L, and relatively small transverse dimension, H, give the reacting fluid a dominant flow direction. This feature makes it virtually impossible for "old" material, near the exit, to come into proximity with "young" material, near the entrance, thus severely limiting longitudinal mixing and thereby precluding backmixing. For the reader unfamiliar with the latter term, we define backmixing in a reactor as mixing of molecules in advanced stages of reaction (low reactant, high product content) with those in early stages (high reactant, low product content). It is obvious that a prerequisite for backmixing is the presence of extensive longitudinal, convective mixing of reacting regions, combined with intermolecular exchange among them by diffusion.

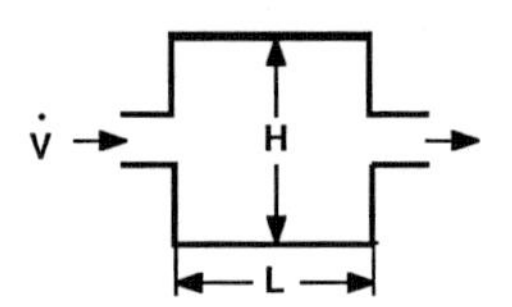

LINEAR : Emulate BR in time
Large aspect ratio ; L/H >> 1
Dominant flow direction
Limited longitudinal mixing
IDEAL : Plug Flow (PF)
No Longitudinal mixing (RTD)
Exit probability of 1 ; all identical ages (RT)

MIXED : Emulate BR in shape
Small aspect ratio ; L/H ≈1
No dominant flow direction
Extensive longitudinal mixing
IDEAL : Ideal Mixer (IM) or "CSTR"
Complete longitudinal mixing
Exit probability independent of age (α) ; random

SIZE : $\lambda_v/\lambda_r \equiv \mathrm{Da} > 1$

SHAPE : $\dot{v}\lambda_v = V \propto LH^2$

Figure 6.1 Schematic representation of general reactor types with distinguishing features

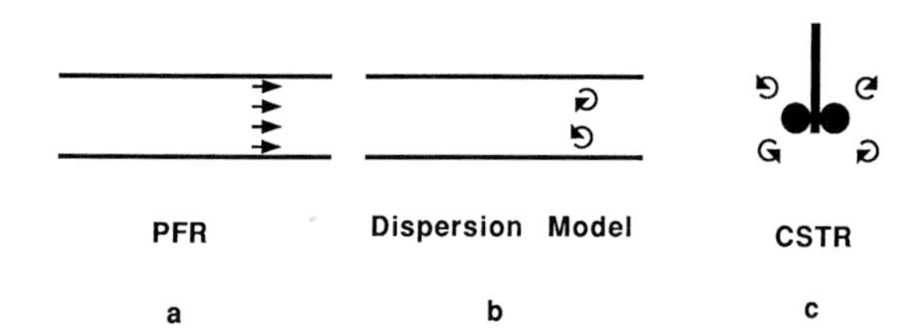

Figure 6.2 Sketch of three primary reactor types

Traditional examples of linear reactors are tubes or pipes, plain or packed with catalyst particles; a less well known one is the extruder. All impart velocity distributions to a reacting fluid, owing to the presence of stationary vessel walls, next to which the fluid is static. The large difference between the residence times of the slow fluid and that of the relatively rapid fluid far from the walls, gives rise to longitudinal mixing of the worst kind, characterized by stagnancy and its compliment, channeling, respectively. This is illustrated in Figs. 3-5, with the aid of tracers, for laminar flow in linear vessels with no moving walls, with one moving wall and with two. The reader will notice that diffusion of tracer has been excluded. This extreme case is sometimes referred to as

Flow

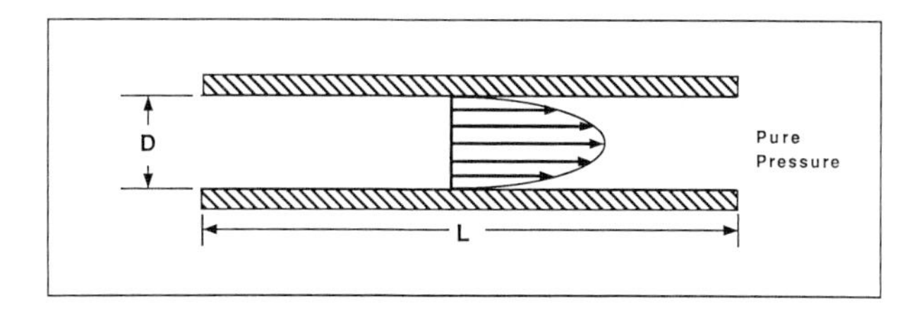

Convective Mixing

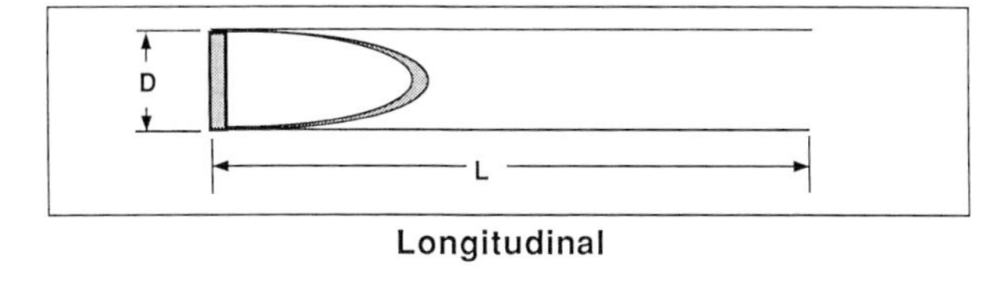

Longitudinal

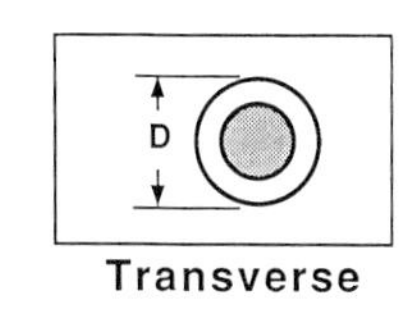

Transverse

Figure 6.3 Laminar flow and mixing patterns in a tube without diffusion

microscopically segregated flow, and is not an unreasonable representation of actual flow of polymer melts.

Biesenberger and Sebastian (1983) have made extensive use of characteristic times to develop dimensionless numbers and criteria for design and scale-up. In particular, they have shown that dimensionless numbers may be expressed as ratios of characteristic times associated with simultaneously occurring rate processes. For example, the Peclet number for mass transfer, whose value reflects the relative importance of diffusion versus convective transport, may be expressed as a ratio of corresponding characteristic times for diffusion and convection, λ_D/λ_C. In particular, for linear vessels in which no transverse convection occurs (Figs. 3-5), the characteristic time for transverse diffusion is

$$\lambda_D = l^2/D \tag{1}$$

where l is a characteristic vessel dimension (H/2 for rectangular slits or radius for cylindrical tubes) and $\lambda_C = \lambda_V$, which is the characteristic time for longitudinal convective transport through the vessel (mean residence time). The phenomenon of microsegregation may now be defined in terms of the inequality

$$\lambda_D \gg \lambda_V \tag{2}$$

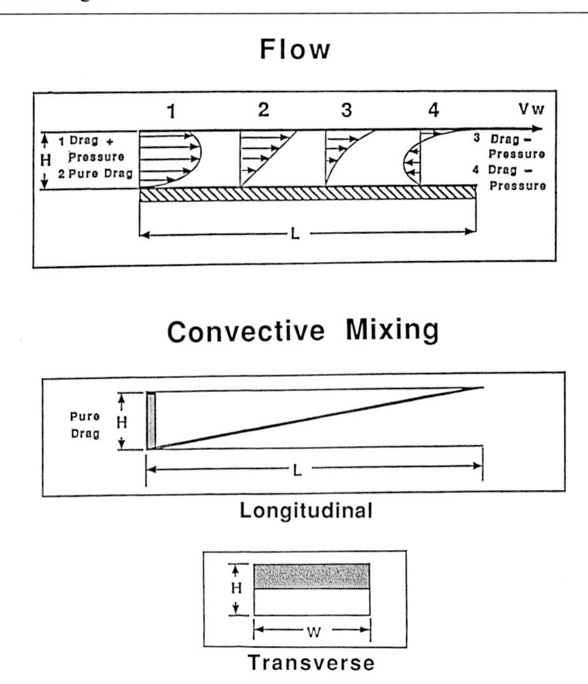

Figure 6.4 Laminar flow and mixing patterns between parallel plates with one plate moving and without diffusion

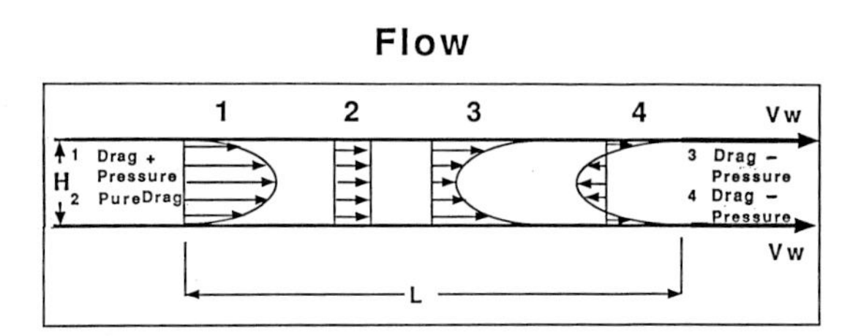

Figure 6.5 Laminar flow patterns between parallel plates with two plates moving and without diffusion

Flow in tubular reactors is generally driven by pressure applied at the inlet. Such vessels are not practical for high viscosity polymer reactions, owing to large pressure drops and plugging due to polymer build-up at the walls. In laminar flow, molecular diffusion is the only mechanism available for transverse mixing, which is therefore virtually absent, as pictured in Fig. 3.

Extruders operate on the principle of drag flow (one moving wall; the screw), and thereby take advantage of high viscosities to generate pressure longitudinally (pressure rise), as illustrated in Fig. 4. Thus they are more suitable for high-viscosity polymer reactions. Unlike the simple parallel plate configuration

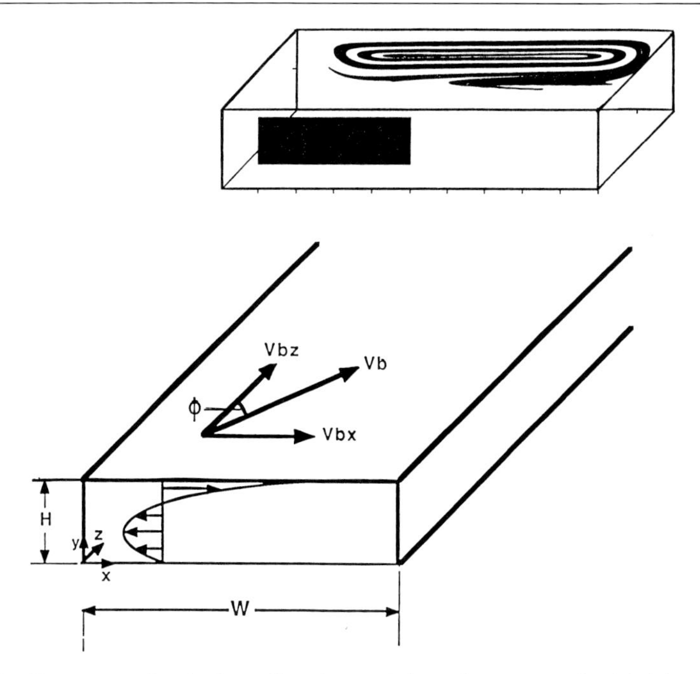

Figure 6.6 Computer simulation of laminar mixing of a rectangular blob in an extruder channel top wall moving - by Prof. F. Ling at Polymer Processing Institute/Stevens Institute of Technology

in Fig. 4, however, they also promote convective transverse mixing, owing to the helix angle of the screw. This is illustrated in Fig. 6.

The ideal linear reactor is the plug-flow reactor (PFR), from which all material exits with identical residence times, either because no transverse velocity distribution exists (Fig. 2a), or because transverse mixing is sufficiently complete to obliterate the residence time differences that such a distribution would produce. In terms of our characteristic times, the latter case may be defined by the inequality

$$\lambda_D \ll \lambda_V \tag{3}$$

as contrasted with Ineq. 2, and is more likely to occur with gases or with liquids in turbulent flow than with liquids in laminar flow, let alone with polymer melts.

This type of characteristic time analysis has been tested against Taylor's criterion for the existence of the so-called "Taylor diffusion" regime (essentially complete transverse mixing) versus the microsegregated regime (essentially no transverse mixing). Since transverse mixing in laminar tubes must occur entirely by diffusion, we set l = R in Eq. 1 and define a dimensionless number, the segregation number (Biesenberger and Sebastian, 1983), in terms of the appropriate characteristic times;

$$Sm \equiv \lambda_D/\lambda_V = R^2 v_{Zav}/DL \quad (4)$$

and require, for the existence of Taylor diffusion, that it be less than some critical value, Sm_{cr}, to be determined either by experimentation or by computation. In the absence of either, one can require, as a first approximation, than $Sm_{cr} \ll 1$. Taylor (1953) computed Sm_{cr} more precisely to be $2/(3.8)^2$. It happens that, to qualify for the Taylor regime, a reactor is, for practical purposes, a PFR to a sufficiently good approximation.

Under steady state conditions, the PFR emulates the perfectly mixed batch reactor (BR) with residence time replacing real time. If we neglect changes in density during reaction, a differential material balance for any reacting species k is:

$$\frac{dC_k}{d\alpha} = r_k \quad (5)$$

where r_k represents the rate function for the formation of k, $\alpha = t$ for the BR and $\alpha = z/v_z$ for the steady state PFR. The solution of Eq. 5 in dimensionless form gives conversion, Φ, as a function of Damkohler number of the first kind, Da,

$$\Phi = f(Da) \quad (6)$$

where

$$Da \equiv \lambda_v/\lambda_r \quad (7)$$

and λ_r is a characteristic time for the reaction, such as half-life, etc. (Biesenberger and Sebastian, 1983).

Actually, the extruder tends to be more plug-like than the tube in laminar flow, owing to the presence of transverse convection in the channels, as mentioned earlier. Self-wiping flights are believed to impart even more plug-like behavior to co-rotating twin screw extruders with tightly intermeshing screws.

In any case, a distinct feature of all linear reactors is a continuous concentration gradient for all reacting species in the longitudinal, or flow, direction, due to the presence of reaction, which renders the exit composition very much different from that at the entrance.

In sharp contrast, the backmix reactor is designed to approach and maintain spatial uniformity throughout its volume with respect to all concentrations as well as temperature. To accomplish this, the vessel must be equipped with a mechanism of some sort to generate thorough convective mixing (Fig. 2c); and, to ensure that such mixing is effective longitudinally, its geometry most likely will demand an aspect ratio L/H that is close to unity. The most well-known example, of course, is the continuous stirred tank reactor (CSTR) in which no dominant flow direction prevails.

Regions consisting of "old" fluid are continuously thrown into contact with regions consisting of "young" fluid: a necessary condition for spatial uniformity. Notwithstanding such intense macroscopic mixing, it may not be sufficient to ensure mixing at the molecular level, which requires exchange of contents among microscopic regions and which must ultimately rely upon molecular diffusion. In the absence of such exchange, we again have microsegregation.

Regardless of its specific geometry, we shall refer to the ideal backmix vessel, in which mixing is complete at all times down to the molecular level (micromixing), as an ideal mixer (IM). In such a vessel the exit probabilities of all molecules are identical and independent of their age α (time spent in the vessel). Thus the steady state material balance for species k, neglecting density changes during reaction, is simply:

$$\frac{\Delta C_k}{\lambda_v} = r_k \tag{8}$$

where ΔC_k represents the difference between the effluent concentration, which is identical to internal concentration, and the feed concentration. The solution of Eq. 8 can also be expressed as conversion as a function of Da, where the functional form is, of course, different from that of Eq. 6.

It should be pointed out that the IM, like the PFR, is an abstraction. Any vessel capable of achieving the requisite spatial uniformity would qualify. In other words, its geometric form need not be that of a CSTR, which is generally fed by pressure-driven flow, or gravity, and emptied via the same means. At the time of this writing, the author is unaware of the existence of a drag flow counterpart of the CSTR, i.e. a vessel whose flow is driven by moving walls and whose mixing dynamics have been proven to be IM-like. A likely candidate, however, in the author's opinion, might be Farrel's Diskpack processor.

The behavior of many actual industrial reactors is not adequately represented by either of the above categories, viz., the PFR with no backmixing and the IM with complete backmixing. That is, their geometry is such that $L \gg H$, yet they exhibit partial backmixing (Fig. 2b). In plain tubes, such mixing could be due to turbulence, or to the combined effects of a laminar velocity profile with rapid transverse diffusion (Taylor "axial" diffusion, as previously described). In packed bed reactors, it could be the result of interparticle eddies.

The steady state differential material balance commonly used by chemical engineers for such "in-between" vessels is the so-called dispersion model

$$0 = -v_{Zav}\frac{dC_k}{dz} + D_L\frac{d^2C_k}{dz^2} + r_k \tag{9}$$

where v_{Zav} is the transverse, area-averaged, longitudinal velocity and D_L represents an apparent or effective longitudinal dispersion coefficient (not diffusivity). The first and second terms on the RHS represent, respectively, longitudinal convection and dispersion. The coefficient must be determined experimentally and is generally correlated and graphed in terms of dimensionless numbers: Peclet number ($Pe_L = Lv_{Zav}/D_L$) vs. Reynolds number. Notice that Pe_L here characterizes longitudinal dispersion, and is in general consistent with our earlier definition, where now $l = L$ in Eq. 1 and $\lambda_c = L/v_{Zav} = \lambda_v$.

The solution of Eq. 9 may be expressed as conversion as a function of Da and Pe_L.

$$\Phi = f(Da, Pe_L) \tag{10}$$

Whereas it may be shown mathematically that the solutions of Eqs. 5 and 8 for the PFR and IM are limiting cases of Eq. 10 as $Pe_L \to \infty$, and $Pe_L \to 0$, respectively, it is impossible, physically, to produce sufficient backmixing in a linear reactor to cause it to approach that in a backmix reactor for reactions whose time frame (λ_r) is of the order of the residence time in the vessel λ_v.

6.1.2 Continuous Drag Flow Reactors

For illustrative purposes, we model simple flow of an incompressible, Newtonian fluid to obtain expressions for mass flow rate in terms of volumetric flow rate.

$$Q = \varrho_p \dot{V} \tag{11}$$

As noted above, flow in conventional reactors is frequently driven by pressure applied to the feed stream. By applying a momentum balance we obtain, for simple pressure flow at steady state,

$$\dot{V} = \dot{V}_p \cong -K \frac{dP}{dz} \tag{12}$$

where:

$$K = \frac{\pi R^4}{8\mu} \text{ for cylindrical tubes} \tag{13}$$

or

$$K = \frac{W_c H^3}{12\mu} \text{ for shallow rectangular channels } (H \ll W_c) \tag{14}$$

The counterpart of Eq. 12 for "pure" (Newtonian) drag flow in shallow rectangular channels driven by a single moving wall is

$$\dot{V}_d \cong A_c v_{Bz}/2 \tag{15}$$

where v_{Bz} is the component of wall velocity in the down-channel direction. Notice that $v_{Bz}/2$ is the average down-channel, Newtonian, fluid velocity. Following convention, wall (screw) velocity is expressed in relative terms as barrel velocity:

$$v_{Bz} = \pi d_B N_R \cos\phi \tag{16}$$

and cross sectional channel area is

$$A_c = \begin{cases} HW_c & \text{for single screw} \quad (17) \\ (A_o/N_c)\sin\phi & \text{for twin screw} \quad (18) \end{cases}$$

The presence of downstream resistance to flow, however, which exists in all practical cases, causes drag flow to be opposed by pressure flow. Thus for full channels, neglecting shape factors, the net flow rate is the algebraic sum of the two.

$$\dot{V} \cong N_c \Big[\underbrace{A_c v_{Bz}/2}_{\dot{V}_d} - \underbrace{K \frac{dP}{dz}}_{\dot{V}_p} \Big] \tag{19}$$

When the helix angle, ϕ, is zero, all wall motion is dedicated to down-channel flow, which is illustrated in Fig. 4; but when $\phi > 0$, as in extruders, a component of wall motion proportional to $\sin\varphi$ is directed toward cross-channel flow, which induces transverse convective mixing, as illustrated in Fig. 6.

Although Eq. 19 is not precise, quantitatively, it is widely used in engineering approximations and for design purposes. It will serve to illustrate the basic aspects of flow in drag flow reactors and, in particular, extruder reactors. For example, together with Eq. 14, it explains how rising viscosity can increase net throughput for a given pressure increase generated per unit distance, z, measured in the down-channel direction; or, alternatively, how it can generate greater pressure per unit distance for a fixed throughput. With the aid of Eq. 16 , the effects of geometry parameters ϕ, H, W_c and L, and RPM (N_R) are also evident.

Various useful geometric expressions are listed in Table 1, such as the relationship between channel length L and distance measured along the screw, or barrel, axis L_B, and the cross sectional area between screws and barrel, A_0, for self-

Table 6.1 Some Geometric Relationships

Twin Screw (Booy, 1978)		Single Screw
$S = (A_0/N_C)\sin\phi$	$A_0 = A_B - A_S$	$L_s = \pi(d_B - 2C)\tan\phi$
$A_B = 2\{R_B^2(\pi - \psi) + (R_B C_L/2)\sin\psi\}$		$N_C(W_C + e) = L_s\cos\phi$
$A_S = 2\{2[\psi C_L^2 - C_L R_B \sin\psi] + \alpha(R_B^2 + (C_L - R_B)^2]\}$		$L_B = L\sin\phi$
$\cos\psi = C_L/d_B$	$\psi = \pi/4 - \alpha/2$	

wiping co-rotating twin screw extruders. The approximate degree of channel fill, g, can be computed from the expression:

$$g \cong \dot{V}_{actual}/\dot{V}_{max} \tag{20}$$

where the denominator represents maximum flow at open discharge via Eq. 15, i.e. drag flow with the channel just full and without pressure generation.

Like most reaction vessels, extruder-reactors are especially suited to certain applications, in particular, high-viscosity reaction media. The ability to pump via internal pressurization, and thereby to overcome the pressure drops incurred in seals, dies, and stages without forwarding capability, drops off appreciably at low viscosities. This is evident from the last term in Eq. 19. Therefore,

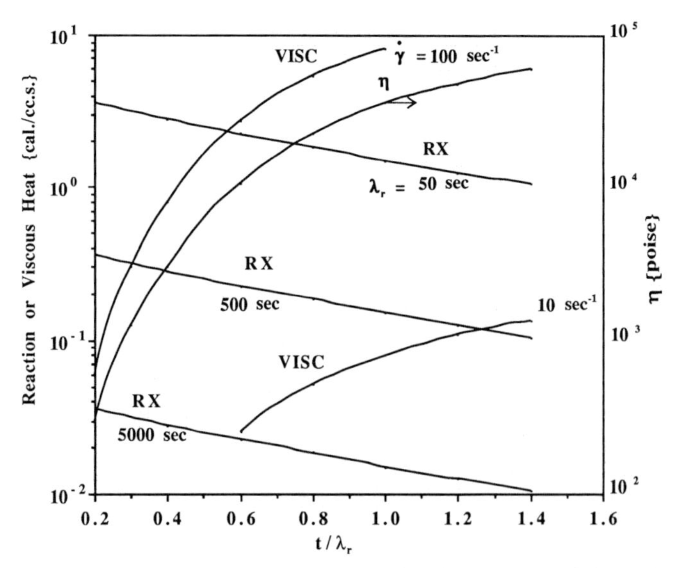

Figure 6.7 Comparison of time-responses to rising viscosity, during three typical free-radical polymerizations (λ_r), of reaction exotherm and viscous heating at three shear rates ($\dot{\gamma}$)

the extruder is not a prudent choice for low-viscosity reaction media. Other disadvantages of extruder-reactors are their relatively short residence times and poor heat transfer characteristics, especially as diameter increases and surface-to-volume ratio decreases concomitantly. The latter is discussed in Chapter 5 and in more detail in Chapter 7.

Typical residence times in extrusion equipment are of the order of minutes, which therefore require that characteristic times for reaction be no greater ($Da > 1$ in Eq. 7) if reasonable conversions are desired. Thus, λ_v is determined by reaction kinetics, and since throughput rate $\dot{V}$ is generally dictated by production requirements, extruder volume (V) may be estimated (Fig. 1).

Regarding shape, channel depth H must be small enough to facilitate the removal of reaction heat, but not too small, or heat generation by viscous dissipation will rise and exacerbate the problem. Thus a long vessel ($L \gg H$) may be needed to satisfy residence time requirements. Fig. 7 is an illustration of the relative importance of reaction exotherm and viscous heating for several shear rates and characteristic reaction times typical for chain reactions and flow in extruders. It is evident that viscous dissipation cannot be ignored out of hand as viscosity grows.

6.1.3 Mixing Mechanisms

The attainment of proper mixing is undoubtedly the single most important consideration when specifying or designing an extruder-reactor. Since chemical reaction is a molecular event, proper mixing in REX means mixing at the molecular level. Selecting the optimum mixing configuration depends upon the nature of the feed stream(s) (solid, liquid, melt or gas) and the specific mixing requirements (dispersive, distributive or both), which are generally not understood precisely and most often represent extrapolations from past experiences. Some fundamental aspects of mixing as they relate to extruder reactions are discussed here, and are amplified in Chapters 1 and 5; but much of the technology available to date, as noted, is based upon empiricism.

For purposes of analysis and comprehension, we focus our attention on a small element or domain within our reactor, illustrated in Fig. 8, which, for lack of a better term, we call a microscopic reaction environment (MRE), and we assign to it a characteristic dimension l. Our MRE could be a dispersed solid particle, a liquid droplet, a blob of melt or a gas bubble, soluble or insoluble in the surrounding continuous medium (e.g., melt); or it could be an imaginary region within the continuous phase itself.

In most situations, the process of mixing the MRE with its surroundings involves two elementary steps: reducing its size from some initial dimension ($l_0 \rightarrow l$) as much as required, and distributing the resulting fragments in space as uniformly as possible. For reactions, the ultimate elements are molecules.

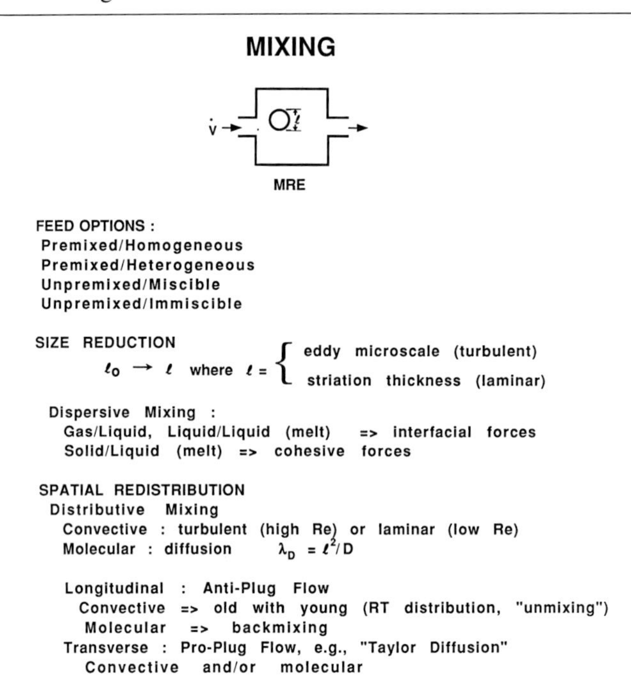

Figure 6.8 Schematic representation of a microscopic reaction environment of characteristic dimension l with various combinations of typical feed modes and mixing mechanisms

Thus we argue that, in general, the process of mixing includes both size reduction and spatial redistribution regardless of whether real interfaces are involved or not. In the language of polymer processing, the former is commonly referred to as dispersive mixing and the latter as distributive mixing.

More commonly, however, dispersive mixing is regarded as the process of breaking up a dispersed phase within a melt via fluid stresses at the interface, which are transmitted throughout the melt from a moving surface (e.g. a screw) and which act against cohesive forces when the dispersed phase is a solid (e.g. a pigment) or interfacial tension when it is a fluid (e.g., another melt). In the latter case, it is well known that the mixing process becomes more difficult as the difference between the viscosities of the two phases becomes greater. Melt elasticity differences also play a role, affecting both the texture of the mixture and the shape of the immiscible domains.

Fluid stresses are, of course, related to local deformation rates by an appropriate constitutive equation. High shear rate is commonly regarded as a requirement for successful dispersive mixing. However, high shear regions in dispersive mixers are invariably preceded by entrance regions in which extensional flow prevails. The relative importance of shear versus extensional stress in dispersive mixing, and the possible dominance of one over the other, is an unresolved issue at this time. In either case, as Chohan et al. (1987) have pointed out, good dispersive mixing requires, in addition to high stress, that the material be repeatedly exposed to such stress (mixing stages), and that the distribution among fluid elements within the fluid experiencing that stress (pass distribution) be as uniform as possible.

Distributive mixing, on the other hand, is actually defined as the spatial redistribution among contiguous regions with "passive" interfaces, i.e. regions within the same phase with imaginary boundaries, or domains consisting of different phases with identical properties. Analytical studies of distributive mixing involve monitoring and modeling the motion of tracers which do not alter the properties of the host fluid. Although real mixtures are generally less than ideal, the knowledge gained in such studies contributes nonetheless to a better understanding of the actual processes that are operative in laminar convective mixing.

As previously noted, to facilitate molecular mixing, we would like to enlist the aid of convective mixing and, in terms of our MRE, reduce l from some initial value, l_0. In order to make use of a convective mixing theory, we assume that our MRE has a passive interface. Thus for turbulent flow fields, l can be associated with the dimensions of a turbulent microscopic eddy (Biesenberger and Sebastian, 1983) and related to Reynolds number

$$l_0/l = \mathrm{f(Re)} \tag{21}$$

where l_0 is a macroscopic dimension associated with the origin of the turbulent field, such as tube radius or turbine blade width. In most REX processes, however, we must rely upon laminar flow to mix, owing to high viscosities and low flow rates, or both (low Re). In such cases, we can interpret l as a "striation thickness".

Laminar mixing theory has shown that reducing l from l_0 is equivalent to increasing "interfacial" (passive) area from an initial value S_0 to S, and that the ratio of these areas is a function of the total shear deformation imparted to the initial area by the prevailing shear field (Mohr et al., 1952):

$$l_0/l \leftarrow S/S_0 = \mathrm{f}(\gamma) \tag{22}$$

Striations in extruder channels are illustrated in Fig. 6.

Erwin (1978a) has pointed out that, in addition to total shear, laminar mixing depends strongly upon orientation of regions, such as the surface of our MRE, relative to the shear field; initial orientation as well as subsequent reorientation. More specifically, he has shown that the ideal laminar mixer exposes each element of surface to many small shear displacements, each of which is followed by optimum reorientation (45°) relative to the shear field. In the limit of an infinite number of infinitesimal shear displacements, this leads to the exponential mixer

$$f(\gamma) = \exp(\gamma/2) \tag{23}$$

Erwin and Mokhtaranian (1983) have also shown that plain extruder channels are merely linear mixers. Therefore, to achieve better than linear mixing, it is essential to reorient streamlines in extrusion equipment. Such reorientation is believed to be the root cause of the successes that some extruder geometries and special mixing devices have had with distributive laminar mixing. These concepts are related to practical situations in more detail in Chapters 1 and 5.

Recently, the mathematical theory of chaos has begun to contribute to our understanding of distributive mixing by formalizing, via Poincaré diagrams, the beneficial effects of subjecting a region to sequential stretching and folding motions (Ottino, 1990). These benefits are well known from experience, in particular from such common operations as kneading dough and mixing rubber in roll mils. In order to make use of these concepts, it is necessary to link criteria for "good" chaotic mixing with actual flow fields prevailing in mixing equipment, as we are attempting here with simple models, such as the MRE.

In pursuit of that goal, we shall apply the concepts of laminar mixing theory to our MRE via characteristic times. To simplify our analysis we begin without reaction. Clearly, our choice of initial dimension l_0 is important, and that choice depends upon the nature of the feed and the particular mixing requirement. In general, however, to achieve "good mixing" in our MRE, it seems reasonable to require, as in Ineq. 3, that

$$\lambda_D \ll \lambda_V \tag{24}$$

with the presumption that the diffusing substance is miscible (soluble).

However, the characteristic length l, which was a fixed vessel dimension in Eq. 1, now refers to our MRE and is therefore a variable function of time if convective mixing occurs, having been reduced from l_0 within the time available in a reactor (λ_V). Thus, to obtain a dimensionless number corresponding to Sm in terms of a fixed characteristic length l_0, we make use of an appropriate relationship between the two lengths, such as Eq. 21 or 22. Obviously, good mixing requires that Sm be "sufficiently" less than some critical value $Sm < Sm_{cr}$ after time λ_V; a large value of Sm would signify microscopic (molecular) segregation. The critical value can be estimated more precisely either by experimentation or via an accurate computer model. In either case its selection will be open to subjective interpretation, since the transition from "poor" to "good" mixing is most likely not a sharp one.

An obvious choice for l_0 would be the characteristic dimension of an inlet stream, in the case of umpremixed feeds, or the mean dimension of a dispersed droplet, in the case of predispersed feed.

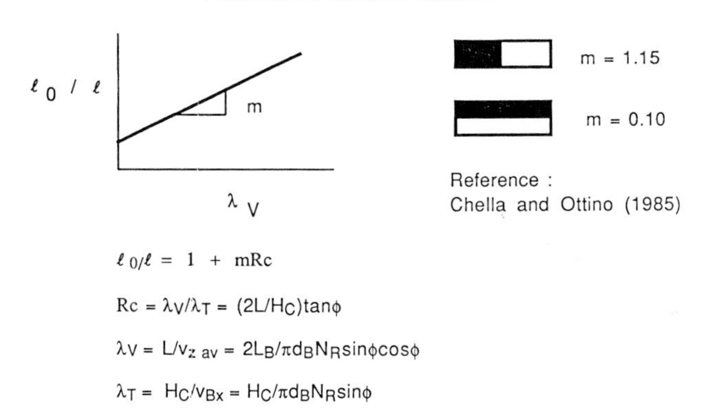

Figure 6.9 Striation thinning in a linear mixer (extruder channel) for two feed configurations of dye blobs

Specifically, for an extruder, l can be equated to the striation thickness after time λ_V resulting from cross-channel convection, as shown in Fig. 6. Expressions for estimating the decline of l as a function of residence time are available for certain flow fields; in fact, the function is linear for plain screw channels as previously noted. The linear functions shown in Fig. 9 (Chella and Ottino, 1985) were rewritten in terms of a characteristic time for transverse flow, l_T, and a dimensionless recirculation number, Rc. Notice that Rc is a function of helix angle, but is independent of screw RPM, since λ_V and λ_T are equally affected by a change in RPM. Notice, also, the effect of initial orientation of the "interface", vertical versus horizontal, shown in the figure in terms of a tracer, and the initial length l_0 which is defined here as one-half the corresponding channel dimension. Thus

$$l_0/l = f(Rc) \cong mRc \tag{25}$$

and the segregation number for a plain single screw channel becomes

$$Sm = l^2/D\lambda_V = l_0^2 v_{Zav}/(mRc)^2 DL \tag{26}$$

which, compared to that for the tube (Eq. 4) with equivalent values for residence time, D and dimension l_0, has been diminished by $(mRc)^2$ in the denominator due to transverse convection.

6.1.4 Mixing and Reaction

Most mixing problems associated with chemical reactions involve three processes which occur simultaneously: convective mixing, diffusion and reaction. Diffusion, which is generally very slow, is the ultimate mechanism for mixing molecular participants in a reaction. However, in the interest of time, it is essential to promote convection as well as to assist the diffusion process by reducing diffusion lengths and/or increasing interfacial areas. To assist us in applying the concepts on mixing briefly reviewed above, we return to our MRE.

Before beginning, it should be pointed out that some reactors remain microsegregated throughout reaction, despite the potential miscibility of their contents. In terms of our MREs, this means that their contents are not mixed with the surroundings at all, or with each other. In some cases this isolation is intentional; in particular if the MREs are self-contained, i.e. each MRE is immiscible with its surroundings, contains all necessary reaction ingredients within its boundaries upon entering the reactor and is quite content to remain microscopically segregated from the others and from its surroundings. Suspension polymerization of an organic monomer in an aqueous medium is a common example. In other cases, microscopic segregation of isolated MREs is the inadvertent result of an inability to provide adequate mixing in spite of a feed stream that is homogeneous, or at least potentially miscible, upon entering the reactor.

In either case the material balance for all such "microsegregated" reactors with self-contained MREs is simply (Biesenberger and Sebastian, 1983)

$$\Delta C_k = \int_0^\infty C_k(\tau)\, e(\tau)\, d\tau \tag{27}$$

which requires for its evaluation a knowledge of the residence time distribution $e(\tau)$ of the particular reactor in question, in addition to the concentration-time profile of the species in question from batch kinetics, $C_k(t)$ (solution of Eq. 5). Residence time distributions are readily accessible experimentally or computationally.

More typically, we presume that the contents of each MRE must be mixed with its surroundings on a molecular level as a prerequisite to reaction. Whereas the three simultaneous processes mentioned above actually occur in parallel, it is reasonable and convenient to postulate that, under ideal conditions, we would prefer them to occur in series: i.e. first, convective mixing to reduce l_0 to l; then, molecular homogenization by diffusion; and, finally, reaction among the well mixed molecules. This "series" approach, which was tacitly applied in the previous treatment of simultaneous convective and diffusive mixing without reaction (Ineq. 24), facilitates the use of characteristic times, which can be readily estimated from kinetic and transport property data. It will also provide us with some guidance in dealing with reactors requiring molecular mixing, whose mixing dynamics are not as simple as that of the PFR, the IM, the dispersion model, or even the microsegregated reactor.

In the presence of reaction, it is not sufficient to achieve molecular mixing within the time frame λ_V, especially when the characteristic time for reaction, λ_r, is smaller (i.e. when $Da > 1$). Therefore, it seems reasonable to compare λ_D

with λ_r, rather than with λ_V. In fact, a dimensionless number was introduced to the chemical reaction engineering literature some time ago for just that purpose. It represents the relative importance of simultaneous diffusion and reaction and is called the Damkohler number of the second kind. Biesenberger and Sebastian (1983) have expressed it as the ratio of appropriate characteristic times

$$Da_{||} = \lambda_D/\lambda_r = l^2/D\lambda_r \tag{28}$$

where l here, as in Pe, generally refers to a fixed vessel dimension.

Two common examples of simultaneous mixing and reaction in reactive extrusion (REX), cited in Chapters 1 - 4, are the cracking of polypropylene, and the maleation of polyolefins and EP elastomers. Both are free radical reactions and both require that small amounts of liquid initiator be mixed with polymer melt as near to molecular uniformity as feasible within times that are short compared to the half-life of the initiator (λ_r). Obviously, to avoid localized, nonuniform reaction in such cases, rapid convective mixing is required to disperse the initiator into tiny droplets for effective diffusive mixing, assuming mutual solubility. Such convective mixing is difficult to achieve in view of the large differences in viscosity, as well as proportions, among the two feeds.

In an effort to develop a criterion for adequate mixing in the presence of reaction utilizing our MRE, it seems reasonable to require, in place of Ineq. 24, that

$$\lambda_D \ll \lambda_r \tag{29}$$

where characteristic length l here refers to our MRE and may therefore again be a variable function of time. Alternatively, by analogy to Eq. 4, we could define a segregation number for reaction

$$Sr = \lambda_D/\lambda_r = l^2/D\lambda_r \tag{30}$$

and consider it to be a generalization of $Da_{||}$, in the same way that we consider the segregation number for mixing Sm to be a generalization of Pe. Thus, for adequate molecular mixing we require that $Sr < Sr_{cr}$, where, again, the critical value must be determined by experimentation or computation. For a plain extruder channel then, by analogy to Eq. 26, we obtain

$$Sr = Da_{||}/(mRc)^2 \tag{31}$$

where the Damkohler number

$$Da_{||} = l_0^2/D\lambda_r \tag{32}$$

is expressed in the conventional manner in terms of a characteristic vessel dimension l_0, and Rc is the recirculation number, which is a measure of the relative rates of cross-channel versus down-channel convection and is defined in Fig. 9. Recall that Rc is a function of helix angle but is independent of screw RPM.

For illustrative purposes, let us consider a polymerization reaction of two monomers, A and B, which has second order kinetics (first order with respect to each monomer) and requires molecular mixing for reaction. Since A and B enter the extruder in separate streams and the primary convective mixing mechanism under examination is due to cross-channel flow, we select the vertical feed configuration in Fig. 10 and set $l_0 = W_c/2$, assuming that A and B enter in stoichiometrically equal amounts.

Our criterion for mixedness, $Sr < Sr_{cr}$, in terms of recirculation number then becomes:

$$Rc > Rc_{cr} = (Da_{\|}/Sr_{cr})^{1/2}/m \tag{33}$$

Notice that, if $Da_{\|}$ for a reaction is a large number, indicating that mixing by diffusion is relatively slow compared to reaction, then the mixing process must be sufficiently enhanced by convection (large mRc) to achieve adequate mixing (small value of Sr).

It is evident from the many cases cited and reviewed by Brown in Chapter 4 that mixing requirements encountered in REX are varied and not as simple as those selected for analytical treatment in our examples. Feed streams may be added separately, or premixed on a macroscopic scale (heterogeneous) or on a molecular scale (homogeneous); they may be solids (pellets, powders), low-viscosity liquids, melts or gases. An attempt has been made in Fig. 8 to categorize and summarize them.

It is also well known from experience that mixing in plain extruder channels is not adequate for most REX applications. Longitudinal mixing, and therefore backmixing, is limited, and has been modeled accordingly by some researchers via Eq. 9 with a concomitantly high value for Pe_L, and transverse mixing, although enhanced somewhat by cross-channel flow, is generally inadequate for conducting reactions. For such purposes, a variety of screw geometries and special mixing devices are available. Selection of the optimum device for dispersive and distributive mixing, which is largely based upon empirical knowledge, depends upon the physical state of the feeds as well as their point of introduction (hopper, side port) and order of addition. Some mixing problems and extruder geometries are reviewed in Chapter 5 by Todd. Our emphasis in Section 6.2 is on the effects of mixing on various types of polymer reactions.

Beyond our heuristic effort in the previous section to relate mixing theory to extrusion equipment, the research literature offers little additional help. Most of the analytical work on mixing with chemical reaction reported in the chemical engineering literature deals with feed streams that are premixed (homogeneous), and examines the effects of reactor dynamics on particular classes of reactions. Furthermore, such effects are traditionally treated in terms of residence time distribution and extent of backmixing, since such reactor characteristics are amenable to mathematical modeling for certain simple reactors. These reactors and their models have been reviewed briefly in the previous section, and the effect of their mixing dynamics on various idealized polymerization schemes have been treated in detail (Biesenberger and Sebastian, 1983).

6.1.5 The Extruder as a Reactor

Several authors in this monograph have tabulated the many virtues and few vices of the extruder as a polymer reactor. Table 2, which represents another attempt, emphasizes those features that affect reactions in particular, viz., side-stream addition, venting and mixing, and Fig. 10 illustrates some of these features and the various machine and process parameters available in the form of a fictitious process configuration.

Certain geometric parameters, such as number and proximity of screws, number of channels N_c, channel depth H and helix angle (or screw lead L_S), have important consequences in REX since they affect key process conditions, such as mixing dynamics, interfacial surface-to-volume ratio, and residence time. For instance, subdividing the melt stream by increasing the number of

Table 6.2 The Extruder as a Reactor

ADVANTAGES

Bulk reaction; no diluents (solvents, water)

Avoid contamination, lower equipment/energy costs

Drag flow; transport high viscosity media] meter feed for
Laminar mixing; mix high viscosity media] independent control

Linear vessel; plug flow

Staging; melting, side-stream addition, mixing, reacting, venting, pressurizing, shaping

Modular, No dead space; twin-screw

DISADVANTAGES

Short residence time

Poor heat transfer

No backmix version

6.2 Polymer Reactions

Various categories of REX processes have been listed in Table 3 with a far less ambitious attempt at completeness than that of Brown (Chapter 4), but rather with a view toward distinguishing between two large classes of polymer reactions, the stepwise and the chain type, which are classified accordingly in Table 4 using characteristic times. In terms of chemical mechanism, reactions that proceed via functional groups are more likely to be stepwise and those that involve active intermediates (free radicals, ions, catalyst complexes) are more likely to be chain-type, although these generalizations are not always true.

We define a polymer reaction as stepwise when the molecules involved in a particular transformation at any time during that reaction are, for the most part, the same molecules as those that are involved in the same transformation at any other time during the reaction. Examples of transformations are: growing polymers or copolymers from monomers by propagation reactions; enlarging pre-existing polymers by chain extension or coupling reactions; modifying polymers by functionalizing or grafting reactions; or, cracking polymers to smaller fragments. They are listed in Table 4.

More specifically, polyamides, polyesters and polyurethanes are formed via stepwise reaction of functional groups; but anionically catalyzed "living polymers" may also be regarded as stepwise reactions, since their active intermediates are deprived of termination. In both cases, the "same" molecules grow in parallel fashion throughout the reaction. Certain chain extension and block copolymer reactions are also stepwise, as shown in the table.

On the other hand, free radical and ionically initiated addition polymerizations with termination are obviously not stepwise reactions, and neither are

Table 6.3 Reactive Extrusion Processes

A. In situ polymerization
  1. Simultaneous polymerization or copolymerization of two or more monomers.
  2. Polymerization of monomer(s) in the presence of polymers.

B. Polymer modification reactions
  1. Functionalizing polymer or oligomer for subsequent reaction or blending with other polymers, fillers, or reinforcing agents.
  2. Grafting polymers, oligomers, or monomers to give: branched or block structures.
  3. Coupling polymers or oligomers of like kind (chain extension) or different kind (copolymerization) to give: block, branched or network structures.

C. Polymer cracking
  Molecular weight reduction: e.g. controlled rheology.

Table 6.4 Reaction Types

Reaction type	Reaction	
Random polymerization		
step	$A_x + B_y \rightarrow A_xB_y$	$\lambda_p \sim \lambda_r$
Addition polymerization		
step	$A_x + A \rightarrow A_{x+1}$	$\lambda_p \sim \lambda_r$
Chain polymerization		
initiation	$A \rightarrow A_1^*$	
propagation	$A_x^* + A \rightarrow A_{x+1}^*$	
or	$A_x^* + A_y \rightarrow A_{x+y}^*$	$\lambda_p \ll \lambda_r$
termination	$A_x^* \rightarrow A_x$	
Chain extension	$2\,A{-}{-}{-}A + BB \rightarrow A{-}{-}{-}ABBA{-}{-}{-}A$	$\lambda_p \sim \lambda_r$
Copolymerization		
propagation	$A_x^* + A \rightarrow A_{x+1}^*$	
and	$A_x^* + B \rightarrow B_{x+1}^*$	
and	$B_x^* + A \rightarrow A_{x+1}^*$	$\lambda_p \ll \lambda_r$
and	$B_x^* + B \rightarrow B_{x+1}^*$	
Block copolymerization		
	$A{-}{-}{-}A^* + xB \rightarrow A{-}{-}{-}AB{-}{-}{-}B^*$	$\lambda_p \ll \lambda_r$
or	$A{-}{-}{-}A + B{-}{-}{-}B \rightarrow A{-}{-}{-}AB{-}{-}{-}B$	$\lambda_p \sim \lambda_r$
Graft Copolymerization		
	$A{-}{-}{*}{-}{-}A + xB \rightarrow A{-}{-}\overset{\substack{B\\ \vert\\ B\\ \vert}}{-}{-}{-}A$	$\lambda_p \ll \lambda_r$
	${-}{-}{-}A{-}{-}{-} + B{-}{-}{-}B \rightarrow {-}{-}{-}\overset{\substack{B\\ \vert\\ B}}{A}{-}{-}{-}$	$\lambda_p \sim \lambda_r$
Depolymerization	$A_x^* \rightarrow A_{x-1}^* + A$	$\lambda_p \ll \lambda_r$
Degradation	$A_x \rightarrow A_y + A_{x-y}$	$\lambda_p \sim \lambda_r$

conventional copolymerizations, since each generation of growing polymers, which is small in number, is short-lived, and is preceded and succeeded by similar generations consisting of different polymers. An interesting hybrid is the graft copolymerization listed in Table 4 in which side chains are grown by a chain addition reaction on pre-existing polymer backbones. Whereas each side chain propagates only once, the same polymer backbone is likely to initiate a graft site at any time throughout the graft reaction.

The distinction between stepwise and chain polymer reaction can be expressed conveniently in terms of characteristic times as

$$\lambda_p \sim \lambda_r \tag{34}$$

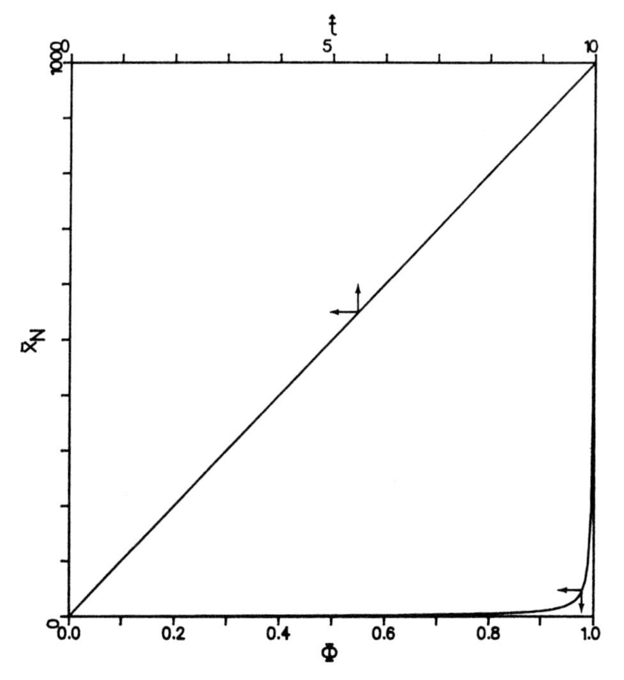

Figure 6.11 Number average degree of polymerization ($\bar{x}_N$) versus dimensionless time ($\hat{t}$) and conversion (Φ) for a random, stepwise polymerization

for stepwise reactions and

$$\lambda_p \ll \lambda_r \tag{35}$$

for chain reactions, where λ_r is the characteristic time associated with the overall reaction from start to finish, and λ_p is the characteristic time associated with the polymer transformation in question (chain propagation, chain coupling, chain functionalization, etc.).

The sharp contrast in behavior between stepwise and chain polymerizations carried out in a batch reactor is evident in Figs. 11-13 (Biesenberger and Sebastian, 1983). Fig. 11 is a graph of average degree of polymerization versus conversion of functional groups for a stepwise polymerization of the "random" type in Table 4. Notice the gelation-like behavior of the DP at large conversions.

In contrast, the average DP of the initial generation of "instantaneous" polymer formed during a chain addition polymerization (Table 4) is high, as shown in Fig. 12, but that of succeeding generations could be either lower or higher, depending upon whether the polymerization is "conventional" or dead-end", respectively, which could cause the overall average DP to drift downward or upward with conversion of monomer, as shown. Similarly, the average copolymer composition during chain addition copolymerization (Table 4) can

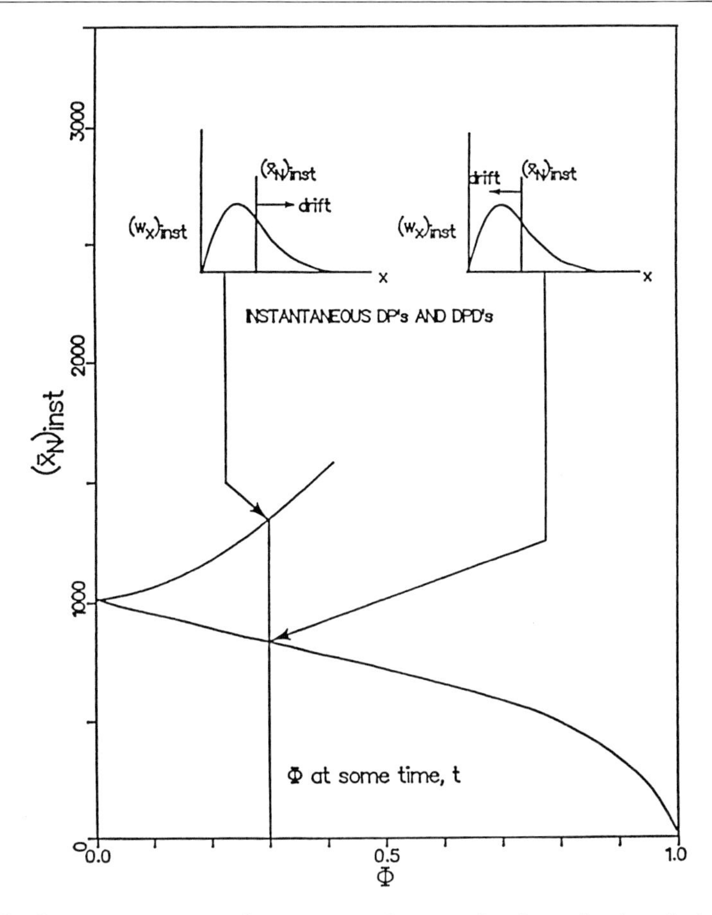

Figure 6.12 Instantaneous number average degree of polymerization $(\bar{x}_N)_{inst}$ versus conversion (Φ) for a chain addition polymerization

drift in either direction with conversion, favoring either monomer A or B as shown in Fig. 13. The behavior of "instantaneous" copolymer versus the corresponding composition of the unreacted co-monomer pool is illustrated in Fig. 14 for various combinations of reactivity ratios. Composition drift can be in either direction depending upon the location of the CC curve relative to the diagonal (Biesenberger and Sebastian, 1983). The basic cause of these drift phenomena in chain reactions carried out in batch reactors is, of course, the different changes in concentration and composition with time among the various reactants: initiator, monomer A, monomer B, etc.

The effect of mixing and reactor type on these distinctly different types of reaction can be analyzed conveniently in terms of our MRE. In a PFR all MREs pass through with identical residence times. Therefore, all polymer reactions conducted therein will proceed precisely as they would in a batch

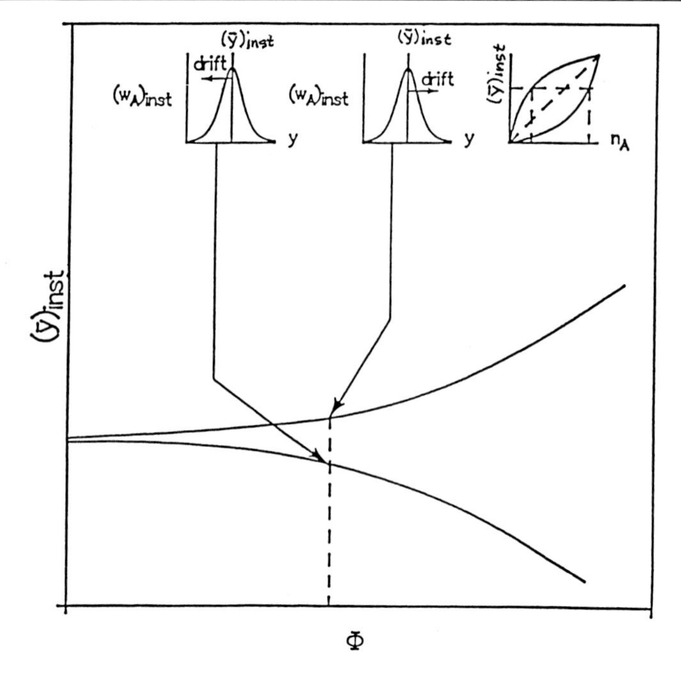

Figure 6.13 Instantaneous copolymer composition $(\bar{y})_{inst}$ versus conversion (Φ) for a chain addition copolymerization

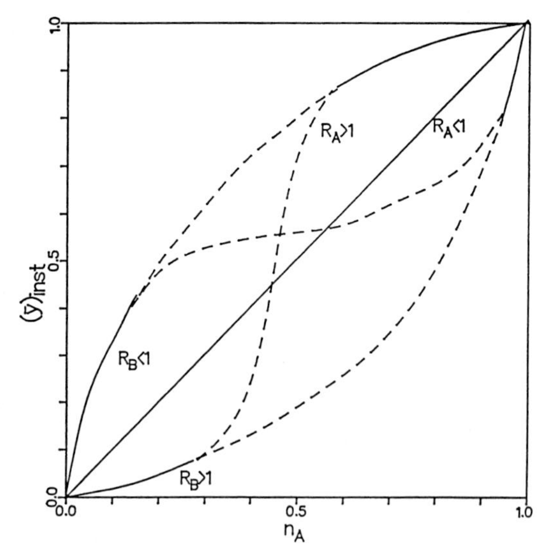

Figure 6.14 Instantaneous copolymer composition $(\bar{y})_{inst}$ versus molar composition of unreacted comonomers (n_A)

reactor, and all molecular characteristics, such as composition and chain length, will be dispersed in accordance with the inherently random nature of the reactions themselves ("statistical dispersion") with additional dispersion in chain type reactions due to the aforementioned drift phenomena ("drift dispersion") (Biesenberger and Sebastian, 1983). In the presence of velocity distributions, however, and their associated residence time distributions, which cannot be avoided in real vessels, all such distributions will be further dispersed.

When MREs interact longitudinally, i.e. when "older" ones and "younger" ones mix on a molecular level (backmix), their contents respond entirely differently for stepwise than for chain type reactions. The process of plunging the molecules participating in a stepwise transformation into varying molecular environments consisting of more and less advanced states of reaction, with their concomitant differences in composition, actually alters the statistical nature of that transformation. In the extreme case of the IM with complete backmixing, the probability of exit from the reactor is available to the same degree for all reacting molecules. Thus, the physical process of exiting is equivalent to the chemical process of termination, which obviously tends to broaden even further the dispersion of the aforementioned distributions. For example, the probabilities of propagation for the first two stepwise polymerizations listed in Table 4 become, respectively

$$\mathrm{Prob} = \mathrm{VkC_p^2/(VkC_p^2 + \dot{V}C_p)} \tag{36}$$

and

$$\mathrm{Prob} = \mathrm{VkC_m C_p/(VkC_m C_p + \dot{V}C_p)} \tag{37}$$

where C_m and C_p are concentrations of monomer and polymer, respectively (Biesenberger and Sebastian, 1983). Notice the exit flow term $\dot{V}$ in each denominator.

The effect of backmixing on chain type reactions, however, is just the opposite. Since $\lambda_p \ll \lambda_v$ for such reactions, the opportunity to exit has virtually no effect on the molecular statistics of such transformations. On the other hand, the homogenization of composition among all MREs with respect to residence time and space has the beneficial effect of reducing drift dispersion.

In the extreme case of IM, drift is eliminated altogether. Consequently, the resulting distributions are the result of random molecular events imbedded in an unchanging environment. Examples are the most Probable DP distribution and the Gaussian CC distribution. Since the latter are quite narrow, the IM provides a means for producing copolymers with very uniform compositions (no drift along the curves in Fig. 14). This probably accounts for the wide use of CSTRs for copolymerizations, as well as the use of diluents - water in emulsions and suspensions, and solvents in solutions - to side-step the mixing problems associated with high-viscosity bulk polymerizations. This observation should provide an incentive for the development of a drag flow version of the IM to deal with those viscosities and thereby avert the numerous economic and environmental problems accompanying the use of diluents delineated by Kowalski in the introduction and in Chapter 1.

Acknowledgment

The author wishes to thank Prof. F. Ling for computing, especially for this chapter, the striation patterns reproduced in Fig. 6.

Symbols

A_0 = free cross-sectional area between barrel and twin screws
A_B = cross-sectional area of barrel
A_C = cross-sectional area of screw channel
A_S = cross-sectional area of twin screws
C = flight clearance
C_k = concentration of substance k
C_L = center-to-center distance between screws
D = molecular diffusivity
Da = dimensionless Damkohler number of the first kind (defined in text)
$Da_{||}$ = dimensionless Damkohler number of the second kind (defined in text)
D_L = effective longitudinal dispersion coefficient
d_B = barrel diameter
e = flight width
g = fractional of channel filled with melt
H = transverse channel dimension (depth in extruder)
L = longitudinal channel dimension (down-channel in extruder)
L_B = longitudinal dimension in direction of screw axis
L_s = screw lead
l = characteristic length (defined in text)
m = slope (defined in text)
n_A = mole fraction of monomer A in a mixture of A and B
N_C = number of channels
N_R = screw RPM
Pe_L = dimensionless longitudinal Peclet number
Q = mass flow rate
R = radius in general

R_B = barrel radius
R_C = dimensionless recirculation number (defined in text)
r = variable radius
r_k = rate of formation of substance k by reaction
S = interfacial area (defined in text)
S_0 = initial interfacial area
S_m = dimensionless segregation number without reaction (defined in text)
S_r = dimensionless segregation number with reaction (defined in text)
t = variable time
V = volume
$\dot{V}$ = volumetric flow rate (total)
$\dot{V}_d$ = volumetric flow rate by drag flow
$\dot{V}_p$ = volumetric flow rate by pressure flow
v_{Bx} = component of barrel velocity in cross-channel direction
v_{Bz} = component of barrel velocity in down-channel direction
v_{Zav} = average down-channel velocity of melt
W_c = channel width (cross-channel)
x_N = number average degree of polymerization
y = number fraction of monomer A in a copolymer chain of A and B
z = variable down-channel distance
α = variable age in vessel or twin-screw tip angle (defined in Table 1)
Φ = conversion
ϕ = screw helix angle
γ = shear
λ_C = $L/v_{Zav} = \lambda_V$
λ_D = characteristic time for diffusion
λ_p = characteristic time for single polymer reaction step
λ_r = characteristic time for overall reaction
λ_V = $V/\dot{V}$ = mean residence time
μ = Newtonian viscosity
ϱ_p = polymer density
τ = residence time
ψ = twin-screw barrel angle (defined in Table 1)

Chapter 7

Heat Transfer in Extruder Reactors

By William M. Davis, Exxon Chemical Co., 1900 East Linden Avenue, P.O. Box 45, Linden, NJ 07036

7.1 Introduction

Control of the chemical reaction in order to achieve the desired product is the objective of the reactive extrusion process. As the rate of reaction approximately doubles with every 10 °C increase in temperature, heat transfer obviously becomes an important consideration. Following are some of the key points that demonstrate the importance of heat transfer in reactive extrusion:

- The kinetics which may include the selectivity and extent of reaction are very sensitive to bulk temperature and its distribution.
- It is important to prevent localized overheating which can result in a high rate of reaction and degradation in the product.
- Since reactive extrusion consists of many stages it may be necessary to have a specific axial temperature profile.
- Heat transfer effects can and often do govern scale-up considerations.
- A stable temperature will be required to maintain good product consistency.

Naturally the screw design will be one that is intended for good mixing. These designs will influence heat generation and transfer. Detailed treatment of these key parameters are beyond the scope of this chapter.

7.2 Model

An approach that has been successfully used in practice is to treat heat transfer in an extruder as that through a cylinder of concentric layers. It is obvious that this may be valid for single screw extruders but questionable for twin screw extruders. Treatment of twin machines via this technique to determine the overall coefficient has shown agreement with more elaborate finite element calculations to within 10%. Of course the internal wetted area is calculated by subtracting the open apex perimeter. As this field is not, analytically, precise it is the writer's opinion that the final 10% accuracy is not worth the expense. Fig. 1 shows the important elements of this model.

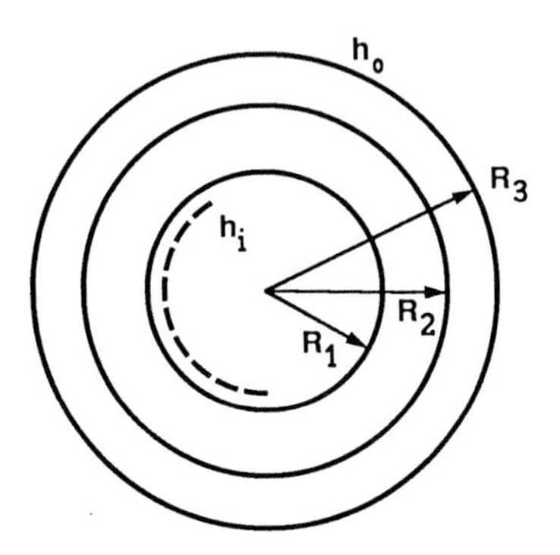

Figure 7.1 Concentric Barrel Model (reproduced by permission of the American Institute of Chemical Engineers. © 1988 AIChE; Davis, 1988)

This model has been constructed as a two piece barrel. The effects of materials of construction along with polymer and external heat transfer coefficients, h_i and h_o, respectively will be discussed in some detail.

Development of the mathematics for concentric cylinder heat transfer is discussed in detail in fundamental heat transfer books, i.e. Kern (1950). This detail will not be presented here.

An overall heat transfer coefficient based on the outer perimeter, U_o, can be defined as:

$$U_o = \frac{q}{\pi D_3 (T_o - T_i) x L}, \quad \frac{\text{BTU/Hr}}{\text{ft}^2 \, ^\circ\text{F}} \tag{1}$$

Expressed as a function of the extruder design components U_o is:

$$U_o = \frac{1}{\dfrac{R_3 \ln(R_3/R_2)}{K_2} + \dfrac{R_3 \ln(R_2/R_1)}{K_1} + \dfrac{R_3}{R_1 h_i} + \dfrac{1}{h_o}} \tag{2}$$

Alternatively the overall heat transfer coefficient can be based on the internal diameter, U_i as:

$$U_i = \frac{q}{\pi D_1 (T_o - T_i) x L}, \quad \frac{BTU/Hr}{ft^2 \, ^\circ F} \tag{3}$$

or:

$$U_i = \frac{1}{\dfrac{R_1 \ln(R_3/R_2)}{K_2} + \dfrac{R_1 \ln(R_2/R_1)}{K_1} + \dfrac{1}{h_i} + \dfrac{R_1}{R_3 h_o}} \tag{4}$$

In theory both of these overall coefficients give the same result for heat transfer. When calculating the heat flux, Q, or ΔT the appropriate area on the ID or OD must be used. However, all heat put into or removed from an extruder must be via jackets or coring. Often times full coverage on the perimeter is not achieved. In these cases it is best to use U_o along with the wetted perimeter to calculate the heat exchange capacity of the extruder.

Equations 2 and 4 are comprised of components from the physical construction of the barrel, i.e. R and K, and process film coefficients h_i and h_o. The heat transfer coefficient of the external fluid, h_o, is easily estimated from fundamental heat transfer sources such as Kern (1950). As a general rule it is sound design practice to size the heat transfer fluid circulation rate so that $h_o > 200$ BTU/Hr ft^2 °F. If this is done, h_o will have a minimal effect on the overall coefficient. At values of $h_o < 100$ the effect on the overall coefficient will be noticeable.

The polymer film coefficient h_i must actually be determined empirically for the system being considered. Typically this value ranges from 30 up to 100 BTU/Hr ft^2 °F as shown by Todd (1988b). This coefficient is usually the limiting component in the expression for the overall heat transfer coefficient. It is a function of mainly RPM and screw to barrel clearance. This will be discussed in more detail.

7.3 Effects of Design Parameters on Heat Transfer

Most commonly the metal of an extruder barrel is believed to have minimal thermal resistance compared to the polymer film coefficient. This can be largely true for small machines especially those made from carbon steel. However, as the size of the extruder grows and if some degree of corrosion protection is desired the total thermal resistance of the barrel can be comparable to the polymer film coefficient. Table 1 (Davis, 1988) gives the calculated components contributing to the overall heat transfer coefficient. The example used is for two piece construction using a liner inserted into a larger outer barrel. Dimensions chosen do not necessarily reflect a "real" system but were chosen so that the effects are large enough to be significant. The expression for U_o is as follows:

$$U_o = \frac{1}{\dfrac{R_3 \ln(R_3/R_2)}{K_2} + \dfrac{R_3/R_2 x \xi}{K_{air}} + \dfrac{R_3 \ln(R_2/R_1)}{K_1} + \dfrac{R_3}{R_1 h_i}} \tag{5}$$

The radial gap between the liner and outer barrel is given as ξ.

The following Table (Davis, 1988) summarizes the components in the denominator of Eq. 5 for the following cases listed.

Table 7.1 The Impact of Barrel Construction on Heat Transfer

U BTU/ft² hr °F	$\frac{D_3 \ln(D_3/D_2)}{2K_2}$	$\frac{D_3 2(2\xi)}{2K_{air}}$	$\frac{D_3 \ln(D_2/D_1)}{2K_1}$	$\frac{D_3}{D_1 h_i}$	Barrel Construction*
12.3	.0079	0	.0074	.066	no air gap all CS
11.8	.0079	.0037	.0074	.066	.001 gap all CS
10.0	.0079	.0186	.0074	.066	.005 gap all CS
9.5	.0079	0	.0317	.066	no gap SS liner
9.2	.0079	.0037	.0317	.066	no gap SS liner & CS barrel
7.6	.0339	0	.0317	.066	no gap SS liner & SS barrel
7.4	.0339	0.0037	.0317	.066	.001 gap SS liner SS barrel

* SS = stainless steel
CS = carbon steel

If K's = ∞ then U = 15.2. This is the overall heat transfer coefficient based only on the polymer film coefficient.

Results obtained from an inspection of this Table show the following:

- The change in materials from all carbon steel to stainless steel results in a significant reduction of heat transfer capacity of about 40%.

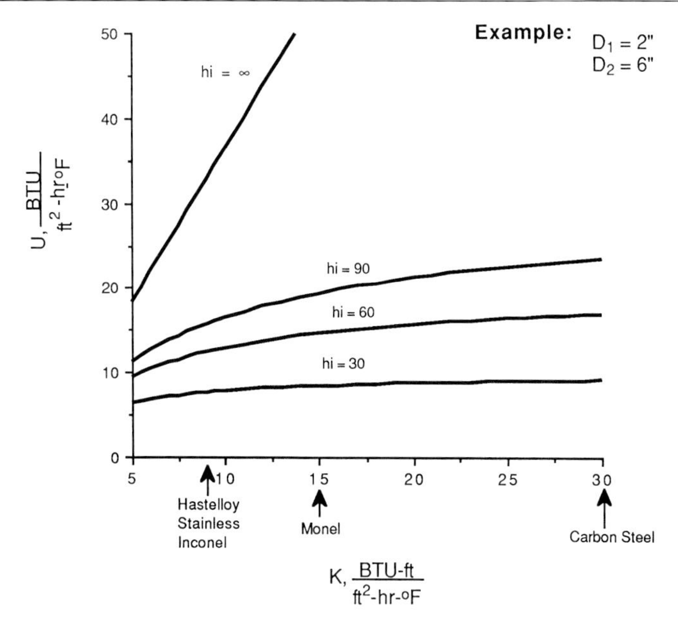

Figure 7.2 Effect of materials of construction on overall heat transfer coefficient (reproduced by permission of the American Institute of Chemical Engineers. © 1988 AIChE; Davis, 1988)

- In each case inclusion of an .0037″ air gap reduced the heat transfer capability from < 5% to 19%. The severity of this effect depends on the size of the air gap. Since there is some metal to metal contact in this area it could be argued that a line contact coefficient of 200-500 BTU/Hr ft^2°F might be more appropriate.

Fig. 2 presents the effect of materials of construction and polymer film coefficient on overall heat transfer. As can be expected the influence of materials is less at lower polymer film coefficients. At film coefficients of 90 the change in materials of construction from carbon to stainless steel can reduce the heat transfer capability by about one half. It is important to remember this effect when doing scale-up. For a thermally sensitive process, pilot plant data obtained on a carbon steel extruder may be very misleading when applied to a larger corrosion resistant machine. The use of nickel based bimetallic coatings with carbon steel backing material is a possible alternative to all stainless steel construction.

Failure to consider the shell side coefficient is a common error as it is usually assumed that the fluid system heat transfer is much higher than that of the polymer and, therefore, negligible.

The effect of the shell side coefficient is shown in Fig. 3. As would be expected there is little effect where the polymer film coefficient is governing. However, at a high h_i of 90 it can be seen that a poorly designed heat transfer fluid

or electrical wattage. In the case of a fluid system the flow rate must be accurately measured as this is the simple greatest error source. The temperature difference is best measured by recording the difference between two instruments and meet the absolute temperature of each.
2. Use the interior as the perimeter for the area as this is the best defined surface.
3. Define the temperature difference to be used. It is probably best to use the film temperature (details to follow) and an average of the heat transfer fluid. For most circulating fluid systems the fluid rate is high and the ΔT small. Therefore, an arithmetic average should be acceptable.
4. Calculate the overall coefficient U_i:

$$U_i = \frac{q_b}{\pi D_1 (T_i - T_o) x L}, \quad \frac{BTU/Hr}{ft^2 {}^\circ F} \tag{6}$$

5. The fluid (shell side) coefficient, h_o, can be estimated from available handbooks or text books such as Kern (1950).
6. Eq. 4 can be used.

$$U_i = \frac{1}{\frac{R_1 \ln(R_2/R_1)}{K} + \frac{1}{h_i} + \frac{R_1}{R_2 h_o}} \tag{7}$$

$$h_i = \left[\frac{1}{U_i} - \frac{R_1 \ln(R_2/R_1)}{K} - \frac{R_1}{R_2 h_o}\right]^{-1} \tag{8}$$

7.5 Temperature Measurement

Accurate temperature measurement is, of course, important for control and analysis of an extruder reactor. The preceding treatment for experimental determination of h_i also requires accurate temperature measurement.

The basic instrument for measurement of the polymer processing temperature is a flush mounted thermocouple shown in Fig. 5. This instrument is essentially surrounded by the metal barrel which is in turn heated or cooled which can cause the indicated temperature to be in error by as much as 40 °C. The error will be biased high or low dependent on whether the barrel is heated or cooled, respectively. Since this type of probe is so affected by the barrel conditions it will be even insensitive to trend indication of the actual polymer temperature.

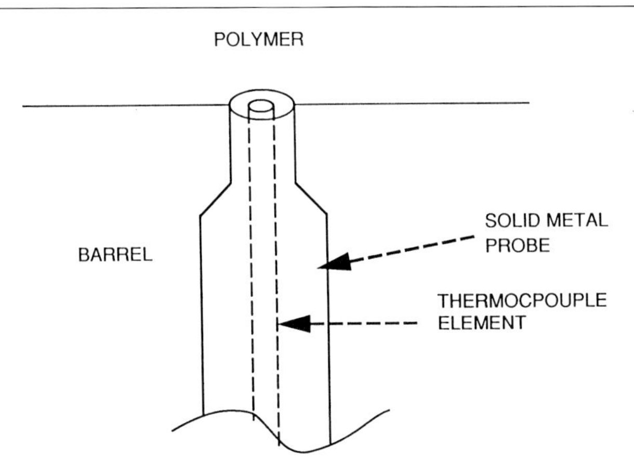

Figure 7.5 Extrusion melt thermocouple (reproduced by permission of the American Institute of Chemical Engineers. © 1988 AIChE; Davis, 1988)

A more accurate probe is the immersed type shown in Fig. 6. This type of probe can be immersed into a melt stream and is best used in a die adaptor at the discharge of the extruder. This type of probe is also subject to error from heating and cooling of the surrounding metal, especially if the annular insulation is inadequate. A more substantial debit is that modification of the screw is required if this probe is used at a mid-point in the process. Also the protruding element is subject to breakage during start-up with cold polymer.

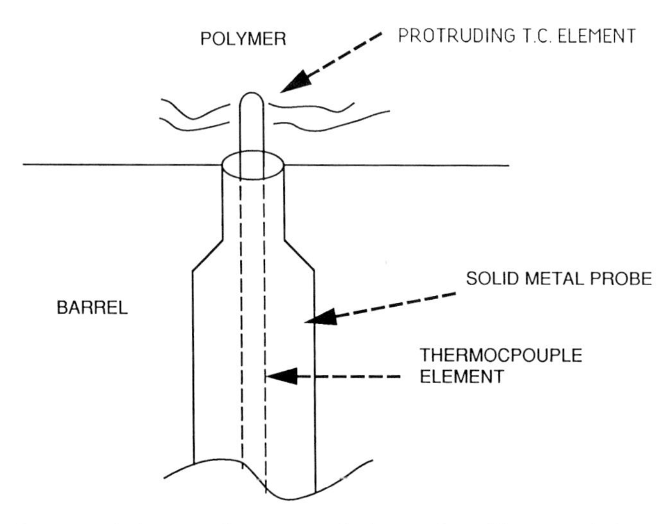

Figure 7.6 Immersed element thermocouple (reproduced by permission of the American Institute of Chemical Engineers. © 1988 AIChE; Davis, 1988)

The most accurate and sensitive instrument is an infrared probe as shown in Fig. 7. This probe has overall dimensions approximately the same as a standard pressure transducer but with a window at the tip.

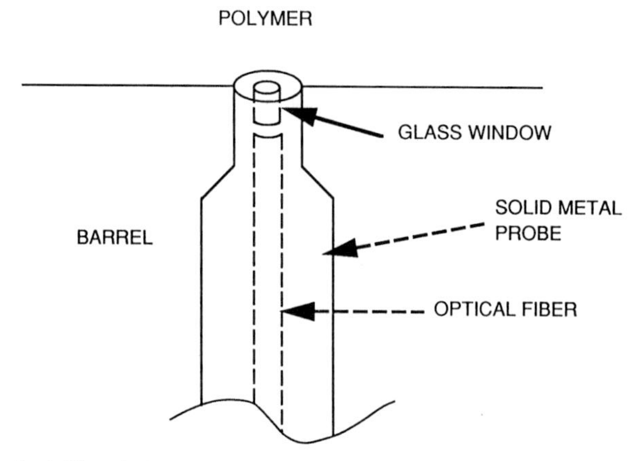

Figure 7.7 Optical fiber infrared temperature probe (reproduced by permission of the American Institute of Chemical Engineers. © 1988 AIChE; Davis, 1988)

An optical fiber is inserted into the probe up to the window. Infrared radiation is then transmitted through the optical fiber to a detector head which translates the signal into a temperature read out.

Fig. 8 gives a comparison of the IR temperature (instrument by Vanzetti Systems, Stoughton, MA) and bulk polymer temperature measured by collecting a sample from an open port at the same axial location. In this example, agreement to within 5 °C has been measured.

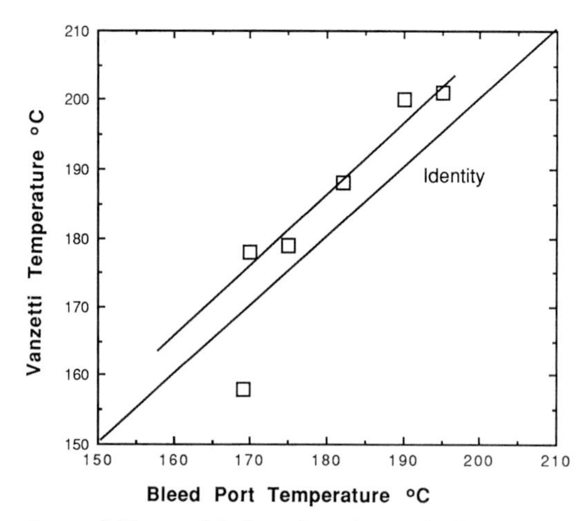

Figure 7.8 Comparison of Vanzetti infrared and true bulk temperatures (reproduced by permission of the American Institute of Chemical Engineers. © 1988 AIChE; Davis, 1988)

A comparison between the Vanzetti and a flush mounted thermocouple is shown in Fig. 9.

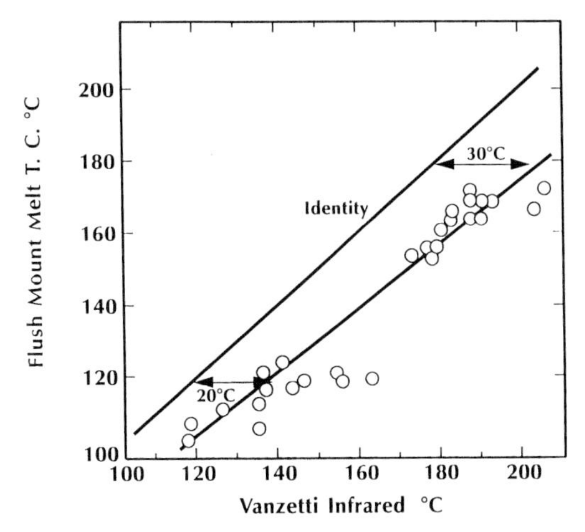

Figure 7.9 Comparison between melt thermocouple and Vanzetti infrared probe measurements (reproduced by permission of the American Institute of Chemical Engineers. © 1988 AIChE; Davis, 1988)

Using the Vanzetti as the basis this shows the melt thermocouple is indicating approximately 20 to 30 °C low. This error would be consistent with a chilled barrel.

The data shown in Fig. 8 gives good agreement between the bulk and Vanzetti temperatures. These experiments were all run with a starve fed screw. In order to assess the performance of the Vanzetti in a full system an experiment was performed by imposing and measuring a temperature profile across a polymer sample as shown in Fig. 10. In this set-up the block containing the polymer sample was placed between heated platens in a press. An emissivity setting of one was used in all these examples. For processes where pigments are added

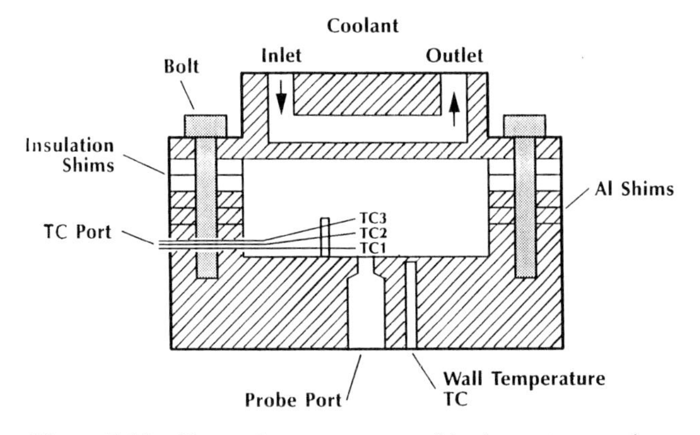

Figure 7.10 Vanzetti apparatus test block - cutaway view

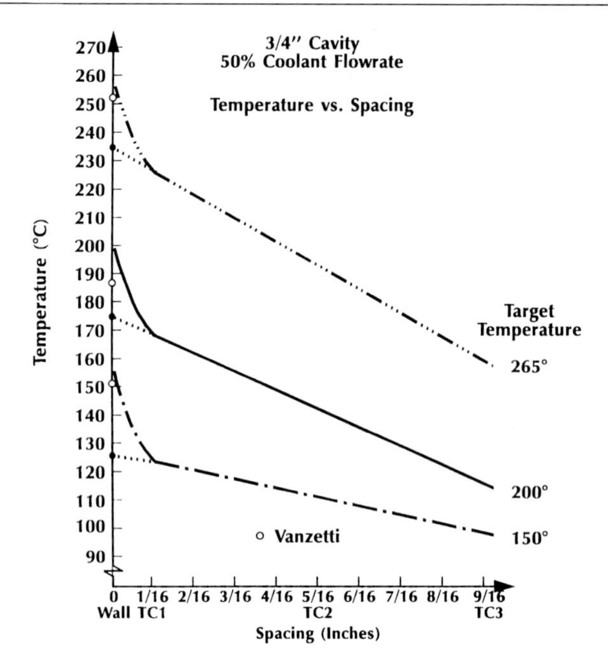

Figure 7.11 Vanzetti apparatus measurements in melted polymer with imposed temperature gradient

this would have to be adjusted. Results of this impaired profile experiment are given in Fig. 11. This shows that the measured Vanzetti temperature more closely agrees with the "Target Temperature" which is the platen set point. The conclusion is that the indicated infrared temperature in an extrusion process is probably an accurate measurement of the barrel film temperature due to its small depth-of-field. Even though the infrared temperature is not a true "bulk" temperature it is the preferred instrument for reactive extrusion because:

1. It is reasonably isolated from conduction effects of the barrel heating/cooling system. The indicated temperature is a true polymer temperature.
2. Process changes and upsets are quickly and accurately detected. Conventional probes can take 30 min to respond to a change due to the dampening effects of the large metal barrel. The infrared system responds in seconds.
3. The maximum process temperature is measured, which occurs due to viscous energy generation between the flight and barrel.

7.6 Temperature Control Schemes

Stable control of the heating/cooling system is important for any reactive extrusion process. Fig. 12 gives the most fundamental control system. In this scheme the metal barrel temperature is measured and controlled. This system is simple and inexpensive. Often the control logic can be simply on-off. Debits are slow response to actual process changes and usually poor relationship to actual polymer temperatures.

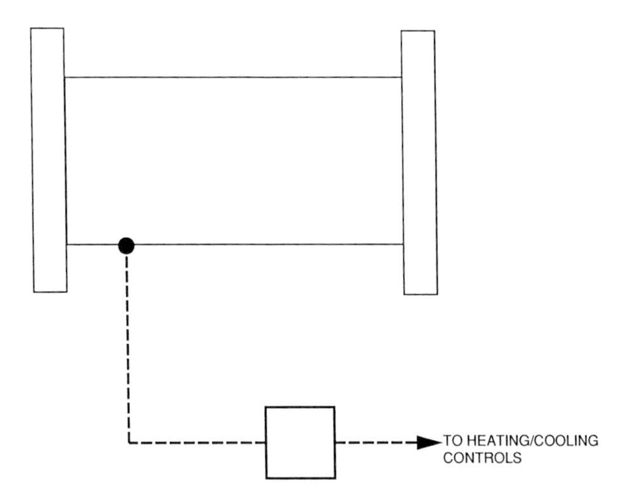

Figure 7.12 Basic extruder temperature control scheme

The next step in refinement for control logic is shown in Fig. 13. Actual polymer temperature is sensed and used for control purposes. Again this system is fairly simple but since the lag time between sensing and response (due to the thermal resistance of the barrel) can be significant, process instabilities can easily result. Again process problems would be more acute with larger machines. Control systems using a predictive logic are possible but require more advanced electronics and must be tuned empirically.

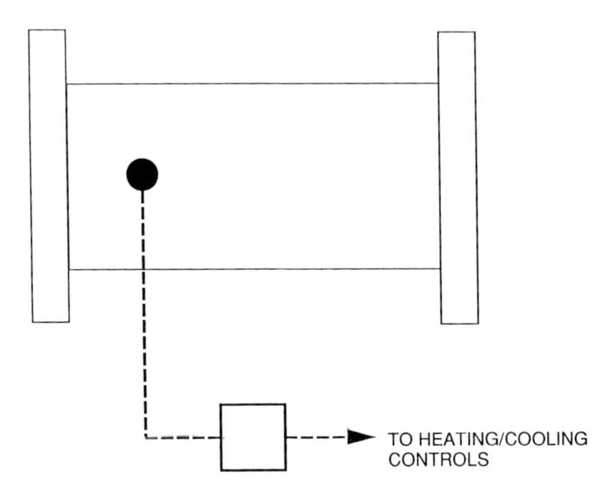

Figure 7.13 Process temperature control scheme

Finally the most practical and reliable control system as shown in Fig. 14 has a cascade logic. In this system the polymer temperature is sensed which provides a set point to the primary control loop which uses the barrel temperature for control. This has the advantage of controlling the actual polymer temperature with the stability of control of the barrel temperature. The required electronics are fairly basic and tuning is easily done. Rauwendaal (1986) has a very comprehensive discussion on control logic as applied to extruders.

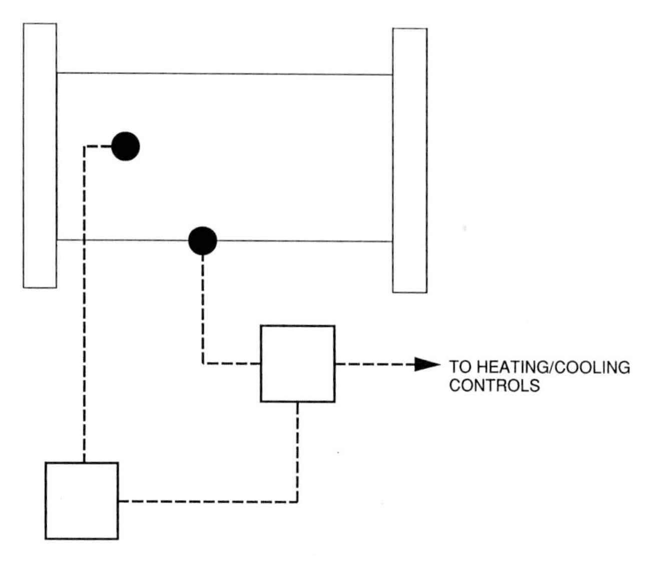

Figure 7.14 Cascade temperature control

7.7 Heat Exchange Barrel Construction

In addition to the items previously discussed attention must be paid to design of the "shell side" of the barrel. The most basic and probably most common is the jacketed barrel shown in Fig. 15. This type of design is best used in smaller extruders and for processes that are not extremely thermally sensitive.

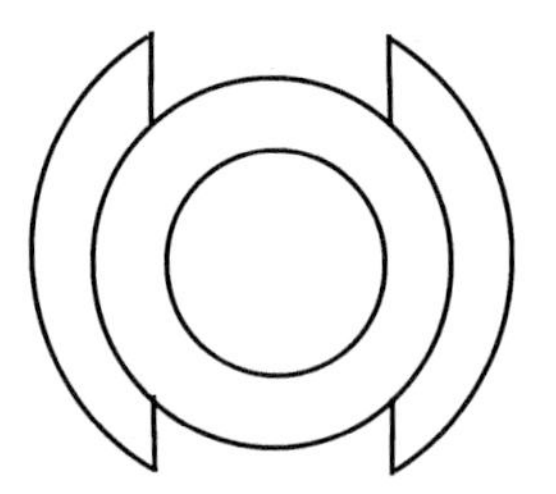

Figure 7.15 Jacketed barrel

Advantages to the jackets are ease of replacement and lowest cost. However, response to process changes can be slow especially for corrosion resistant barrel materials.

A "wet liner" approach to barrel design is shown in Fig. 16. In this system a thin liner is inserted into a larger outer barrel with a space allowed for coolant flow between the two. This has the advantage of minimizing the distance between the process and heat transfer fluid which should in practice maximize heat transfer capability and minimize response time. The largest debit to this design is the potential for coolant fluid leaks through seal welds and instrument ports. Also differences in thermal expansion can cause cracks if the barrel and liner are made from different materials.

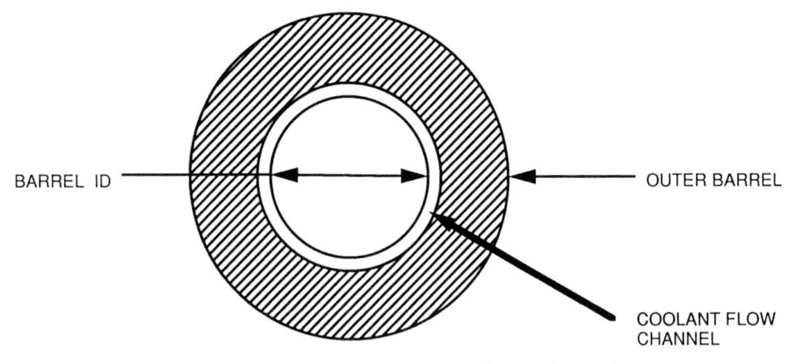

Figure 7.16 "Wet Liner" barrel

The cored barrel as shown in Fig. 17 is probably the most common and mechanically most reliable where enhanced heat transfer is sought. The cores can be located very close ($\sim$ 3/4 inch) to the barrel ID. In this type of system several cores are often piped in series to maximize the fluid velocity and hence h_0. For longer barrels gun drilling to close tolerances can be expensive especially for alloy steels. Also both the barrel and heat transfer fluid system must be carefully designed to avoid cracking due to thermal shock. This is probably more extreme for the case of the "wet liner".

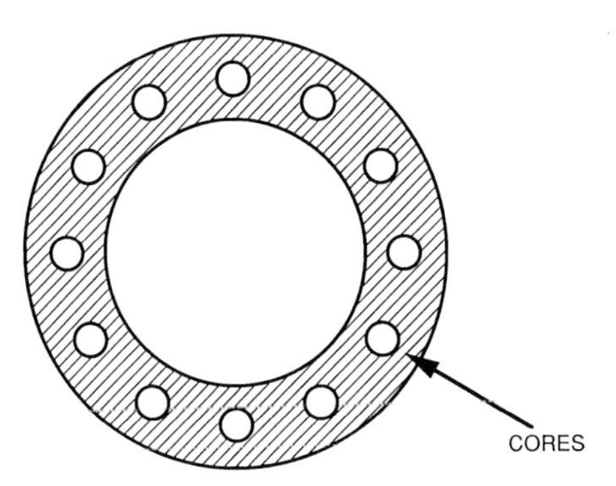

Figure 7.17 Cored barrel

In summary, for a process where heat input is primarily required the standard jacket design should be sufficient. However, where heat removal must be maximized either the "wet liner" or cored barrel must be used. These designs are more expensive and have some mechanical risks which can be minimized with proper attention to details. As polymer mixing or reaction exotherms both result in adiabatic temperature increases the need to remove heat is usually much more acute than heat infusion for extruder reactors.

7.8 Sources of Heat Generation

Using an overall energy balance as a starting point the amount of heat that needs to be transferred through the barrel can be expressed as

$$q_b = MC_p(T_p - T_f) + \Delta H_t - E - q_r \tag{9}$$

E = mechanical energy input from the extruder drive
q_r = heat of reaction

where $+ q_b$ indicates a net energy input from the heat transfer system and $- q_b$ means a net energy removal. Typically for reactive extrusion $(E + q_r) > MC_p(T_p - T_f) + \Delta H_t$. Therefore, there is a need to remove heat. The term E denotes the amount of net power inputted from the extruder drive. This is primarily a function of polymer rheology and screw design. For Newtonian fluid the relationship for E is given in Tadmor and Klein, p. 235 (1970) as:

$$E = P \mu N^2 (\pi D)^2 \frac{W}{H} \frac{L}{\sin\theta} f \tag{10}$$

Inspection of this equation shows that viscous energy increases as the square of the screw speed and increase to the channel depth, H. Since a usual response to the need to improve mixing is to increase the screw speed and to use a more restrictive screw element, both of these changes will result in a temperature rise and, as a result, an increase in the rate of reaction. The end result would be overreaction and/or product degradation. Another factor which is evident from Eq. 10 is the increase of viscous energy with the cube of the dimension scale (i.e. $D^2 \times L$). This can impact scale-up which will be discussed in more detail later. Another good reference for fundamental energy balance in an extruder is given by Stevens (1988).

Obviously there is a need to optimize the screw design for mixing while minimizing the energy required. This is an important consideration and beyond the scope of this chapter.

Another important source of temperature generation is in the screw flight to barrel clearance. As this is the point in the process with the maximum shear rate it is also where the temperature is a maximum. Temperature generation in this clearance can be modelled as couette flow where:

$$E_f = V_f \times \tau x \dot{\gamma} \tag{11}$$

The amount of material dragged between the flight and barrel can be modelled as drag flow under a sliding plate. Applied to an extruder the resulting relationship is:

$$Q_f = \frac{N(\pi D)^2}{2} \delta \tan\theta \tag{12}$$

An adiabatic temperature rise can be calculated as:

$$\Delta T = \frac{E_f}{Q_f \times \varrho \times C_p} \tag{13}$$

with a final relationship for a power low fluid expressed as:

$$\Delta T(^\circ C) = \frac{2\eta(\pi D N)^n e}{\delta^{n+1} \varrho C_p} \times (2.4 \times 10^{-8}) \tag{14}$$

where

$$\begin{array}{ll} e\,[=]\,\text{cm} & C_p\,[=]\,\text{kcal/kg °C} \\ D\,[=]\,\text{cm} & N\,[=]\,\text{rev/sec} \\ \delta\,[=]\,\text{cm} & \eta \propto \text{poise} \\ \varrho\,[=]\,\text{g/cc} & \mu = \eta\dot{\gamma}^{n-1} \end{array} \tag{15}$$

Rauwendaal and Ingen Housz (1988) give a detailed discussion of leakage flow in the screw to barrel clearance.

Upon inspection of Eq. 14 it becomes obvious that the viscous energy input generated in this clearance can be reduced by minimizing the flight width, e, and screw to barrel clearance, δ. However, as the following example shows, increasing the clearance will not only decrease the temperature rise but also the film heat transfer coefficient h_i. In this example h_i is estimated as the conduction through a stagnant film.

Calculations for a typical radial clearance for a 3 inch extruder with a radial clearance of .003 inch are indicated in bold in Table 2. These calculations show that increasing the clearance, δ, from .003 to .015 inch reduces the temperature rise by about 93% but the film coefficient, h_i, is reduced by a lesser amount of 80%. Therefore, based only on the clearance temperature generation there

Table 7.2 Temperature Generation in Flight / Barrel Clearance

Extruder dimensions	Polymer Properties
$D = 3'' = 7.62\,cm$	$k = .96 \frac{BTU\text{-}in}{ft^2\text{-}Hr\text{-}°F}$
$N = 60$ RPM $= 1$ Rev/sec	$n = .66$
$e = .25'' = .635$ cm	$\eta = 15{,}063$
$\theta = 17.7$	$C_p = .55$ kcal/kg-°C $\varrho = .75$ g/cc

δ (in)	$\dot{\gamma}$ (sec^{-1})	η (poise)	ΔT, °C	h_i
.003	**3,141**	**1,056**	**30.0**	**320**
.005	1,885	1,245	12.9	192
.010	942	1,565	4.1	96
.015	628	1,790	2.1	64
.020	471	1,968	1.3	48

should be a net temperature reduction by increasing the clearance. However, this does not address the heat removal from the main part of the channel in the screw. As previously stated, extruders, particularly for low viscosity polymers, are designed with small clearances to enhance heat transfer by increasing h_i. Optimization of the overall heat generation and transfer is very complex. It involves the polymer rheology, reaction details, and screw design. The overall problem, to this writer's knowledge, has not yet been numerically accomplished.

7.9 Scale-Up

When scaling-up from laboratory or pilot plant heat transfer can often be the limiting factor. The following equations summarize scaling factors:

Capacity:
$$\frac{Q_2}{Q_1} \propto \left(\frac{N_2}{N_1}\right) \times \left(\frac{D_2}{D_1}\right)^3 \tag{16}$$

Power:
$$\frac{E_2}{E_1} \propto \left(\frac{N_2}{N_1}\right)^2 \times \left(\frac{D_2}{D_1}\right)^3 \tag{17}$$

Heat Transfer Area:
$$\frac{A_2}{A_1} \propto \left(\frac{D_2}{D_1}\right)^2 \tag{18}$$

An inspection of these equations shows that the conveying capacity and energy generation increase with D^3 (for $N_2 = N_1$) whereas the ability to remove the generated heat increases with only D^2. It is for this reason that manufacturers often give $D^{2.5}$ as the scale-up rule since the screw speed is usually reduced with increasing size. A common rule for adjusting the screw speed on scale-up is when this rule is used the energy generation to surface area remains constant on scale-up basis.

$$N_2 = N_1 \left(\frac{D_2}{D_1}\right)^{-.5} \tag{19}$$

Using Eq. 16 will almost always give an overly optimistic projection of capacity for the larger machine. In fact, the more conservative use of area can also be misleading and optimistic if the barrel heat exchange designs are radically different. When scaling-up a thermally sensitive process it is necessary to analyze the overall heat transfer capabilities of the extruders.

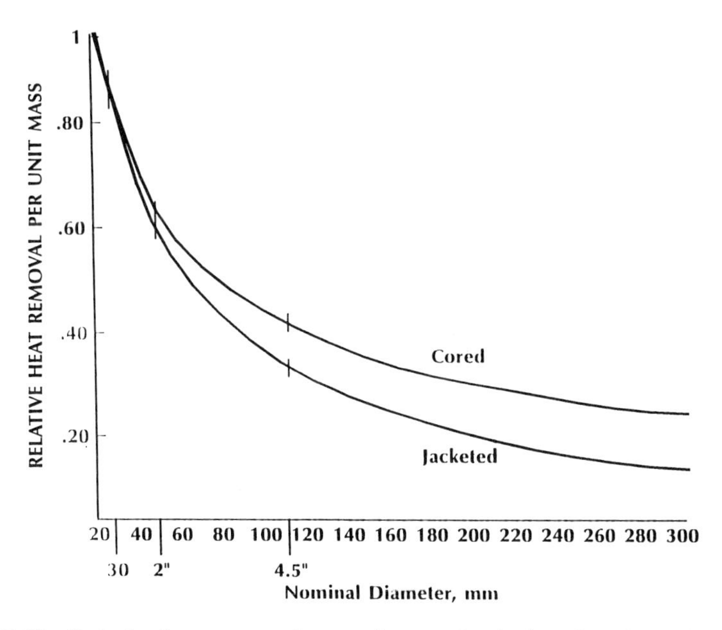

Figure 7.18 Relative heat removal per unit mass for jacketed and cored extruders

Fig. 18 gives the calculated relative heat removal per unit mass using a 20 mm laboratory scale extruder as a basis. As can be seen the difference in heat removal capability from 20 mm to 50 mm is about 40%. Attempting to scale-up from a 20 or 30 mm extruder to commercial sizes is not sound practice as the capacity of the larger extruder will usually be overstated. Often times reactions which are feasible in a small laboratory scale extruder may prove to be not so in a larger machine because of the greater temperature distribution and resulting higher bulk temperatures in the larger machine. It is probably best to use a 50 mm extruder as the smallest size from which one should scale-up to a commercial process. Also shown in Fig. 18 is the difference in heat removal between cored and jacketed barrels which is most significant for the larger sized machine.

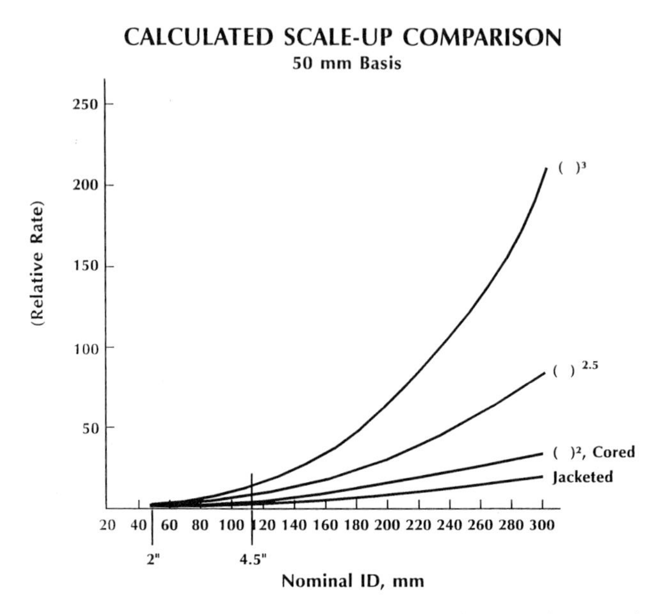

Figure 7.19 Projected capacities of extruders for various scale-up rules and barrel construction

The difference in projected capacities is best shown in Fig. 19 which uses a 50 mm extruder as a basis. This shows that use of the diameter cubed will grossly overestimate the capacity of a commercial scale extruder in comparison to the other rules. As previously stated a detailed analysis of the heat transfer capability is usually needed for any scale-up projection for extruder reactors.

Scale-Up Example

Following is an example which has been inspired by reality but is in fact hypothetical. The situation being addressed is scale-up from a 2½ inch pilot extruder to a 6 inch commercial extruder for a temperature sensitive process.

Pilot Line Information:
Dia. $2^1/_2''$
L/D = 15
Capacity = 250 Ibs/Hr
Polymer feed temperature = 80 °F
Process temperature = 400 °F
Power requirement = 9 lbs/Hr/HP

Calculate the cooling/heating load for the pilot line situation.

Heat Capacity = .6 BTU/lb-°F

Adiabatic temperature rise:

$$\Delta T = \frac{42.44 \frac{\text{BTU/min}}{\text{HP}} \times 60}{9 \frac{\text{lb/Hr}}{\text{HP}} \times .6 \frac{\text{BTU}}{\text{IB-°F}}}$$

Therefore energy must be removed through the extruder walls and design of an efficient heat removal system will probably be especially crucial for the scale-up process. On a unit mass basis the energy to be removed is

$$\Delta H = 250 \frac{\text{lbs}}{\text{Hr}} \times .6 \frac{\text{BTU}}{\text{lb-°F}} \times (552 - 400)\,°\text{F}$$

$$= 22{,}800 \text{ BTU/Hr}$$

or

$$\Rightarrow 91.2 \frac{\text{BTU}}{\text{lb}}$$

Using a scale-up exponent of 2.5 the projected capacity of a 6 inch extruder is

$$Q = 250 \times (6/2.5)^{2.5} = 2{,}231 \text{ lb/Hr}$$

Therefore, the required heat removal from the process is

$$\Delta H = 91.2 \frac{\text{BTU}}{\text{lb}} \times 2{,}231 \frac{\text{lb}}{\text{Hr}} = 203{,}452 \frac{\text{BTU}}{\text{Hr}}$$

The initial extruder design proposed by the extruder vendor is

Stainless steel liner: $1^1/_2''$ thick $K_1 = 7 \dfrac{\text{BTU-ft}}{\text{ft}^2\text{-Hr-°F}}$

Carbon Steel Barrel: $2''$ thick $K_3 = 30 \dfrac{\text{BTU-ft}}{\text{ft}^2\text{-Hr-°F}}$

Therefore, the various dimensions to be used for heat transfer calculations are:

$$D_1 = 6'' = .5$$

$$D_2 = 6'' + 2 \times 1.5'' = 9'' = .75'$$

$$D_3 = 9'' + 2 \times 2'' = 13'' = 1.08'$$

$$L = 15 \times {}^{6''}/_{12''} = 7.5'$$

Since the initial design is for a jacketed extruder and in any case part of the objective is to design the heat removal system duty, the overall heat transfer coefficient will be calculated on the OD:

$$U = \frac{1}{\dfrac{D_3 \ln(D_3/D_2)}{2K_3} + \dfrac{D_3 \ln(D_2/D_1)}{2K_1} + \dfrac{D_3}{D_1 h_i} + \dfrac{1}{h_o}}$$

Based on pilot line measurements a film coefficient of 60 will be used. It will be assumed that the coolant circulation system can be properly designed so that $h_o \geq 200$.

Then the overall coefficient for this design becomes

$$U = \frac{1}{\dfrac{1.08 \ln(1.08/.75)}{2 \times 30} + \dfrac{1.08 \ln(.75/.5)}{2 \times 7} + \dfrac{1.08}{.5 \times 60} + \dfrac{1}{200}}$$

$$= 12.7 \frac{\text{BTU}}{\text{ft}^2\text{-Hr-°F}}$$

Calculate the necessary coolant temperature:

$$\Delta H = U \times (\pi D_3) \times L \times (T_{process} - T_{coolant})$$

where $\Delta T = T_{process} - T_{coolant}$

$$\Delta T = 630\,°F$$

Therefore the required coolant temperature is

$$T_{coolant} = 400\,°F - 630\,°F = -230\,°F$$

This is impractical from the viewpoint of cost of the refrigeration system, installing operating cost, and process.

In order to move the heat transfer fluid in closer to the barrel ID jackets will be replaced by a set of cores in the outer barrel spaced on a 10.5 inch diameter.

The inner perimeter of the cores will be used as the diametrical basis of U:

$$U = \cfrac{1}{\cfrac{.875 \ln(.875/.75)}{2 \times 30} + \cfrac{.875 \ln(.75/.5)}{2 \times 7} + \cfrac{.875}{.5 \times 60} + \cfrac{1}{200}}$$

$$= \frac{1}{.00225 + .0253 + .0291 + .0050}$$

$$= 16.2\ \text{BTU/ft}^2\text{-Hr-°F}$$

It should be noted from the expression for U that the thermal resistance of the outer barrel is the smallest component. Therefore, the exact diameter assumed for the coring is not extremely important. The ΔT between the coolant and process required for this barrel design is:

$$\Delta T = \frac{203{,}452}{\pi \times .875 \times 16.2 \times 7.5} = 609\,°F$$

$$T_{coolant} = (400 - 609)\,°F = -209\,°F$$

This is still much too cold to be practical.

An inspection of the expression for U shows that the most significant components are the thermal resistance from the stainless steel liner and film coefficient. A next approach could be to make the liner as thin as possible. Suppose a mechanical strength analysis shows that the liner thickness could be reduced to $^7/_8$ inch and the coring in the outer barrel maintained with a $^3/_4$ inch spacing between the barrel OD and the nearest core location.

Then the pertinent dimensions become:

$$D_1 = 6'' = .5'$$

$$D_2 = 6'' + 2 \times {}^7/_8'' = 7.75'' = .65'$$

$$D_3 = 7.75'' + 2 \times 75'' = 9.25'' = .77'$$

The heat transfer coefficient is then:

$$U = \frac{1}{\frac{.77\ln(.77/.65)}{2 \times 30} + \frac{.77\ln(.65/.5)}{2 \times 7} + \frac{.77}{.5 \times 60} + \frac{1}{200}}$$

$$= 21.2 \frac{\text{BTU}}{\text{ft}^2\text{-Hr-}^\circ\text{F}}$$

Therefore,

$$\Delta T = \frac{203{,}452}{\pi \times .77 \times 7.5 \times 21.2} = 529\,^\circ\text{F}$$

$$T_{coolant} = 400 - 529 = -129\,^\circ\text{F}$$

This temperature is still too low to be practical.

As has been previously discussed carbon steel has much better conductivity than stainless. Therefore, the next logical construction alternative is to substitute a carbon steel liner for the stainless steel one. A bimetallic coating can be used on the liner ID for corrosion and wear protection. Since these coatings are only about $^1/_{16}$ to $^1/_8$ inch thick their effect on overall conductivity is negligible. If the air gap is neglected the expression for U becomes:

$$U = \frac{1}{\frac{.77\ln(.77/.5)}{2 \times 30} + \frac{.77}{.5 \times 60} + \frac{1}{200}}$$

$$= 27.6 \frac{\text{BTU}}{\text{ft}^2\text{-Hr-}^\circ\text{F}}$$

$$\Delta T = \frac{203{,}452}{\pi \times .77 \times 7.5 \times 27.6} = 406\,^\circ\text{F}$$

$$T_{coolant} = (400 - 406)\,^\circ\text{F} = -6\,^\circ\text{F}$$

This temperature is achievable with a chiller. However, chillers can be expensive to purchase and operate. Suppose the pilot plant results show that the film coefficient, h_i, ranges up to 80. It may then be worthwhile to calculate U with this h_i:

$$U = \frac{1}{\frac{.77\ln(.77/.5)}{2 \times 30} + \frac{.77}{.5 \times 80} + \frac{1}{200}}$$

$$= 33.6 \frac{\text{BTU}}{\text{ft}^2\text{-Hr-}^\circ\text{F}}$$

$$\Delta T = \frac{203{,}452}{\pi \times .77 \times 7.5 \times 33.6} = 334\,^\circ\text{F}$$

$$T_{coolant} = 400 - 334 = -66\,^\circ\text{F}$$

This is a reasonable temperature for the coolant but still requires a chiller.

Suppose a chilled system of 50 °F is available; calculations can then be done for the two cases of h_i to evaluate the effect on capacity the results of which are given in the following table.

Table 7.3 Capacity Calculations for the Two h_i's Considered for 50 °F Chilled Coolant.

h_i	U	ΔH	Capacity, lbs/Hr
60	27.6	175,258	1,921
80	33.6	213,358	2,339

Therefore, if the 50 °F chiller is available, it seems reasonable to use this as a starting point.

If the capacity proves to be deficient the 50 °F chiller can be replaced or debottlenecked. Of course, the heat transfer fluid must be carefully selected to avoid significant decreases to the shell side coefficient, h_o. Decreases in temperature will increase the visocity of the heat transfer fluid and can significantly decrease U.

Symbols

C_p = Heat capacity
D = Nominal diameter
D_1 = Inner diameter
D_2 = Mid diameter
D_3 = Outer diameter
E = Drive motor energy
e = Flight width
E_f = Energy generation in flight OD to barrel ID clearance
f = Geometrically determined factor
H = Channel depth
ΔH_t = Heat of transition
h_i = Polymer film heat transfer coefficient
h_o = Shell side heat transfer coefficient
K_1 = Thermal conductivity of inner barrel
K_2 = Thermal conductivity of outer barrel
K_{air} = Thermal conductivity of air
L = Length
M = Mass
N = Screw RPM
Q = Volumetric flow rate
q = Heat flux
q_b = Heat transferred through barrel
Q_f = Volumetric rate of material dragged under flight
q_r = Heat of reaction
T_f = Feed temperature
T_i = Film temperature
T_o = "Shell" side temperature
T_p = Product temperature
U_i = Overall heat transfer coefficient based on the ID
U_o = Overall heat transfer coefficient based on the OD
V_f = Sliding velocity of flight
W = Channel width
$\dot{\gamma}$ = Shear rate
δ = Screw OD to barrel ID radial clearance
η = Power law viscosity proportionality constant
θ = Helix angle
ϱ = Density
τ = Shear stress
μ = Viscosity

References

Abdou-Sabet, S., and M.A. Fath, US 4,311,628, Monsanto, (1982).
Abdou-Sabet, S., and K.-S. Shen, US 4,594,390, Monsanto, (1986).
Abe, K., S. Yamauchi, and A. Ohkubo, US 4,460,743, Mitsubishi Petrochemical, (1984).
Abe, H., T. Nishio, Y. Suzuki, T. Sanada, S. Hosoda, and T. Okada, EPA 295,103, Sumitomo Chemical, (1988).
Abe, H., T. Nishio, Y. Suzuki, and T. Sanada, EPA 308,255, Sumitomo Chemical, (1989).
Adachi, T., and S. Mizuno, US 4,305,863, Dainippon Ink and Chemicals, (1981).
Agarwal, P.K., I. Duvdevani, D.G. Peiffer, and R.D. Lundberg, *J. Polym. Sci.: Part B: Polym. Phys. Ed., 25,* 839 (1987).
Aharoni, S.M., and T. Largman, US 4,417,031, Allied, (1983a).
Aharoni, S.M., and T. Largman, US 4,390,667, Allied, (1983b).
Aharoni, S.M., W.B. Hammond, J.S. Szobota, and D. Masilamani, *J. Polym. Sci., Polym. Chem. Ed., 22,* 2567 (1984a).
Aharoni, S.M., W.B. Hammond, J.S. Szobota, and D. Masilamani, *J. Polym. Sci., Polym. Chem. Ed., 22,* 2579 (1984b).
Aharoni, S.M., and D. Masilamani, US 4,568,720, Allied, (1986).
Aharoni, S.M., C.E. Forbes, W.B. Hammond, D.M. Hindenlang, F. Mares, K. O'Brien, and R.D. Sedgwick, *J. Polym. Sci.: Part A: Polym. Chem. Ed., 24,* 1281 (1986).
Ajiboye, O., and G. Scott, *Polym. Degradation Stab., 4,* 415 (1982).
Akaboshi, M., K. Uragami, E. Fukita, and A. Fukai, US 3,072,624, Kurashiki Rayon, (1963).
Akkapeddi, M.K., and J. Gervasi, *Polym. Prepr. Am. Chem. Soc. Div. Polym. Chem., 29 (1),* 567 (1988).
Akkapeddi, M.K., B. Van Buskirk, and A.C. Brown, PCT Intl. Appl. WO 88/08433, Allied-Signal, (1988).
Akkapeddi, M.K., J.C. Haylock, and J.A. Gervasi, US 4,847,322, Allied-Signal, (1989).
Aldred, A.C.G., and S.G. Fogg, US 4,591,615, British Petroleum, (1986).
Al-Malaika, S., *Polym. Prepr. Am. Chem. Soc. Div. Polym. Chem., 29 (1),* 555 (1988).
Al-Malaika, S., S. Honggokusumo, and G. Scott, *Polym. Degradation Stab., 16,* 25 (1986).
Andersen, P.G., US 4,476,283, Uniroyal, (1984).
Andersen, P.G., and M.J. Kenny, *Soc. Plast. Eng. RETEC "Polyolefins VI" Tech. Pap.,* p.156, Houston, Texas, Febr. 26-28, 1989.
Anonymous, GB 998,439, Du Pont, (1965).
Anonymous, GB 1,043,082, Chemische Werke Hüls, (1966).
Anonymous, GB 1,168,959, Allied Chemical, (1969a).
Anonymous, GB 1,141,118, Toyo Rayon, (1969b).
Anonymous, GB 1,217,231, Asahi, (1970).
Anonymous, GB 1,253,632, Toray Industries, (1971).
Anonymous, GB 1,442,681, Chemie Linz, (1976).
Anonymous, GB 1,506,430, BASF, (1978).
Anonymous, *Plast. Technol., 31,* 8, 17 (1985a).
Anonymous, *Plast. World, 43,* 9, 85 (1985b).
Anonymous, *Mod. Plast., 63,* 6, 15 (1986).
Aoyama, T., H. Okasaka, and H. Kodama, US 4,010,219, Toray Industries, (1977).

Arai, Y., US 4,459,391, Unitika, (1984).
Arai, Y., and T. Tanaka, US 4,459,392, Unitika, (1984a).
Arai, Y., and T. Tanaka, US 4,459,390, Unitika, (1984b).
Arai, Y., and T. Tanaka, US 4,520,175, Unitika, (1985a).
Arai, Y., and T. Tanaka, US 4,520,174, Unitika, (1985b).
Arai, Y., T. Tanaka, and K. Kamiyama, US 4,529,779, Unitika, (1985).
Armstrong, R.G., US 3,373,222, Continental Can, (1968a).
Armstrong, R.G., US 3,373,223, Continental Can, (1968b).
Atochem, Product Bulletin "Half-life: Peroxide Selection Based on Half-Life" , Buffalo, N.Y., 1990.
Aycock, D.F., and S.-P. Ting, US 4,600,741, General Electric, (1986).
Aycock, D.F., and S.-P. Ting, US 4,642,358, General Electric, (1987).
Baba, K., T. Shiota, K. Murakami, and K. Ono, US 3,887,534, Sumitomo, (1975).
Baker, W.E., and M. Saleem, *Polym. Eng. Sci., 27,* 1634 (1987a).
Baker, W.E., and M. Saleem, *Polymer, 28,* 2057 (1987b).
Baldwin, F. P., *Rubber Chem. Technol., 52,* 677 (1979).
Ban, L.L., M.J. Doyle, M.M. Disko, and G.R. Smith, *Polym. Commun., 29,* 163 (1988).
Ban, L.L., M.J. Doyle, M.M. Disko, G. Braun, and G.R. Smith in *Integration of Fundamental Polymer Science and Technology,* P.J. Lemstra and L.A. Kleintjens, eds., Elsevier Applied Science, New York, 1989, p. 117.
Banucci, E.C., and G.A. Mellinger, US 4,073,773, General Electric, (1978).
Barnabeo, A.E., US 4,551,504, Union Carbide, (1985).
Barnewall, J.M., and A.S. Scheibelhoffer, US 3,975,329, Goodyear Tire and Rubber, (1976).
Bartilla, T., D. Kirch, J. Nordmeier, E. Prömper, and Th. Strauch, *Adv. Polym. Technol., 6,* 339 (1986).
Beauxis, J. and R.C. Kowalski, US 3,593,011, Esso Res. & Eng. Co. (1971).
Becker, R., and W. Rohr, EPA 289,925, BASF, (1988).
Beers, M.D., H. Chura, and R.J. Robilard, US 3,960,802, General Electric, (1976).
Belz, R.K., EPA 143,894, (1985).
Belz, R.K., US 4,612,355, RB Kunststoffpatent-Verwertungs, (1986).
Benedetti, E., F. Posar, A. D'Alessio, P. Vergamini, M. Aglietto, G. Ruggeri, and F. Ciardelli, *Brit. Polym., J., 17,* 34 (1985).
Benedetti, E., A. D'Alessio, M. Aglietto, G. Ruggeri, P. Vergamini, and F. Ciardelli, *Polym. Eng. Sci., 26,* 9 (1986).
Berardinelli, F.M., and R. Edelman, US 4,052,360, Celanese, (1977).
Berghaus, U., and W. Michaeli, *Soc. Plast. Eng. ANTEC Tech. Pap., 36,* 1929 (1990).
Bergström, C., and T.H. Palmgren, EPA 247,877, Neste Oy, (1987).
Bergström, C., J. Brenner, and P. Stenius, *J. Appl. Polym. Sci., 23,* 3653 (1979).
Bergström, C., B. Björkvall, B. Stenlund, J. Huttunen, and C.-J. Johansson, US 4,746,704, Neste Oy, (1988).
Bernhardt, E. C., *Processing of Thermoplastic Materials,* Reinhold, NY (1959).
Biensan, M., and P. Potin, US 4,067,861, Ato Chimie, (1978).
Biesenberger, J. A., Ed., *Devolatilization of Polymers,* Hanser Publ., Munich (1983).
Biesenberger,J.A., and D.H. Sebastian, *Principles of Polymerization Engineering,* John Wiley, New York (1983).
Biesenberger, J.A., S.K. Dey, and J. Brizzolara, *Polym. Eng. Sci., 30,* 1493 (1990).
Bigio, D., and L. Erwin, *Soc. Plast. Eng. ANTEC Tech. Pap., 30,* 45, (1984).
Bigio, D., and S. Zerafadi, Proc. 4th Ann. *Polym. Processing Soc.* Meeting, section 4, p. 6, Osaka, Japan, May 1988.
Binsack, R., D. Rempel, H. Korber, and D. Neuray, US 4,260,690, Bayer, (1981).

Blondel, P., and C. Jungblut, US 4,595,730, Atochem, (1986).
Bloor, R., *Plast. Techol., 27,* 2, 83 (1981).
Bodolus, C.L., and W.J. Miloscia, US 4,463,137, Standard Oil, (1984).
Bodolus, C.L., and D.A. Woodhead, US 4,542,189, Standard Oil, (1985).
Bodolus, C.L., and D.A. Woodhead, US 4,692,498, Standard Oil, (1987).
Boivin, D.W., and R.A. Zelonka, US 4,678,834, Du Pont Canada, (1987).
Boocock, J.R., US 4,801,656, Du Pont Canada, (1989).
Booy, M.L., *Polym. Eng. Sci., 18,* 973 (1978).
Borggreve, R.J.M., and R.J. Gaymans, *Polymer 30,* 63 (1989a).
Borggreve, R.J.M., and R.J. Gaymans, *Polymer 30,* 71 (1989b).
Borggreve, R.J.M., and R.J. Gaymans, *Polymer 30,* 78 (1989c).
Borggreve, R.J.M., R.J. Gaymans, and A.R. Luttmer, *Makromol. Chem., Makromol. Symp., 16,* 195 (1988a).
Borggreve, R.J.M., R.J. Gaymans, J. Schuijer, and J.F. Ingen Housz, *Polymer 28,* 1489 (1988b).
Borman, W.F.H., and J.A. Rock, US 4,020,122, General Electric (1977).
Bouilloux, A., J. Druz, and M. Lambla, *Polym. Process Eng.,3,* 235 (1986).
Bourland, L.G., M.E. London, and T.A. Cooper, *Proc. RAPRA Tech. Semin. "Reactive Processing - Practice and Possibilities",* Shawbury, England, Sept 11-12, 1989.
Bratawidjaja, A.S., I. Gitopadmoyo, Y. Watanabe, and T. Hatakeyama, *J. Appl. Polym. Sci., 37,* 1141 (1989).
Brandstetter, F., J. Hambrecht, H. Reimann, R. Pflueger and K. Ruppmich, German 3,120,803, B.A.S.F., A.G. (1982a).
Brandstetter, F., J. Hambrecht and B. Scharf, EP 062,838, B.A.S.F., A.G. (1982b).
Braun, D., and W. Illing, *Angew. Makromol. Chem., 154,* 179 (1987).
Brown, G.D., US 4,564,349, Union Carbide, (1986).
Brown, S.B., and D.J. McFay, *Polym. Prepr. Am. Chem. Soc. Div. Polym. Chem., 27* (1), 333 (1986).
Brown, S.B., and D.J. McFay, US 4,680,329, General Electric, (1987).
Brown, S.B., and C.M. Orlando, "Reactive Extrusion" in *Encycl. Polym. Sci. Eng.,* J.I. Kroschwitz, ed., Wiley, New York, *14,* 169 (1988).
Bruzzone, M., S. Gordini, and K. Wyllie, US 4,714,747, Enichem Elastomeri, (1987).
Bucknall, C.B., *Toughened Plastics,* Applied Science Publishers, London, 1977.
Buntin, R.R., J.P. Keller, and J.W. Harding, US 3,849,241, Exxon, (1974).
Buntin, R.R., J.P. Keller, and J.W. Harding, US 3,978,185, Exxon, (1976).
Bush, J.L., and C.W. Milligan, US 3,649,578, Du Pont, (1972).
Callais, P.A., *Proc. First Intern. Congress "Compalloy '89",* p. 247, New Orleans, La, April 5-7, 1989.
Campbell, J.R., P.M. Conroy, and R.A. Florence, *Polym. Prepr. Am. Chem. Soc. Div. Polym. Chem., 27* (1), 331 (1986).
Campbell, J.R., R.E. Williams, S.B. Brown, P.M. Conroy, and R.A. Florence, US 4,840,982, General Electric (1989).
Carbonaro, A., S. Gordini, and S. Cucinella, EPA 127,236, Enichem Polimeri, (1984).
Carlsson, D.J., and D.M. Wiles, "Degradation" , in *Encycl. Polym. Sci. Eng.,* J.I. Kroschwitz, ed.,Wiley, New York, *4,* 630 (1986).
Cartasegna, S., *Rubber Chem. Technol., 49,* 722 (1986).
Casale, A., and R.S. Porter, *Adv. Polym. Sci., 17,* 1 (1975).
Casale, A., and R.S. Porter, *Polymer Stress Reactions,* Academic Press, New York, vol. 1, 1978, p. 226 ff.
Castagna, E.G., A Schrage, and M.F. Repiscak, US 3,940,379, Dart Industries, (1976).
Caywood, jr., S.W., US 3,884,882, Du Pont, (1975).

Caywood, jr., S.W., US 4,010,223, Du Pont, (1977).
Chan, C.M., and A.C. Tanous, US 4,616,056, Raychem, (1986).
Chan, C.-M., and S. Venkatraman, *J. Appl. Polym. Sci., 32,* 5933 (1986).
Charles, J.J., and R.C. Gasman, US 4,246,377, GAF, (1981).
Chavan, V. V., *AIChE J., 29,* 177 (1983).
Chella, R., and J.M. Ottino, *ACS Symp. Ser., 196,* 567 (1981).
Chella, R., and J.M. Ottino, *I.E.C. Fund., 24,* 170 (1985).
Chen, I. M., and C. M. Shiah, *Plast. Eng., 45,* 10, 33 (1989).
Cheung, M.-F., A. Golovoy, R.O. Carter, and H. van Oene, *Ind. Eng. Chem. Res., 28,* 476 (1989).
Chiang, W.-Y., and W.-D. Yang, *J. Appl. Polym. Sci., 35,* 807 (1988).
Chohan. R.K., B. David, A. Nir and Z. Tadmor, *Intern. Polym. Processing, 2,* 1 (1987).
Choudhury, N.R., and A.K. Bhowmick, *J. Appl. Polym. Sci., 38,* 1091 (1989).
Christensen, D.C., and R.G. Voss, US 4,532,310, (1985).
Cimmino, S., L. D'Orazio, R. Greco, G. Maglio, M. Malinconico, C. Mancarella, E. Matuscelli, R. Palumbo, and G. Ragosta, *Polym. Eng. Sci., 24,* 48 (1984).
Cimmino, S., F. Coppola, L. D'Orazio, R. Greco, G. Maglio, M. Malinconico, C. Mancarella, E. Matuscelli, and G. Ragosta, *Polymer, 27,* 1874 (1986).
Clarke, C.M., US 3,318,848, Celanese, (1967).
Clementini, L., and L. Spagnoli, US 4,578,428, Montedison, (1986).
Coleman, E.A., US 4,195,134, GAF, (1980).
Colombo, E.A., T.H. Kwack, and T.-K. Su, US 4,614,764, Mobil Oil, (1986).
Coran, A.Y., and R.P. Patel, US 4,104,210, Monsanto, (1978a).
Coran, A.Y., and R.P. Patel, US 4,116,914, Monsanto, (1978b).
Coran, A.Y., and R.P. Patel, US 4,130,534, Monsanto, (1978c).
Coran, A.Y., B. Das, and R.P. Patel, US 4,130,535, Monsanto, (1978).
Coran, A.Y., and R.P. Patel, US 4,141,863, Monsanto, (1979a).
Coran, A.Y., and R.P. Patel, US 4,141,878, Monsanto, (1979b).
Coran, A.Y., and R.P. Patel, US 4,173,556, Monsanto, (1979c).
Coran, A.Y., and R. Patel, US 4,207,404, Monsanto, (1980a).
Coran, A.Y., and R. Patel, *Rubber Chem. Technol., 53,* 141 (1980b).
Coran, A.Y., and R. Patel, *Rubber Chem. Technol., 53,* 781 (1980c).
Coran, A.Y., and R. Patel, US 4,299,931, Monsanto, (1981a).
Coran, A.Y., and R.P. Patel, US 4,271,049, Monsanto, (1981b).
Coran, A.Y., and R.P. Patel, US 4,287,324, Monsanto, (1981c).
Coran, A.Y., and R.P. Patel, US 4,288,570, Monsanto, (1981d).
Coran, A.Y., and R. Patel, *Rubber Chem. Technol., 54,* 892 (1981e).
Coran, A.Y., and R. Patel, US 4,338,413, Monsanto, (1982a).
Coran, A.Y., and R. Patel, US 4,355,139, Monsanto, (1982b).
Coran, A.Y., R.P. Patel, and D. Williams, *Rubber Chem. Technol., 55,* 116 (1982).
Coran, A.Y., and R. Patel, *Rubber Chem. Technol., 56,* 210 (1983a).
Coran, A.Y., and R. Patel, *Rubber Chem. Technol., 56,* 1045 (1983b).
Coran, A.Y., R. Patel, and D. Williams-Headd, *Rubber Chem. Technol., 58,* 1014 (1985).
Corbett, J.M., and C.R. Bearden, US 4,454,086, Dow, (1984).
Cordes, C., and H.-J. Sterzel, US 4,064,103, BASF, (1977).
Costanza, J.B., and F.M. Berardinelli, GB 1,578,472, Celanese, (1980).
Crespy, A., B. Joncour, J.P. Prevost, J.P. Cavrot, and C. Caze, *Eur. Polym. J., 22,* 505 (1986).
Curry J., S. Jackson, B. Stoehrer and A. Van der Veen, *Chem. Eng. Progr., 84,* 11, 43 (1988).
Curto, D., A. Valenza, and F.P. La Mantia, *J. Appl. Polym. Sci., 39,* 865 (1990).

Dagli, S.S., M. Xanthos,and J.A. Biesenberger, *Proc. 4th Intern. Congress "Compalloy '91"*, p. 255, New Orleans, La., Jan. 30- Febr. 1, 1991.
Davis, J.H., GB 1,403,797, Imperial Chemical Industries, (1975).
Davis, W.M., *Chem. Eng. Progr.*, 84, 11, 35 (1988).
Davison, S., US 4,578,430, Shell Oil Co., (1986).
Dean, B.D., *J. Elastomers Plast., 17,* 55 (1985).
Dean, B.D., US 4,568,724, Atlantic Richfield, (1986).
Deibig, H., and R.K. Belz, EPA 143,935, Belland and R.K. Belz, (1985).
den Otter, J.L., U.S. Pat. 4,483,257, Nederlandse Centrale Organisatie voor Toegepast-Natuurwetenschappelijk Onderzoek, (1984).
Dey, S.K., and J.A. Biesenberger, *Soc. Plast. Eng. ANTEC Tech. Pap., 33,* 133 (1987)
Dijkstra, A.J., I. Goodman, and J.A.W. Reid, US 3,533,157, Imperial Chemical Industries, (1971).
D'Orazio, L., C. Mancarella, and E. Martuscelli, *J. Mater. Sci., 23,* 161 (1988).
Dorn, M., *Adv. Polym. Technol., 5,* 87 (1985).
Dow Chemical Co., Technical Bulletin, "Reactive Polystyrene" , (1985).
Dreiblatt, A., H. Herrmann, and H.J. Nettelnbreker, *Plast. Eng.,43,* 10, 31, (1987).
Droescher, M., S. Mumcu, K. Burzin, C. Gerth, and H. Heuer, US 4,555,550, Chemische Werke Huels, (1985a).
Droescher, M., K. Burzin, C. Gerth, and H. Heuer, US 4,550,148, Chemische Werke Huels, (1985b).
Du, C.C., Ph.D. Dissertation, Stevens Institute of Technology, Hoboken, N.J. (1987).
Dunphy, J.F., US 4,851,473, Du Pont, (1989).
Dyer, R.F., and A.K. Meyer, US 3,808,302, Eastman Kodak, (1974).
Edgar, O.B., GB 1,162,691, Imperial Chemical Industries, (1969).
Edwards, D.C., and D. Padliya, EPA 180,444, Polysar, (1986).
Eguiazaba, J.I., and J. Nazabal, *Makromol. Chem., Macromol. Symp. 20/21,* 255 (1988).
Ehrig, R.J., and R.C. Weil, US 4,707,524, Aristech, (1987).
Eise, K., *Plast. Compounding,* January/February, 1986, p. 44.
Engler, D.A., and A.R. Maistrovich, EPA 160,397, Minnesota Mining and Manufacturing, (1985).
Epstein, B.N., US 4,174,358, Du Pont, (1979a).
Epstein, B.N., US 4,172,859, Du Pont, (1979b).
Epstein, B.N., and R.U. Pagilagan, US 4,410,661, Du Pont, (1983).
Erwin, L, *Soc. Plast. Eng. ANTEC Tech. Pap., 24,* 488, (1978a).
Erwin, L. *Polym. Eng. Sci., 18,* 738 (1978b).
Erwin, L. and F. Mokhtarian, *Polym. Eng. Sci., 23,* 49 (1983).
Fisher, G.J., F. Brown, and W.E. Heinz, US 3,254,053, Celanese, (1966).
Flood, J.C., and D.A. Plank, US 3,996,310, Exxon, (1976).
Follows, G.W., C.G. Hart, and J. Massey, US 4,591,468, Imperial Chemical Industries, (1986).
Fowler, M.W., and W.E. Baker, *Polym. Eng. Sci., 28,* 1427 (1988)
Fox, D.W., and R.B. Allen, "Compatibility" in *Encycl. Polym. Sci. Eng.*, J.I. Kroschwitz, ed.,Wiley, New York, *3,* 772 (1985).
Freitag, D., L. Bottenbruch, and M. Schmidt, US 4,533,702, Bayer, (1985).
Fritz, H.-G., and B. Stöhrer, *Int. Polym. Process., 1,* 31 (1986).
Frund, jr., Z.N., *Plast. Compounding,* September/October, 1986, p. 24.
Fryc, B.F., K.A. Pigott, and J.H. Saunders, US 3,233,025, Mobay, (1966).
Fujii, S., and S.-P. Ting, US 4,728,461, General Electric, (1988).
Fujii, S., H. Ishida, M. Morioka, A. Saito, and R. van der Meer, US 4,873,276, General Electric, (1989).

Fujimoto, I., S. Isshiki, Y. Kurita, and Y. Sato, US 4,228,255, Fujikura Cable Works, (1980).
Fujita, Y., M. Sakuma, K. Kitano, M. Sakaizawa, Y. Yagi, N. Yamamoto, and T. Yokokura, EPA 235,876, Tonen Sekiyu, (1987).
Fukuda, K., and H. Kasahara, US 4,386,176, Asahi-Dow, (1983).
Fukui, O., Y. Inuizawa, S. Hinenoya, and Y. Takasaki, US 4,562,230, Ube Industries, (1985).
Fulger, C.V., H.D. Stahl, E.J. Turek, and R. Bayha, US 4,508,745, General Foods, (1985).
Gailus, D. W., and L. Erwin, *Soc. Plast. Eng. ANTEC Tech. Pap., 27,* 639, (1981).
Gale, G.M., and A.A. Sorio, EPA 163,865, Union Carbide, (1985).
Gallucci, R.R., US 4,632,962, General Electric, (1986).
Gallucci, R.R., US 4,654,401, General Electric, (1987).
Gallucci, R.R., and R.C. Going, *J. Appl. Polym. Sci., 27,* 425 (1982).
Gallucci, R.R., R. van der Meer, and R.W. Avakian, US 4,873,286, General Electric, (1989).
Garagnani, E., A. Moro, and R. Marzola, US 4,548,993, Montedison, (1985).
Gardner, I. J., J. V. Fusco, N. F. Newman, R. C. Kowalski, W. M. Davis, and F. P. Baldwin, US 4,702,901, Exxon Chemical Co. (1987).
Gattiglia, E., F.P. La Mantia, A. Tirturro, and A. Valenza, *Polym. Bull., 21,* 47 (1989).
Gaylord, N. G., 3, 708, 555, Gaylord Research Institute (1973).
Gaylord, N. G., *J. Macromol. Sci.-Rev. Macromol, Chem., C13,* 235 (1975).
Gaylord, N. G., 3, 956, 230, Champion International Corp. (1976).
Gaylord, N.G., US 4,071,494, Champion International, (1978).
Gaylord, N.G., US 4,506,056, Gaylord Research Institute, (1985).
Gaylord, N.G., *J. Macromol. Sci. Chem., A26,* 1211 (1989).
Gaylord, N.G., and S. Maiti, *J. Polym. Sci., Polym. Lett. Ed., 11,* 253 (1973).
Gaylord, N.G., H. Ender, L. Davis, and A. Takahashi in *Modification of Polymers,* C.E. Caraher and M. Tsuda, eds., ACS Symp. Ser. *121,* American Chemical Society, Washington, D.C., 1980, p. 469.
Gaylord, N. G., and J. Y. Koo, *J. Polym. Sci., Polym. Lett. Ed., 19,* 107 (1981).
Gaylord, N.G., and M. Mehta, *J. Polym. Sci.: Polym. Letters Ed., 20,* 481 (1982).
Gaylord, N.G., and R. Mehta, *J. Polym. Sci.: Part A: Polym. Chem., 26,* 1189 (1988).
Gaylord, N.G., M. Mehta and V. Kumar in *Modification of Polymers,* C.E. Caraher and J.A. Moore, eds., Plenum Press, New York, 1983, p. 171.
Gaylord, N. G., M. Mehta, V. Kumar and M. Tazi, *J. Appl. Polym. Sci., 38,* 359 (1989).
Gaylord, N.G., M. Mehta, and R. Mehta, *J. Appl. Polym. Sci., 33,* 2549 (1987).
Gaylord, N.G., and M.K. Mishra, *J. Polym. Sci.: Polym. Letters Ed., 21,* 23 (1983).
Gaylord, N.G., and A. Takahashi in *Modification of Polymers,* C.E. Caraher and J.A. Moore, eds., Plenum Press, New York, 1983, p. 183.
Gebauer, P., W. Käufer, H. Klinkenberg, and H. Söntgerath, US 4,178,277, Dynamit Nobel, (1979).
Gergen, W.P., and S. Davison, U.S. 4, 119, 607, Shell Oil (1978).
Gergen, W.P., R.G. Lutz, and R. Gelles, US 4,578,429, Shell Oil, (1986).
Ghaemy, M., and G. Scott, *Polym. Degradation Stab., 3,* 405 (1981).
Gillette, P.C., US 4,812,519, Hercules, (1989).
Gilliam, K.D., and E.E. Paschke, US 4,166,873, Standard Oil, (1979).
Gimpel, F., US 4,529,750, (1985).
Giroud-Abel, B., and J. Goletto, EPA 109,342, Rhone-Poulenc Specialites Chimiques, (1984).

Glander, F., and H.U. Voigt, US 4,289,860, Kabel- und Metallwerke Gutehoffnungshütte, (1981).
Golba, jr., J.C., and G.T. Seeger, *Plast. Eng., 43,* 3, 57 (1987).
Golder, M.D., US 3,839,267, Celanese, (1974a).
Golder, M.D., US 3,853,806, (1974b).
Golovoy, A., M.F. Cheung, and H. van Oene, *Polym. Eng. Sci., 27,* 1642 (1987).
Golovoy, A., M.-F. Cheung, K.R. Carduner, and M.J. Rokosz, *Polym. Eng. Sci., 29,* 1226 (1989).
Gorman, J.E., and J.A. Morris, US 4,822,857, Shell Oil, (1989).
Gouinlock, E.V., H.W. Marciniak, M.H. Shatz, E.J. Quinn, and R.R. Hindersinn, *J. Appl. Polym. Sci.,12,* 2403 (1968).
Goyert, W., E. Meisert, W. Grimm, A. Eitel, H. Wagner, G. Niederdellmann, and B. Quiring, US 4,261,946, Bayer, (1981).
Goyert, W., A. Awater, W. Grimm, K.-H. Ott, W. Oberkirch, and H. Wagner, US 4,317,890, Bayer, (1982).
Goyert, W., W. Grimm, A. Awater, H. Wagner, and B. Krüger, US 4,500,671, Bayer, (1985).
Goyert, W., J. Winkler, H. Perrey, and H. Heidingsfeld, US 4,762,884, Bayer, (1988).
Gozdz, A.S., and W. Trochimczuk, *J. Appl. Polym. Sci., 25,* 947 (1980).
Grabowski, T. S., U.S. 3, 130, 177, Borg-Warner (1964a).
Grabowski, T. S., U.S. 3, 134, 746, Borg-Warner (1964b).
Grant, T.S., and D.V. Howe, US 4,740,552, Borg-Warner, (1988).
Grant, T.S., R.L. Jalbert, and D. Whalen, US 4,732,938, Borg-Warner, (1988).
Greco, R., M. Malinconico, E. Matuscelli, G. Ragosta, and G. Scarinzi, *Polymer, 28,,* 1185 (1987).
Greco, R., P. Musto, F. Riva, and G. Maglio, *J. Appl. Polym. Sci., 37,* 789 (1989).
Greene, R.E., and E.T. Pieski, US 3,144,436, Du Pont, (1964).
Gregorian, R.S., US 3,182,033, W.R. Grace, (1965).
Grenci, J., D. Rosendale, and M. Xanthos, Unpublished Data, Polymer Processing Institute, Hoboken, N.J., (1989)
Hagger, J.M.R., B.G. Howell, and M.J. Poole, GB 2,202,537, BICC, (1988).
Hallden-Abberton, M.P., N.M. Bortnick, L.A. Cohen, W.T. Freed, and H.C. Fromuth, US 4,727,117, Rohm and Haas, (1988).
Hamersma, W.J.L.A., R.W. Avakian, and C.M.E. Bailly, EPA 274,140, General Electric, (1988).
Hammer, C.F., and H.K. Sinclair, US 3,972,961, Du Pont, (1976).
Hammer, C.F., and H.K. Sinclair, US 4,017,557, Du Pont, (1977).
Han, C.Y., and W.L. Gately, US 4,689,372, General Electric, (1987).
Harayama, H., K. Shinkai, H. Takahashi, and M. Mizusako, US 4,320,214, Sekisui, (1982).
Hartman, P.F., US 3,909,463, Allied Chemical, (1975).
Hathaway, S.J., and R.A. Pyles, US 4,732,934, General Electric, (1988).
Hathaway, S.J., and R.A. Pyles, US 4,800,218, General Electric, (1989).
Heinemeyer, B.W., and S.D. Tatum, US 4,612,156, Dow, (1986).
Henman, T.J., in *Developments in Polymer Stabilization-1,* G. Scott, ed., Applied Science Publishers, London, 1979, p.39.
Henman, T.J., in *Degradation and Stabilization of Polyolefins,* N.S. Allen, ed., Applied Science Publishers, New York, 1983, p. 29.
Hepp, L.R., EPA 149,192, General Electric, (1985).
Herberg, M.J., R.F. Macander, and T.R. Stegman, US 4,551,515, General Electric, (1985).

Hergenrother, W.L., and A.W. Greenstreet, EPA 155,995, Firestone Tire & Rubber, (1985).
Hergenrother, W.L., M.G. Matlock, and R.J. Ambrose, US 4,427,828, Firestone Tire & Rubber, (1984).
Hergenrother, W.L., M.G. Matlock, and R.J. Ambrose, US 4,508,874, Firestone Tire & Rubber, (1985).
Herman, J.T., K.L. Bryce, and G.M. Lancaster, US 4,420,580, Dow, (1983).
Herten, J.F., and B.D. Louies, EPA 170,790, Allied, (1986).
Hertlein, T., and H.-G. Fritz, *Kunstst.-Ger. Plast., 78,* 606 (1988).
Hobbs, S.Y., R.C. Bopp, and V.H. Watkins, *Polym. Eng. Sci., 23,* 380 (1983).
Hochstrasser, U., and E. Kertscher, US 4,351,790, Maillefer, (1982).
Hohlfeld, R.W., US 4,590,241, Dow, (1986).
Hold, P., D.H. Sebastian and R. Shamar, *Adv. Polym. Technol., 2,* 259 (1982).
Homma, S., M. Kashiwagi, and M. Tanaka, US 4,775,713, Mitsui Petrochemical Industries, (1988).
Howland, C., and L. Erwin, *Soc. Plast. Eng. ANTEC Tech. Pap., 29,* 113 (1983).
Huber, G.R., L.G. Wenger, B.W. Hauck, G.J. Rokey, L.E. Schmelzie, and T.R. Hartter, US 4,728,367, Wenger, (1988).
Hudec, P., and L. Obdrzalek, *Angew. Macromol. Chem., 89,* 41 (1980).
Hunt, P.A., and J. Maslen, GB 1,437,176, Imperial Chemical Industries, (1976).
Hyun, M.E., and S.C. Kim, *Polym. Eng. Sci., 28,* 743 (1988).
Ide, F., K. Kamada and A. Hasegawa, *Kobunshi Kagaku, 25,* 107, 167, 298 (1968a).
Ide, F., K. Kamada and A. Hasegawa, *Kobunshi Kagaku, 29,* 259, 264 (1968b).
Ide, F., and A. Hasegawa, *J. Appl. Polym. Sci., 18,* 963 (1974).
Ide, F., and I. Sasaki, US 4,003,874, Mitsubishi Rayon, (1977).
Illing, G., *Mod. Plast., 46,* 8, 70 (1969).
Illing, G., US 3,536,680, Werner and Pfleiderer, (1970).
Inata, H., and S. Matsumura, *J. Appl. Polym. Sci., 32,* 5193 (1986).
Inata, H., M. Ogasawara, T. Morinaga, and A. Norike, US 4,196,066, Teijin, (1980).
Inata, H., M. Ogasawara, T. Morinaga, and A. Norike, US 4,269,947, Teijin, (1981).
Inoue, T., M. Hattori, and K. Hayama, US 4,698,395, Mitsubishi Petrochemical, (1987).
Izawa, S., J. Ohzeki, T. Yahata, and A. Nakanishi, US 4,132,684, Asahi-Dow, (1979).
Jalbert, R.L., and T.S. Grant, US 4,654,405, Borg-Warner Chemicals, (1987).
Janeschitz-Kriegel, H. and J. Schiff, *Plast. and Polym.,* Dec. 1969, p. 523.
Jepson, C.H., *Ind. Eng. Chem., 45,* 992 (1953).
Johnson, B.C., T.W. Hovatter, S.T. Rice, and H.S. Chao, US 4,808,674, General Electric, (1989).
Jones, G.D., and R.M. Nowak, US 3,177,270, Dow, (1965).
Jones, W.J., and R. A. Mendelson, U.S. 4, 569, 969, Monsanto (1986).
Kampouris, E.M., and A.G. Andreopoulos, *J. Appl. Polym. Sci., 34,* 1209 (1987).
Kasahara, H., K. Fukuda and H. Suzuki, US 4,339,376, Asahi-Dow, (1982).
Kasahara, H., K. Tazaki, K. Fukuda, and H. Suzuki, US 4,421,892, Asahi-Dow, (1983).
Kato, Y., M. Yuyama, M. Moriya, H. Matsuura, S. Iijima, and T. Hashimoto, US 4,789,709, Sumitomo, (1988).
Kauth, H., K. Reinking, and D. Freitag, US 4,782,123, Bayer, (1988).
Kawada, T., Y. Matsuo, K. Makino, and N. Oshima, US 4,739,011, Japan Synthetic Rubber, (1988).
Keller, J.P., J.S. Prentice, and J.W. Harding, US 3,755,527, Esso, (1973).
Keogh, M.J., US 4,291,136, Union Carbide, (1981).
Keogh, M.J., US 4,328,323, Union Carbide, (1982).
Keogh, M.J., US 4,526,930, Union Carbide, (1985).

Keogh, M.J., US 4,593,071, Union Carbide, (1986).

Keogh, M.J., S.L. Wallace, and G.D. Brown, US 4,489,029, Union Carbide, (1984).

Kern, D.Q., *Process Heat Transfer,* McGraw-Hill, New York, 1950.

Kerschbaum, K., and A. Konicka, US 4,716,000, Rosendahl Maschinen, (1987).

Khanna, Y.P., E.A. Turi, S.M. Aharoni, and T. Largman, US 4,417,032, Allied, (1983).

Kircher K., *Chemical Reactions in Plastics Processing,* pp.80-81, Hanser Publishers, Munich, 1987.

Kobayashi, T., H. Kitagawa, C. Sugitawa, and S. Kobayashi, US 4,584,353, Toyo Boseki, (1986).

Kobayashi, Y., T. Itou, and T. Inoue, EPA 282,052, Mitsubishi Petrochemical, (1988).

Kodama, T., I. Sasaki, and H. Mori, US 4,141,882, Mitsubishi Rayon, (1979).

Kohan, M.I., W.H. Martin, and C.K. Rosenbaum, US 3,676,400, Du Pont, (1972).

Kolouch, R.J., and R.H. Michel, US 4,409,167, Du Pont, (1983).

Kopchik, R.M., US 4,246,374, Rohm and Haas, (1981).

Korber, H., EPA 2,760, Bayer, (1979).

Korber, H., P. Tacke, F. Fahnler, D. Neuray, and F. Heydenreich, US 4,362,846, Bayer, (1982).

Korver, G.L., US 4,071,504, Goodyear Tire and Rubber, (1978).

Kosaka, N., S. Izawa, and J. Sugiyama, US 4,483,958, Asahi-Dow, (1984).

Kosanovich, G.M., and G. Salee, US 4,415,721, Occidental Chemical, (1983).

Kosanovich, G.M., and G. Salee, US 4,465,819, Occidental Chemical, (1984a).

Kosanovich, G.M., and G. Salee, US 4,490,519, Occidental Chemical, (1984b).

Kotliar, A.M., *J. Polym. Sci., Macromol. Rev., 16,* 367 (1981)

Kotnour, T., R.L. Barber, and W.L. Krueger, EPA 160,394, Minnesota Mining and Manufacturing, (1985).

Kotnour, T. A., R.L. Barber and W.L. Krueger, US 4,619,979, 3M (1986).

Kowalski, R. C., Ph.D. Thesis, Polytechnic Institute of Brooklyn (1963).

Kowalski, R. C., Proc. Fifth International Congress of Rheology, Kyoto, Japan, October, 1968.

Kowalski, R.C., US 3,563,972, Esso, (1971).

Kowalski, R.C., American Chemical Society Polym. Div. Topical Workshop on Polymerization and Polymer Modification by Reactive Processing, May 21-24, 1985a, Bermuda.

Kowalski, R. C., US 4,508,592, Exxon Chemical Co. (1985b).

Kowalski, R.C., in *History of Polyolefins,* R.B. Seymour and T. Cheng, eds., D. Reidel Publishing Co., Boston, 1986, p. 307.

Kowalski, R.C., J.W. Harrison, J.C. Staton, and J.P. Keller, US 3,608,001, Esso, (1971).

Kowalski, R.C., W.M. Davis, N.F. Newman, and L. Erwin, US 4,513,116, Exxon, (1985a).

Kowalski, R.C., W.M. Davis, N.F. Newman, Z.A. Foroulis, and F.P. Baldwin, US 4,548,995, Exxon, (1985b).

Kowalski, R.C., W.M. Davis, N.F. Newman, Z.A. Foroulis, and F.P. Baldwin, US 4,554,326, Exxon, (1985c).

Kowalski, R.C., W.M. Davis, N.F. Newman, Z.A. Foroulis, and F.P. Baldwin, US 4,563,506, Exxon, (1986).

Kowalski, R. C., Proc. First Intern. Congress "Compalloy '89" , New Orleans, La, April 5-7, 1989a.

Kowalski, R. C., *Chem. Eng. Progr., 85,* 5, 67 (1989b).

Krabbenhoft, H.O., US 4,579,905, General Electric, (1986).

Krebaum, L.J., W.C.L. Wu, and J. Machonis, jr., US 3,882,194, Chemplex, (1975).

Kresge, E. N., R. H. Schatz, and H-C. Wang in *Encycl. Polym.Sci. Eng.,* J.I. Kroschwitz, ed., Wiley, New York, *8,* 423, (1987).
Kriek, G.R., J.Y.J. Chung, G.E. Reinert, and D. Neuray, US 4,542,177, Mobay, (1985).
Kubanek, V., Z. Sterbacek, J. Kralicek, J. Marik,and B. Casensky, GB 2,096,155, Chemopetrol (1982).
Lambla, M., *Polym. Process Eng., 5,* 297 (1988).
Lambla, M., J. Druz, and A. Bouilloux, *Polym. Eng. Sci., 27,* 1221 (1987a).
Lambla, M., J. Druz, and A. Bouilloux in *New Polymeric Materials. Reactive Processing and Physical Properties,* E. Martuscelli and C. Marchetta, eds., VNU Science Press, Utrecht, 1987b, p. 33.
Largman, T., and S.M. Aharoni, US 4,433,116, Allied, (1984).
Lazarus, S.D., and K. Chakravarti, US 4,115,350, Allied, (1978).
Lazarus, S.D., and R.A. Lofquist, US 4,348,314, Allied, (1982).
Lee, R.W., and W.J. Miloscia, US 4,410,659, Standard Oil, (1983).
Lindt, J.T., *Polym. Process Eng.,1,* 37 (1983).
Lo, J.-D., and W.R. Schlich, US 4,585,852, General Electric (1986).
Lucas, J.M., and A.C. Perricone, GB 2,177,410, Milchem, (1987).
Luijk, P., E. Van Gelderen, and G.P. Schipper, US 3,644,248, Shell Oil, (1972).
Lundberg, R.D., "Ionic Polymers" in *Encycl. Polym. Sci. Eng.,* J.I. Kroschwitz, ed., Wiley, New York, *8,* 393 (1987).
Lundberg, R.D., I. Duvdevani, D.G. Peiffer, and P.K. Agarwal in *Advances in Polymer Blends and Alloys Technology,* vol. 1, M.A. Kohudic ed., Technomic Publishing Co., Lancaster, 1987, p. 131.
Lutz, R.G., R. Gelles, and W.P. Gergen, US 4,795,782, Shell Oil, (1989).
Mack, W.A., *Chem. Eng.*, 99 (1972).
Mack, W.A., and R. Herter, *Chem. Eng. Progr.,72,* 1, 64 (1976).
Maresca, L.M., US 4,877,848, General Electric, (1989).
Martuscelli, E., F. Riva, C. Sellitti, and C. Silvestre, *Polymer 26,* 270 (1985).
Mashita, K., T. Fujii, and T. Oomae, EPA 177,151, Sumitomo Chemical, (1986).
Mashita, K., J. Nambu, and S. Ishii, EPA 234,819, Sumitomo Chemical, (1987).
Mashita, K., T. Fujii, and T. Omae, US 4,780,505, Sumitomo Chemical, (1988).
Matsumura, S., H. Inata, and T. Morinaga, US 4,351,936, Teijin, (1982).
Matsuoka, N., H. Matsumoto, Y. Hori, Y. Miki, K. Sano, and I. Ijichi, US 4,487,897, Nitto Electric Industrial, (1984).
Matsuura, K., N. Yamaoka, and M. Miyoshi, US 4,412,042, Nippon Oil (1983).
Mawatari, M., T. Itoh, S. Tsuchikawa, and S. Kimura, US 4,839,425, Japan Synthetic Rubber, (1989).
McKnight, W.J., R.W. Lenz, P.V. Musto, and R.J. Somani, *Polym. Eng. Sci., 25,* 1124 (1985).
McNally, D., H.L. La Nieve, and J.L. Costanzo, EPA 180,471, Celanese, (1986).
Menges, G., *Macromol. Chem., Macromol. Symp., 23,* 13 (1989)
Menges, G., and T. Bartilla, *Polym. Eng. Sci., 27,* 1216 (1987).
Mesrobian, R.B., P.E. Sellers, and D. Adomaitis, US 3,373,224, Continental Can, (1968).
Meyuhas, G.S., A. Moses, Y. Reibenbach, and Z. Tadmor, *J. Polym. Sci., Polym. Lett. Ed.,11,* 103 (1973).
Michel, A., US 4,443,584, CNRS, (1984).
Mijangos, C., G. Martinez, A. Michel, J. Millan, and A. Guyot, *Eur. Polym. J., 20,* 1 (1984).
Mijangos, C., A. Martinez, and A. Michel, *Eur. Polym. J., 22,* 417 (1986).

Mijangos, C., J.M. Gomez-Elvira, G. Martinez, and J. Millan, *Macromol. Chem., Macromol. Symp., 25,* 209 (1989).
Mizuno, S., and T. Adachi, US 4,254,010, Dainippon Ink and Chemicals, (1981).
Mizuno, S., R. Ishikawa, and M. Miyazaki, US 4,085,086, Dainippon Ink and Chemicals, (1978).
Mizuno, S., and T. Sugie, US 4,165,307, Dainippon Ink and Chemicals, (1979).
Moberly, C.W., US 4,510, 297, (1985).
Mohr, W.D., R.L. Saxton and C.H. Jepson, *Ind. Eng. Chem., 49,* 1855 (1957).
Moncur, M.V., US 4,350,794, Monsanto, (1982).
Mondragon, I., and J. Nazabal, *Polym. Eng. Sci., 25,* 178 (1985).
Mondragon, I., and J. Nazabal, *J. Appl. Polym. Sci., 32,* 6191 (1986).
Mondragon, I., and J. Nazabal, *J. Mater. Sci. Lett., 6,* 698 (1987).
Mondragon, I., M. Gaztelumendi, and J. Nazabal, *Polym. Eng. Sci., 26,* 1478 (1986a).
Mondragon, I., M. Gaztelumendi, and J. Nazabal, *Polym. Eng. Sci., 28,* 1126 (1986b).
Morgan, R.A., and W.H. Sloan, EPA 150,953, Du Pont, (1985).
Moriya, Y., N. Suzuki, and Y. Okada, US 4,753,990, Nippon Oil and Fats, (1988).
Moriya, Y., N. Suzuki, and H. Goto, US 4,839,423, Nippon Oil and Fats, (1989).
Morman, M.T., and T.J. Wisneski, US 4,451,589, Kimberly-Clark, (1984).
Motooka, M., and H. Mantoku, US 4,616,059, Mitsui Petrochemical Industries, (1986).
Mueller-Tamm, H., W. Immel, H. Mohr, and K.-H. Fauth, EPA 35,677, BASF, (1981).
Munteanu, D., in *Metal-Containing Polymeric Systems,* J.E. Sheats, C.E. Carraher, and C.U. Pittman, eds., Plenum Press, New York, 1985, p.479.
Munteanu, D., in *Developments in Polymer Stabilization-8,* G. Scott, ed., Elsevier Applied Science, New York, 1987, p. 179.
Murch, L.E., US 3,845,163, Du Pont, (1974).
Müssig, B., R.-V. Meyer, B. Brassat, and R. Dhein, US 4,820,771, Bayer, (1989).
Nakazima, O., and S. Izawa, EPA 282,664, Asahi, (1988).
Nakazima, O., and S. Izawa, US 4,863,996, Asahi, (1989).
Nangeroni, J. F., K. Eise and D.S. Kidwell, *Polym. Proc. Eng., 3,* 85 (1985).
Narkis, M., and R. Wallerstein, *Polym. Commun., 27,* 314 (1986).
Neilinger, W., D. Wittman, U. Westeppe, L. Bottenbruch, J. Kirsch, and H.-J. Füllmann, EPA 291,796, Bayer, (1988).
Neill, P.L., K.L. Bryce, and G.M. Lancaster, US 4,666,988, Dow, (1987).
Neill, P.L., G.M. Lancaster, and K.L. Bryce, US 4,774,290, Dow, (1988).
Nelb, R.G., K. Onder, K.W. Rausch, and J.A. Vanderlip, US 4,672,094, Dow, (1987).
Nettelnbrecker, H.-J., and B. Stoehrer, VDI-Gessellschaft Kunststofftechnik, p. 225, VDI-Verlag, Düsseldorf, 1988.
Newman, N.F., and R.C. Kowalski, US 4,384,072, Exxon, (1983).
Newman, N.F., and R.C. Kowalski, US 4,486,575, Exxon, (1984).
Newman, N.F., and R.C. Kowalski, US 4,501,859, Exxon, (1985).
Nichols, R.J., *Mod. Plast., 63,* 9, 90 (1986).
Nishio, T., T. Sanada, and T. Okada, EPA 270,247, Sumitomo, (1988).
Nogues, P., US 4,727,120, Societe Atochem, (1988a).
Nogues, P., US 4,735,992, Societe Atochem, (1988b).
Nowak, R.M., US 3,270,090, Dow, (1966).
Nowak, R.M., and G.D. Jones, US 3,177,269, Dow, (1965).
Ogihara, S., Y. Nakamura, and O. Fukui, US 4,032,592, Ube Industries, (1977).
Ohmae, T., N. Yamaguchi, Y. Toyoshima, T. Kawakita, K. Mashita, and J. Nambu, EPA 258,040, Sumitomo Chemical, (1988).
Ohmura, Y., S. Maruyama, and H. Kawasaki, US 4,339,555, Mitsubishi Chemical Industries, (1982).

Okamoto, T., K. Yasue, T. Marutani, and Y. Fukushima, US 4,804,707, Unitika, (1989).
Okuzumi, Y., US 4,017,463, Goodyear Tire and Rubber, (1977).
Olivier, E.J., US 4,594,386, Copolymer Rubber & Chemical, (1986a).
Olivier, E.J., PCT Intl. Appl. WO 86/04076, Copolymer Rubber and Chemical, (1986b).
Oostenbrink, A.J., R.J.M. Borggreve, and R.J. Gaymans in *Integration of Fundamental Polymer Science and Technology,* P.J. Lemstra and L.A. Kleintjens, eds., Elsevier Applied Science, New York, 1989, p. 123.
Orgen, D.E., US 3,418,280, Celanese, (1968).
Otawa, Y., A. Uchiyama, K. Hiraoka, K. Okamoto, and S. Shimizu, EPA 269,274, Mitsui Petrochemical Industries, (1987).
Ottino, J.M., *Annu. Rev. Fluid Mech., 22,* 207 (1990).
Owens, F.H., and J.S. Clovis, US 3,668,274, Rohm and Haas, (1972).
Pabedinskas, A., W.R. Cluett, and S.T. Balke, *Polym. Eng. Sci., 29,* 993 (1989).
Park, C.P., US 4,554,293, Dow, (1985).
Paul, D.R., and S. Newman, *Polymer Blends,* Academic Press, New York, vols. 1 & 2, 1978.
Paul, D.R., J.W. Barlow, and H. Keskkula, "Polymer Blends" in *Encycl. Polym. Sci. Eng.*, J.I. Kroschwitz, ed., Wiley, New York, *12,* 417 (1988).
Peiffer, D.G., I. Duvdevani, P.K. Agarwal, and R.D. Lundberg, *J. Polym. Sci.: Polym. Lett. Ed., 24,* 581 (1986).
Pelrin, J.E., GB 1,601,063, Chromerics, (1981).
Pennwalt Corp., Lucidol Div., Product Bulletin "Dialkyl Peroxides" , Buffalo, N.Y., 1989a.
Pennwalt Corp., Lucidol Div., Bulletin 16.1274 "Chemical Curing of Elastomers and Crosslinking of Thermoplastics" , Buffalo, N.Y., 1989b.
Peroxid-Chemie GmbH, Technical Bulletin "PP Degradation with Organic Peroxides" , Hoellriegelskreuth, Germany, July, 1980.
Perron, P.J., and E.A. Bourbonais, PCT Intl. Appl. WO 88/06174, Dexter, (1988a).
Perron, P.J., and E.A. Bourbonais, US 4,782,114, Dexter, (1988b).
Phadke, S.V., EPA 274,744, General Electric, (1987).
Phadke, S.V., US 4,757,112, Copolymer Rubber & Chemical, (1988a).
Phadke, S.V., EPA 279,502, General Electric, (1988b).
Pillon, L.Z., and L.A. Utracki, *Polym. Eng. Sci., 24,* 1300 (1984).
Pillon, L.Z., L.A. Utracki, and D.W. Pillon, *Polym. Eng. Sci., 27,* 562 (1987a).
Pillon, L.Z., J. Lara, and D.W. Pillon, *Polym. Eng. Sci., 27,* 984 (1987b).
Piotrowski, B., R. Buening, H. Hanisch, and B. Janser, DE 3,327,596, Dynamit Nobel, (1985).
Poole, M.J., US 4,136,132, BICC, (1979).
Porter, R.S., and A. Casale, *Polym. Eng. Sci., 25,* 129 (1985).
Porter, R.S., J.M. Jonza, M. Kimura, C.R. Desper, and E.R. George, *Polym. Eng. Sci., 29,* 55 (1989).
Powell, R.J., and G.W. Prejean, US 3,969,434, Du Pont, (1976).
Pratt, C.F., PCT Intl. Appl. WO 88/07065, General Electric, (1988).
Pratt, C.F., S.V. Phadke, and E.J. Olivier, PCT Intl. Appl. WO 88/05452, General Electric, (1988).
Quiring, B., G. Niederdellmann, W. Goyert, and H. Wagner, US 4,245,081, Bayer, (1981a).
Quiring, B., W. Wenzel, G. Niederdellmann, H. Wagner, and W. Goyert, US 4,286,080, Bayer, (1981b).
Ratzsch, M., *Makromol. Chem. Macromol. Symp., 12,* 165 (1987).

Ratzsch, M., U. Hofmann, M. Gebauer, G. Hoffman, G. Bergmann, and H. Schade, GB 2,116,981, VEB Leuna-Werke, (1983).
Rausch, jr., K.W., and T.R. McClellan, US 3,642,964, Upjohn, (1972).
Rauwendaal, C., *Polymer Extrusion,* Carl Hanser Verlag, Munich (1986).
Rauwendaal, C. and J.F. Ingen Housz, *Adv. Polym. Technol., 8,* 289 (1988).
Rees, R.W., US 3,404,134, Du Pont, (1968).
Rees, R.W., US 3,471,460, Du Pont, (1969).
Rees, R.W., "Reversible Crosslinking" in *Encycl. Polym. Sci. Eng.*, J.I. Kroschwitz, ed., Wiley, New York, *4,* 395 (1986).
Reischl, A., US 4,595,709, Bayer, (1986).
Richardson, P.N., US 4,283,502, Du Pont, (1981).
Roberts, R., US 4,055,549, Union Carbide, (1977).
Robeson, L.M., *J. Appl. Polym. Sci., 30,* 4081 (1985).
Romanini, D., E. Garagnani, and E. Marchetti in *New Polymeric Materials. Reactive Processing and Physical Properties,* E. Martuscelli and C. Marchetta, eds., VNU Science Press, Utrecht, 1987, p. 56.
Rosendale, D.,and J.A. Biesenberger, *Adv. Chem. Ser., 227,* pp. 267 -283, ACS, Washington, D.C., 1990.
Rotermund, U., *Polyurethanes World Congr.* FSK/SPI, 632 (1987).
Rothwell, R.E., H.H. Rowan, and J.J. Dunbar, US 4,374,960, Allied, (1983).
Roura, M.J., US 4,346,194, Du Pont, (1982).
Roura, M.J., US 4,478,978, Du Pont, (1984).
Rudin, A., *J. Macromol. Chem. - Rev. Macromol. Chem., C19* 267 (1980)
Rugg, B.A., and W. Brenner, US 4,316,747, New York Univ., (1982).
Rugg, B.A., and W. Brenner, US 4,368,079, New York Univ., (1983).
Rugg, B.A., and R. Stanton, US 4,316,748, New York Univ., (1982a).
Rugg, B.A., and R. Stanton, US 4,363,671, New York Univ., (1982b).
Rugg, B.A., and R. Stanton, US 4,390,375, New York Univ., (1983).
Rugg, B.A., and R. Stanton, US 4,591,386, New York Univ., (1986).
Ryason, P.R., US 4,206,713, NASA, (1980).
Ryu, S.H., C.G Gogos, and M. Xanthos, *Soc. Plast. Eng. ANTEC Tech. Pap., 35,* 879 (1989).
St. Clair, D.J., and S.S. Chin, US 4,783,504, Shell Oil, (1988).
Saito, A., A. Yamori, and T. Ibaragi, US 4,292,414, Asahi, (1981a).
Saito, A., A. Yamori, and H. Morita, US 4,308,353, Asahi, (1981b).
Sakai, T., *Soc. Plast. Eng. ANTEC Tech. Pap., 34,* 1853 (1988).
Saleem, M., and W.E. Baker, *J. Appl. Polym Sci., 39,* 655 (1990).
Sano, H., and K. Ohno, EPA 268,280, Mitsubishi Petrochemical, (1988).
Sawden, F.H., EPA 274,424, Du Pont Canada, (1988).
Saxton, R.L., US 4,338,405, Du Pont, (1982).
Schaufelberger, G. F., French 1, 346, 533, Union Carbide (1963).
Schleese, E., H.M. Schmidtchen, H.U. Voigt, F. Patzke, and G. Klein, GB 2,061,967, Kabel- und Metallwerke Gutehoffnungshutte, (1981).
Schmid, E., and M. Hoppe, GB 2,131,037, EMS-Inventa, (1984).
Schmidt, L.R., and E.M. Lovgren, US 4,421,907, General Electric, (1983).
Schmidt, L.R., and E.M. Lovgren, US 4,443,591, General Electric, (1984).
Schmidt, L.R., and E.M. Lovgren, US 4,511,535, General Electric, (1985).
Schmidt, L.R., E.M. Lovgren, and P.G. Meissner, US 4,443,592, General Electric, (1984).
Schmukler, S., M. Shida, and J. Machonis, jr., US 4,575,532, Norchem, (1986a).
Schmukler, S., M. Shida, and J. Machonis, jr., US 4,600,746, Norchem, (1986b).

Schollenberger, C.S., K. Dinsberg, and F.D. Stewart, *Rubber Chem. Technol., 55,* 137 (1982).
Schollenberger, C.S., K. Dinsberg, and F.D. Stewart in *Urethane Chemistry and Applications,* K.N. Edwards, ed., ACS Symp. Ser. 172, American Chemical Society, Washington, D.C., 1981, p. 433.
Schott, N.R., and B. Sanderford, *Coat. Plast. Prepr. Am. Chem. Soc. Div. Org. Coat. Plast. Chem., 37* (2), 73 (1977).
Schuddemage, H.-D., H. Jastrow, and H. Barth, US 4,007,234, Hoechst, (1977).
Schuetz, J.E., R.W. Hohlfeld, and B.C. Meridith, US 4,864,002, Dow, (1989).
Schwarz, E.C.A., US 3,627,867, Du Pont, (1971).
Scott, C.E., and C.W. Macosko, *Polym. Prepr. Am. Chem. Soc., Div. Polym. Chem., 29,* 1, 561 (1988).
Scott, H.G., US 3,646,155, Midland Silicones, (1972).
Scott, G., US 4,213,892, (1980).
Scott, G., in *Developments in Polymer Stabilization-8,* G. Scott, ed., Elsevier Applied Science, New York, 1987, p. 209.
Scott, H.G., and J.F. Humphries, *Mod. Plast. 50,* 3, 82 (1973).
Scott, G., and E. Setoudeh, *Polym. Degradation Stab., 5,* 1 (1983).
Scott, G., and S.M. Tavakoli, *Polym. Degradation Stab.,4,* 343 (1982).
Seddon, R.M., W.H. Russell, and K.B. Rollins, US 3,253,818, Celanese, (1966).
Seddon, R.M., and L.D. Scarbrough, US 3,442,866, Celanese, (1969).
Semanchik, M., and D.M. Braunstein, US 4,105,637, Celanese, (1978).
Sextro, G., K.F. Mueck, K. Berg, and E. Fischer, EPA 279,289, Hoechst, (1988).
Shibuya, N., and K. Kosegaki, EPA 248,526, Mitsubishi Petrochemical, (1987).
Shibuya, N., Y. Sobajima, and H. Sano,US 4,743,651, Mitsubishi Petrochemical, (1988a).
Shibuya, N., K. Takagi, S. Hattori, T. Kobayashi, and H. Sano, EPA 270,796, Mitsubishi Petrochemical, (1988b).
Shida, M., J. Machonis, Jr., S. Schmukler and R. J. Zeitlin, U.S. 4, 087, 587, Chemplex (1978).
Shiraki, T., F. Hayano, and H. Morita, US 4,628,072, Asahi, (1986).
Shiraki, T., F. Hayano, and H. Morita, US 4,657,970, Asahi, (1987a).
Shiraki, T., F. Hayano, and H. Morita, US 4,657,971, Asahi, (1987b).
Shuttleworth, R., and W.F. Watson, *Macromol. Syn., 5,* 65 (1974).
Shyu, W.B., and D.A. Woodhead, US 4,753,997, Standard Oil, (1988).
Siadat, B., M. Malone, and S. Middleman, *Polym. Eng. Sci., 19,* 787 (1979).
Siadat, B., R.D. Lundberg, and R.W. Lenz, *Polym. Eng. Sci., 20,* 530 (1980).
Simmons, A., and W.E. Baker, *Polym. Eng. Sci., 29,* 1117 (1989).
Simmons, A., and W.E. Baker, *Polym. Commun., 31,* 20 (1990).
Sims, W.M., US 3,966,839, Foster Grant, (1976).
Sivavec, T.M., US 4,808,671, General Electric, (1989).
Smith, J.J., W.A. Miller, and F.P. Reding, US 3,412,080, Union Carbide, (1968).
Solc, K., ed., *Polymer Compatibility and Incompatibility. Principles and Practice,* Harwood Academic Publishers, New York, 1981.
Sopko, T.M., and R.E. Lorentz, US 4,812,544, Lubrizol, (1989).
Sperling, L.H., "Microphase Structure" in *Encycl. Polym. Sci. Eng.*, J.I. Kroschwitz, ed., Wiley, New York, *9,* 770 (1987).
Spielau, P., W. Kühnel, K. Klaar, B. Gaspar, R. Weiss, H. Ulb, H.-U. Breitscheidel, G. Klingberg, and J. Fenske, US 4,652,326, Dynamit Nobel, (1987).
Staas, W.H., EPA 33,220, Rohm and Haas, (1981).
Staton, J.C., J.P. Keller, R.C. Kowalski, and J.W. Harrison, US 3,551,943, Esso, (1971).

Statz, R.J., US 4,766,174, Du Pont, (1988).
Steinkamp, R.A., and T.J. Grail, US 3,862,265, Exxon Research and Engineering, (1975).
Steinkamp, R. A., and T. J. Grail, U.S. 3,953,655, Exxon Research & Engineering, (1976).
Stenmark, D.G., and R.L. Heinrich, US 3,884,451, Exxon Research and Engineering, (1975).
Stevens, M.J., *Extruder Principles and Operation,* Elsevier Applied Science, New York 1988.
Stober, K.E., and J.L. Amos, US 2,530,409, Dow (1950).
Strait, C.A., G.M. Lancaster, and R.L. Tabor, US 4,762,890, Dow, (1988).
Strauss, H.W., US 4,716,202, Du Pont, (1987).
Streetman, W.E., US 4,497,934, American Cyanamid, (1985).
Stuber, N.P., and M. Tirrell, *Polym. Process Eng., 3,* 71 (1985).
Subramanian, P.M., and V. Mehra, *Polym. Eng. Sci., 27,* 663 (1987).
Sugio, A., T. Furusawa, K. Tanaka, T, Umemura, and H. Urabe, US 4,115,369, Mitsubishi Gas Chemical, (1978).
Sugio, A., M. Okabe, and A. Amagai, EPA 148,774, Mitsubishi Gas Chemical, (1985).
Sumimura, S., and A. Tazawa, US 4,696,970, Toray Silicone, (1987).
Sutter, H., M. Beck, F. Haas, and G. Marwede, GB 1,347,088, Bayer, (1974).
Sutter, H., K. Nothen, and F. Haas, US 3,780,139, Bayer, (1973).
Sutter, H., and R. Peuker, US 4,058,654, Bayer, (1977).
Suwanda, D., R. Lew, and S.T. Balke, *J. Appl. Polym. Sci., 35,* 1019 (1988a).
Suwanda, D., R. Lew, and S.T. Balke, *J. Appl. Polym. Sci., 35,* 1033 (1988b).
Swarbrick, P., W.J. Green, and C. Maillefer, US 4,117,195, BICC and Establissements Maillefer, (1978).
Swiger, R.T., and P.C. Juliano, US 4,147,740, General Electric, (1979).
Swiger, R.T., and L.A. Mango, DE 2,722,270, General Electric, (1977).
Sybert, P.D., C.Y. Han, S.B. Brown, D.J. McFay, W.L Gately, and J.A. Tyrell, PCT Intl. Appl. WO 87/07279, General Electric, (1987).
Tabor, R.L., and J.A. Allen, US 4,684,576, Dow, (1987).
Tabor, R.L., P.L. Neill, and B.L. Davis, US 4,739,017, Dow, (1988).
Tadmor, Z., and C. G. Gogos, *Principles of Polymer Processing,* John Wiley, New York, 1979.
Tadmor, Z. and I. Klein, *Engineering Principles of Plasticating Extrusion,* Van Nostrand Reinhold, New York, 1970.
Takekoshi, T., and J.E. Kochanowski, US 3,833,546, General Electric, (1974).
Takekoshi, T., and J.E. Kochanowski, US 4,011,198, General Electric, (1977).
Tanaka, M., and S. Honma, EPA 246,729, Mitsui Petrochemical Industries, (1987).
Taubitz, C., E. Seiler, and L. Schlemmer, EPA 223,115, BASF, (1987a).
Taubitz, C., E. Seiler, and L. Schlemmer, EPA 226,002, BASF, (1987b).
Taubitz, C., E. Seiler, and L. Schlemmer, EPA 222,246, BASF, (1987c).
Taubitz, C., E. Seiler, J. Hambrecht, K. Mitulla, and K. Boehlke, EPA 237,710, BASF, (1987d).
Taubitz, C., E. Seiler, R. Bruessau, and D. Wagner, EPA 285,968, BASF, (1988a).
Taubitz, C., E. Seiler, and L. Schlemmer, US 4,780,509, BASF, (1988b).
Taubitz, C., E. Seiler, and L. Schlemmer, US 4,751,268, BASF, (1988c).
Taubitz, C., E. Seiler, K. Boehlke, B. Ostermayer, H. Gausepohl, and V. Muench, EPA 285,976, BASF (1988d).
Taubitz, C., E. Seiler, H. Gausepohl, K. Boehlke, and R. Bueschl, EPA 289,780, BASF, (1988e).

Taubitz, C., L. Schlemmer, E. Seiler, K. Boehlke, and H. Gausepohl, EPA 301,404, BASF, (1989a).
Taubitz, C., E. Seiler, K. Boehlke, K. Bronstert, and D. Wagner, EPA 298,365, BASF, (1989b).
Taubitz, C., H. Gausepohl, E. Seiler, and L. Schlemmer, US 4,797,453, BASF, (1989c).
Taylor, G.I., *Proc. R. Soc. London,* Ser.A, *219,* 186 (1953).
Teyssie, Ph., R. Fayt, and R. Jerome, *Macromol. Chem., Macromol. Symp., 16,* 41 (1988).
Thomas, N.W., US 4,101,601, Celanese, (1978).
Thomas, N.W., F.M. Berardinelli, and R. Edelman, US 4,128,599, Celanese, (1978a).
Thomas, N.W., F.M. Berardinelli, and R. Edelman, US 4,071,503, Celanese, (1978b).
Thomas, N.W., F.M. Berardinelli, and R. Edelman, US 4,110,302, Celanese, (1978c).
Tintel, C., *Integr. Fundam. Polym. Sci. Technol.*--2, [Proc. Int. Meet. Polym. Sci. Technol. Rolduc Polym. Meet.--2] 1987, 64 (1988).
Titzmann, R., H. Thaler, and J. Walter, US 3,657,191, Farbwerke Hoechst, (1972).
Todd, D.B., *Polym. Process Eng. 6,* 15 (1988a).
Todd, D.B., *Soc. Plast. Eng. ANTEC Tech. Pap., 34,* 54 (1988b).
Todd, D.B., *Polym.-Plast. Technol. Eng., 28,* 2, 123 (1989).
Togo, S., A. Amagai, Y. Kondo, and T. Yamada, EPA 268,486, Mitsubishi Gas Chemical, (1988).
Toyama, K., I. Shimizu, T. Imamura, and A. Nakanishi, US 4,097,556, Asahi-Dow, (1978).
Trieschmann, H.G., H. Moeller, G. Schmidtthomee, C. Alt, and R. Herbeck, GB 1,042,178, BASF, (1966).
Trivedi, B.C., and B.M. Culbertson, *Maleic Anhydride,* Plenum Press, New York, 1982, p. 459.
Trolez, Y., C. Macosko, and A. Bouilloux, Proc. Sixth Ann. *Polym. Processing Soc.* Meeting, Paper 01-16, Nice, France, April 17-20, 1990.
Tucker, C.S., and R.J. Nichols, *Soc. Plast. Eng. ANTEC Tech. Pap., 33,* 117 (1987a)
Tucker, C.S, and R.J. Nichols, *Plast. Eng., 43,* 5, 27 (1987b).
Tzoganakis, C., *Adv. Polym. Technol., 9,* 321 (1989).
Tzoganakis, C., J. Vlachopoulos, and A.E. Hamielec, *Int. Polym. Process. 3,* 141 (1988a).
Tzoganakis, C., J. Vlachopoulos, and A.E. Hamielec, *Polym. Eng. Sci., 28,* 170 (1988b).
Tzoganakis C., J. Vlachopoulos, A.E. Hamielec, and D. M. Shinozaki, *Polym. Eng. Sci., 29,* 390, (1989c).
Tzoganakis, C., Y. Tang, J. Vlachopoulos, and A.E. Hamielec, *Polym. Process Eng., 6,* 29 (1988c).
Tzoganakis, C., Y. Tang, J. Vlachopoulos, and A.E. Hamielec, *J. Appl. Polym. Sci., 37,* 681 (1989a).
Tzoganakis, C., Y. Tang, J. Vlachopoulos, and A.E. Hamielec, *Polym.-Plast. Technol. Eng., 28,* 319, (1989b).
Udding, A.C., GB 2,205,103, Shell International Research, (1988).
Ueda, S., H. Harada, K. Yoshida, and K. Kasei, US 4,772,664, Asahi, (1988).
Ueno, K., and T. Maruyama, US 4,315,086, Sumitomo Chemical, (1982a).
Ueno, K., and T. Maruyama, US 4,338,410, Sumitomo Chemical, (1982b). Ullrich, M., E. Meisert, and A. Eitel, US 3,963,679, Bayer, (1976).
Umpleby, J.D., EPA 181,735, BP Chemicals, (1986).
Valsamis, L.N., and E.L. Canedo, *Int. Polym. Process., 4,* 247 (1989).
Van Ballegooie, P., and A. Rudin, *Polym. Eng. Sci., 28,* 1434 (1988).

Van Buskirk, B., and M.K. Akkapeddi, *Polym. Prepr. Am. Chem. Soc. Div. Polym. Chem., 29* (1), 333 (1988).
van der Meer, R., and J.B. Yates, PCT Intl. Appl. WO 87/00540, General Electric, (1987).
Voigt, H.-U., H.-P. Stehmann, M. Völker, and D. Keuper, GB 1,495,850, Kabel- und Metallwerke Gutehoffnungshutte, (1977).
Voigt, H.U., M. Völker, and H.-P. Stehmann, US 4,117,063, Kabel- und Metallwerke Gutehoffnungshutte, (1978).
Vollmert B., *Polymer Chemistry,* pp. 245-246, Springer-Verlag, New York, 1973.
Vroomans, H.J., EPA 286,734, Stamicarbon, (1988).
Waggoner, M.G., US 4,639,495, Du Pont, (1987).
Waniczek, H., and H. Bartl, EPA 128,443, Bayer, (1984).
Waniczek, H., G. Hohmann, H. Bartl, and L. Mott, US 4,602,056, Bayer, (1986).
Watkins, K.R., and L.R. Dean, US 4,359,557, Eastman Kodak, (1982).
Watson, A.T., H.L. Wilder, K.W. Bartz, and R.A. Steinkamp, US 3,898,209, Exxon, (1975).
Weiss, K.A., US 4,816,515, General Electric, (1989).
Wheeler, J.R., US 4,595,546, Crompton & Knowles, (1986).
White, G., US 4,737,547, Du Pont Canada, (1988).
White, J. L., W. Szydlowski, K. Min, and M. H. Kim, *Adv. Polym. Technol., 7,* 295 (1987).
Wielgolinski, L., and J. Nangeroni, *Adv. Polym. Technol., 3,* 99 (1983).
Wong, C.S., US 4,857,254, Du Pont Canada, (1989).
Wong, C.S., and R.A. Zelonka, US 4,612,155, Du Pont Canada, (1986).
Wong, P.C., EPA 266,994, Du Pont Canada, (1988).
Woodbrey, J.C., and M.V. Moncur, US 4,320,213, Monsanto, (1982).
Wu, S., Polym. Eng. Sci. 27, 335 (1987)
Wu, W.C.L., L.J. Krebaum, and J. Machonis, jr., US 3,873,643, Chemplex, (1975).
Xanthos, M., C.G. Gogos, and S.H. Ryu, Proc. Summer *Polym. Processing Soc.* Meeting, Paper 7F, Amherst, Mass., August 16-17, 1989.
Yamamoto, N., M. Isoi, M. Yoda, and S. Wada, US 4,146,529, Toa Nenryo, (1979).
Yates, J.B., US 4,755,566, General Electric, (1988).
Yates, J.B., and T.J. Ullman, US 4,745,157, General Electric, (1988).
Yates, J.B., and D.M. White, US 4,859,739, General Electric, (1989).
Yonekura, K., A. Uchiyama, and A. Matsuda, US 4,785,045, Mitsui Petrochemical Industries, (1988).
Yu, D.W., M. Xanthos, and C.G. Gogos, *Soc. Plast. Eng. ANTEC Tech. Pap., 36,* 1917, (1990).
Ziegler, W., F. Brandstetter, F. Weiss, and L. Schlemmer, EPA 193,110, BASF, (1986).
Zeitler, G., H. Mueller-Tamm, and F. Urban, US 3,949,019, BASF, (1976).
Zeitler, G., F. Werner, G. Bittner, K.-H. Baumann, A. Roeber, L. Metzinger, R. Ohlinger, and H. D. Zettler, US 4,597,927, BASF, (1986).
Zeitlin, R.J., US 3,267,173, Allied Chemical, (1966).
Zerafadi, S., and D. Bigio, Proc. 4th Ann. *Polym. Processing Soc.* Meeting, section 6, p. 10, May 1988.

Index